METHODS OF FRACTURE MECHANICS:
SOLID MATTER PHYSICS

SOLID MECHANICS AND ITS APPLICATIONS
Volume 51

Aims and Scope of the Series

The fundamental questions arising in mechanics are: *Why?*, *How?*, and *How much?* The aim of this series is to provide lucid accounts written by authoritative researchers giving vision and insight in answering these questions on the subject of mechanics as it relates to solids.

The scope of the series covers the entire spectrum of solid mechanics. Thus it includes the foundation of mechanics; variational formulations; computational mechanics; statics, kinematics and dynamics of rigid and elastic bodies; vibrations of solids and structures; dynamical systems and chaos; the theories of elasticity, plasticity and viscoelasticity; composite materials; rods, beams, shells and membranes; structural control and stability; soils, rocks and geomechanics; fracture; tribology; experimental mechanics; biomechanics and machine design.

The median level of presentation is the first year graduate student. Some texts are monographs defining the current state of the field; others are accessible to final year undergraduates; but essentially the emphasis is on readability and clarity.

For a list of related mechanics titles, see final pages.

Methods of Fracture Mechanics: Solid Matter Physics

by

G.P. CHEREPANOV
College of Engineering and Design,
Florida International University,
Miami, U.S.A.

KLUWER ACADEMIC PUBLISHERS
DORDRECHT / BOSTON / LONDON

A C.I.P. Catalogue record for this book is available from the Library of Congress.

ISBN 0-7923-4408-1

Published by Kluwer Academic Publishers,
P.O. Box 17, 3300 AA Dordrecht, The Netherlands.

Kluwer Academic Publishers incorporates
the publishing programmes of
D. Reidel, Martinus Nijhoff, Dr W. Junk and MTP Press.

Sold and distributed in the U.S.A. and Canada
by Kluwer Academic Publishers,
101 Philip Drive, Norwell, MA 02061, U.S.A.

In all other countries, sold and distributed
by Kluwer Academic Publishers Group,
P.O. Box 322, 3300 AH Dordrecht, The Netherlands.

Printed on acid-free paper

Printed in the Netherlands

To the memory

of my mother

Alexandra Petrovna Cherepanova

TABLE OF CONTENTS

AUTHOR'S PREFACE

A method is a way of achieving a result. Without mastering a method, one can only hope to memorize the result. Using the result without proper knowledge of the underlying method and hence neglecting the limitations of this result usually leads to terrible mistakes. Therefore, methods form a most substantial part of a discipline. This book is for those hard-working students who desire to master the methods for the shortest time. Look at the problems placed at the end of a Chapter and try to solve them. If you can solve these problems without reading the Chapter, you do not need to scrutinize it.

Your skills in mastering the methods of the Chapter are proportional to the number of problems you can solve. This is the main indicator of a successful study of the Chapter. The highest praise of the book would be an acquired ability of the reader to solve all these problems. The methods are treated on some particular exemplar cases, which are usually simpler than the corresponding problems at the end of Chapters. To help to solve the latter, hints, sometimes in many details, are provided. So, the Chapter problems constitute an inalienable part and play a decisive role in the book. This approach allows one to get into the complicated methods in a straightforward manner, but it requires an intense work of the reader to solve each Chapter' s problems. As an example of the similar approach or style, the well-known series of books on theoretical physics by Landau and Lifshitz can be mentioned. A systematic treatment of all these methods would necessitate dozens of volumes. The solutions manual is available to qualified users. The book is rich in topics for MS and PhD theses in the area of solid and fracture mechanics and physics. Therefore, it will be useful for a teacher and one who seeks a degree in this field.

Every discipline uses the known methods accommodated to its specific demands and develops its own methods that penetrate and influence other disciplines. The present volume is dedicated to some applications of the methods of fracture mechanics in solid matter physics. It is planned that the following volumes will be dedicated to some applications of these methods in mechanics of solids, theory of materials, structural integrity, material processing and manufacturing, optimal design, technology, and geophysics.

The topics are selected from the most challenging problems that are interesting for applications. As a supplementary text, the book can be used by teachers and graduate students in core courses on fracture mechanics, solid matter physics, and mechanics of solids, or in a special course on the topic of a Chapter. For convenience of reading, pertinent references are placed at the end of every Chapter. Every Chapter begins with a short preamble indicating the main methods and procedures treated in the Chapter. For those who are interested mostly in methods, it is recommended that, at first, they look through the preambles.

The basic methods used and developed in the book are as follows:

- **Mathematical modeling** methods including identification methods, the method of analogy, trial-and-error methods, and Mach's "thought economy" principle;
- Methods of solving **ordinary and partial differential equations, functional and finite difference equations of discrete mathematics, singular and regular integral equations, and integro-differential** equations which appear in various boundary value problems of fracture mechanics ;

- Methods of solving miscellaneous boundary value problems of fracture mechanics including:asymptotic **methods of boundary layer, the method of self-similar - solutions, the matched asymptotic expansions, the method of characteristics, integral transform methods, the Wiener-Hopf and Noble-Jones methods of reduction to the Riemann and Hilbert problems** of the theory of functions of a complex variable, and the **perturbation method**;
- Method of **invariant integrals,** method of **singular and generalized solutions,** and method of **driving or configurational forces**, which were first developed in fracture mechanics;
- Methods of **fractal geometry, catastrophe theory or singular mappings;**
- **Numerical methods** including finite element and finite difference methods.

The most important stage in the study of a problem is the formulation of a predictive model; it should properly address the main challenges of the problem. From an explorer, this stage requires the knowledge of a certain critical amount of specific basic information about the underlying phenomenon (original data base), a logical comprehensive analysis of the information (data analysis), and as a result, an intuitive finding of the mathematical model which can properly describe the main features of the phenomenon. Depending on the goals of an investigation one model can fit better than another one. Inventing and using different models make up the arts and skills of experts in any field of knowledge. In brief, a good model should be:

- Logically self-consistent, that is, not contradicting itself;
- Adequate to a corresponding real phenomenon;
- Predictive, that is, capable of predicting events and situations outside an original data basis, within the limits of necessary accuracy;
- Consistent, that is, not contradicting the approved models in a common area of their action;
- Verifiable, that is, capable of being checked by test or mind;
- Practical, that is, simple and convenient for use;
- Economical, that is, containing a necessary minimum of new constants; and last but not the least
- Beautiful, that is, delivering aesthetic emotions to the inventor and users.

As a rule, independent explorers studying the same problem invent different models, and only one of them survives. A model can be powerful or less powerful depending on the breadth and depth of its predictions. It is recognized that many models can possibly describe a phenomenon, but there is only one which is best. Creation and search for the best models is the final goal of any science or knowledge. Those who know methods better are more successful in the model invention.

Fracture mechanics has influenced the fundamentals of solid matter physics. In this book we consider the application of basic methods and ideas of fracture mechanics to the following problems of solid matter physics.

— Surface energy of solids (Chapter 1);
— Fluctuations and the kinetic theory of fracture (Chapter 2);
— Crack nucleation (Chapter 3);
— Physics of sintering (Chapter 4);

— Point defects in solids (Chapter 5);
— Dislocation emission (Chapter 6);
— Relativistic electron beams in a solid (Chapter 7);
— Fractals in fracture of solids (Chapter 8).

More detailed information about the Chapter is given in Table of Contents.

Many people have, directly or indirectly, been involved in the creation of the book.

My wife Larisa Cherepanov typed most part of the book and prepared figures. Without her devotion to the project, the book could not have been written. My senior colleagues and former friends, first of all Lev Galin, Vadim Sokolovsky, Grigory Barenblatt and Leonid Sedov, influenced my way of thinking and style of work. Jock Eshelby, John Gilman, Jim Rice, Bernie Budiansky, John Hutchinson and others offered constructive criticism of some topics of the book. It is my duty to express my sincere gratitute to all of them .

My special thanks go to Professor Gladwell, Editor of this series, who generously read a draft of the entire manuscript. The help of Helen Rooney, Cathy Rustad, Maggie Zabalo and Caroline Liberty, is gratefully acknowledged. Many other individuals involved in the project are identified on the relevant pages of the book or in the bibliography.

Some results of investigations treated in the book were found thanks to NSF Grant MSS 9224936, NASA Grant 541803700, AFOSR Grant 59107, and U. S. Army SSDC Grant DASG 60-94-C-0032 funding author's research during 1991-1995.

CHAPTER 1

SURFACE ENERGY OF SOLIDS

The methods of fracture mechanics treated here are designed to define, calculate, and measure the surface energy of a solid, which generally depends on how the surface and material were manufactured, and on many other factors. Some new examples of the derivation of self-similar singular solutions to some non-linear partial differential equation systems of hyperbolic and elliptic type are given (Section 1.6 and 1.7). The work with the energy flow rate expressed by the invariant contour integral is shown for most important practical situations (Sections 1.3 - 1.10).

1.1. General Definition

Surface energy is the energy or work expended to create a surface in an inert environment or vacuum. (Interface surface energy is characterized by the work spent to create an interface). Generally, this work depends on the way the surface is created. Consider, for example, the following ways:

Transforming a liquid or gas into a solid (Chapter 1);
Sintering (Chapter 4);
Cracking a solid (Chapters 1-3, and 6);
Cutting a solid by a thin blade (Chapter 9);
Chemical or electrochemical deposition or dissolution of a solid; and
Cutting a solid by a thin beam of superluminal electrons (Chapter 7).

Other processes and ways of surface creation or manufacturing are possible; surface energy of the same chemical material can be thousand times greater for one way than that for another one.

It is a widespread belief that the surface energy (per unit area) is a physical material constant. In fact, it is a function of many factors including the manufacturing process. However, there is one important particular case in which the surface energy can be considered a physical material constant: when the surface was created in a reversible, or close to reversible, thermodynamic process. Many manufacturing technologies dealing with the transformation of a liquid or gas into a solid, cracking an ideally brittle material, sintering, or liquid metal embrittlement apply such processes.

1.2. Surface Energy As a Physical Constant

If a surface is a result of a reversible thermodynamic process, the surface energy of a solid matter is the portion of free energy of the solid accounting for a reversible restructuring of electron structure near a surface. This case is the one which has been treated most often in the physical literature (see, e.g., references[1-5]). In this case, the concepts of "surface and interfacial tensions" are useful. These concepts make sense when a matter is a liquid or a creeping solid, so that atoms possess a mobility and can create a thin membrane shell on the surface of the solid, which is extended by in-plane atomic forces. The principal tensile forces, S_1 and S_2, acting in the tangent plane of the shell along principal directions, 1 and 2, can be represented by the first terms of the Taylor series:

$$S_1 = \gamma_1^1 + \varepsilon_1(d\gamma / d\varepsilon_1), \quad S_2 = \gamma_2^1 + \varepsilon_2(d\gamma / d\varepsilon_2) \tag{1.1}$$

Here, ε_1 and ε_2 are the corresponding strains, and γ_1 and γ_2 are the forces of surface tension (per unit length) independent of strains. The second terms in (1.1) are negligibly small, and atomic forces in the shell are the same along all in-plane directions; hence

$$S_1 = S_2 = \gamma_1^1 = \gamma_2^1 = \gamma \qquad \text{(surface tension)} \tag{1.2}$$

To increase the surface area of the shell by dA, it is necessary to produce the reversible work, γdA; therefore, the surface energy is identically equal to the surface tension.

Interfacial tension and interface surface energy are introduced analogously. Adhesion energy expended to create an interface between materials 1 and 2 is equal to

$$\Gamma_{12} = \gamma_1 + \gamma_2 - \gamma_{12} \tag{1.3}$$

based on the energy conservation law. Here, γ_1 and γ_2 are surface energies of the materials, γ_{12} is the interface surface energy, and Γ_{12} is the adhesion energy (per unit area). The adhesion energy Γ_{12} is the main physical property characterizing the bonding strength of solid materials. When $\gamma_1 = \gamma_2$, and $\gamma_{12} = 0$, Γ_{12} is equal to the cohesion energy, $\Gamma_{11} = 2\gamma_1$. When material 1 is a liquid, Γ_{12} characterizes wetting. In this case, we have (Figure 1.1)

$$\gamma_{2v} = \gamma_{12} + \gamma_1 \cos\theta \tag{1.4}$$

Here, γ_1 and γ_{2v} is the surface tension of the liquid and solid respectively (along the border with a vapor), and θ is the contact angle between solid and liquid surfaces at the point where the solid, liquid, and vapor meet. Equation (1.4) is the equation stating the balance of tension forces at the contact point.

Substituting γ_{12} from (1.4) into (1.3) gives

$$\Gamma_{12} = \gamma_1 (1 + \cos\theta) + \gamma_2 - \gamma_{2v} \tag{1.5}$$

for the adhesion energy of a liquid-solid interface.

Equation (1.4) shows that, if $\gamma_{2v} > \gamma_{12} + \gamma_1$, the balance of tension forces at the contact point is impossible. In this case, a liquid drop spreads over a solid surface until the film thickness achieves the atomic size (ideal wetting). Equation (1.4) also shows that, if $\gamma_{2v} < \gamma_{12} - \gamma_1$, balance is again impossible; in this case, a liquid drop tends to form a ball (or balls) having a one-point contact with the solid (ideal unwetting).

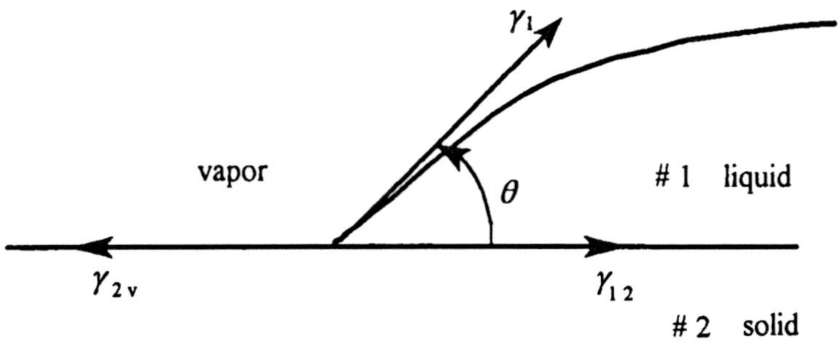

Figure 1.1. A contact of vapor, liquid and solid phases

When

$$\gamma_{12} - \gamma_1 < \gamma_{2v} < \gamma_{12} + \gamma_1, \tag{1.6}$$

the local balance of tension forces at the contact line is possible ($2\pi < \theta < 0$). In this intermediate case a liquid drop takes a certain shape with a finite contact area along the solid surface (wetting).

Surface tension, γ, and cohesion energy, 2γ, can be easily calculated for many materials based on their interatomic potentials and structures[1-6]. For example, the value of γ for a silicate glass of a given chemical composition with ionic bonds can be predicted[3] with an error less than 4% as compared to exact experimental measurements of γ by means of fracture mechanics.

The methods of fracture mechanics include creation of a cracked specimen with a crack, loading the specimen, and measurement of the load at which the crack initiates or grows. The stress intensity factor, K_I, is determined from the solution of the corresponding problem of the theory of elasticity, in terms of loads and geometrical parameters of the specimen and crack; this is sufficient in the case of local symmetry when the crack is of opening mode.

The stress intensity factor K_I is defined as

$$K_I = \lim_{\varepsilon \to 0} \left[\sigma_y \ (2\pi\varepsilon)^{1/2} \right], \tag{1.7}$$

where σ_y is the normal stress on the crack continuation, and ε is a small distance from the crack tip. Equation (1.7) is a corollary of the result obtained from the mathematical theory of elasticity, that the stress is infinite at the crack tip, proportionally to $1/\sqrt{\varepsilon}$, so that K_I characterizes the proportionality factor. A tremendous album of such solutions is available in the literature on fracture mechanics (see, e.g., Appendix in reference[3]). Using these solutions the critical value of K_I at the crack growth, referred to as fracture toughness, K_{IC}, can be found from the measurement of the critical load.

Surface energy is expressed through fracture toughness as[7]

$$\gamma = \frac{(1-\nu^2)K_{IC}^{\ 2}}{2E} \tag{1.8}$$

Here E is Young's modulus and ν is Poisson's ratio. From (1.8), one may easily find the surface tension, γ.

Specimens of any form and crack configuration can be used. Results of measurements of γ by the methods of fracture mechanics do not depend on the geometry of specimens and cracks and agree well with estimates and measurements of γ that are generally recognized in the physical literature[1-6], in the case if cracking is a reversible process. The value of γ in thermodynamically reversible processes of surface formation varies[1-6] from 1,000 to 3,000 ergs/cm^2 for metals, alloys, and glass; from 700 to 5,000 ergs/cm^2 for ceramics and silicon; from 100 to 1,000 ergs/cm^2 for ice, metal oxides, and salts; and from 5,000 to 7,000 ergs/cm^2 for diamond. The latter is probably the maximum possible value of surface tension for brittle solids. One should be aware that the value of γ may be very sensible to small variations of chemical composition of the material and the gaseous or liquid environment. Therefore, using the exact values of γ reported in numerous physical sources makes sense only if the conditions that existed during measurements are reproduced. Unfortunately, many authors do not pay proper attention to these effects and do not report the actual chemical compositions and environmental conditions.

Due to thermal expansion, surface tension of a condensed matter grows by cooling as a linear function of temperature[5]. This property has been used to predict the surface tension of a solid material by its surface tension at liquid state, where it is easily measured. The value of $d\gamma/dT$ has been measured and tabulated for many materials in the liquid state; it varies mostly from -0.1 to -0.5 ergs/(°/C·cm^2). If these data are not available, one can use the estimate,

$$\frac{d\gamma}{dT} = -\frac{(\gamma)_{T=T_M}}{T_C - T_M}, \tag{1.9}$$

because surface tension at critical temperature T_C when liquid and gas are indistinguishable equals zero (T_M is the melting temperature).

Adhesion energy characterizes the work spent to adhere or separate two materials. (Cohesion energy, 2γ, is the work to cohere or disconnect two pieces of the same material). The work expended to separate these materials may be absolutely different from that spent to adhere them. Also, cohesion energy may differ many times from fracture energy. It is only for ideally brittle materials that they coincide. Amorphous solids like glass, ice, ceramics, and some rocks are close to ideally brittle. In ideally brittle materials, cracks and fractures should be reversible, that is, they should disappear when all loads vanish. Practically, it is difficult to observe this even in glass because a crack surface is far from flat, so that small debris and projections prevent the crack from closing.

1.3. Surface Energy As a Process Dependent Property

Most materials including metals, alloys, composites and polymers are not brittle. The work spent to crack a material is much greater than that to cohere two pieces of the material. Many experiments pioneered by Orowan, Irwin, Zener, and Hollomon in the 40's and 50's showed that the surface energy measured by methods of fracture mechanics was usually 100 to 100,000 times greater than the cohesion energy. It appeared that the separation of a material by a crack and hence formation of a surface involves a great amount of plastic strain related to dislocation emission and yielding (in crystalline and polycrystalline materials[3]), transverse shear interface slips (in composites[6]), and restructuring of the molecular chain structure (in polymers[3]).

Investigation of surface energy of a solid in many different processes of surface creation accompanied by large amounts of plastic strain has required an approach independent of constitutive equations of a solid, that is, valid for any solids, elastic or inelastic. Such an approach to surface energy suggested by the present author[8] in 1967 has provided the following general equation for the surface energy of a solid:

$$\gamma = \frac{1}{2}\lim_{\varepsilon\to 0}\oint_{\Sigma_\varepsilon}\left[n_1(K+W) - T_1\frac{\partial u_1}{\partial x_1} - T_2\frac{\partial u_2}{\partial x_1} - T_3\frac{\partial u_3}{\partial x_1}\right]ds \tag{1.10}$$

Here:

$K = \frac{1}{2}\rho\left(v_1^2 + v_2^2 + v_3^2\right) \equiv \frac{1}{2}\rho v_i v_i$, kinetic energy per unit volume;

$W = \int \sigma_{ik} d\varepsilon_{ik} \equiv \int(\sigma_{11}d\varepsilon_{11} + 2\sigma_{12}d\varepsilon_{12} + \sigma_{22}d\varepsilon_{22} + 2\sigma_{23}d\varepsilon_{23} + 2\sigma_{13}d\varepsilon_{13} + \sigma_{33}d\varepsilon_{33})$,

deformation work per unit volume;

$T_i = \sigma_{ij}n_j \equiv \sigma_{i1}n_1 + \sigma_{i2}n_2$ (where i = 1,2,3), tractions on Σ_ε;

Σ_ε, an arbitrary closed path encircling a point 0 at the moving front of a crack in the plane, $x_3 = 0$, which is perpendicular to the crack front at point 0 (Figure 1.2);

ds, an arc element of Σ_ε;

n_1 and n_2, coordinates of unit normal vector to Σ_ε in the plane, $x_3 = 0$;

x_1, x_2, and x_3, Cartesian coordinates with the origin, point 0, the x_1 axis being a crack continuation and the x_3 axis along the crack front;

u_1, u_2, and u_3, displacement components;

v_1, v_2, and v_3, material velocity components;

σ_{11}, σ_{12}, σ_{22}, σ_{13}, σ_{23}, and σ_{33}, Cartesian stress tensor components;

ε_{11}, ε_{12}, ε_{22}, ε_{13}, ε_{23}, and ε_{33}, Cartesian strain tensor components; and

$\lim_{\varepsilon\to 0}$ means that Σ_ε shrinks to point 0.

The Equation (1.10) is valid if the crack grows along its continuation, that is, along the x_1 axis of Figure 1.2. For other paths, the equation of surface energy can be found in Ref.[3]. The latter case is important for mixed mode fracture.

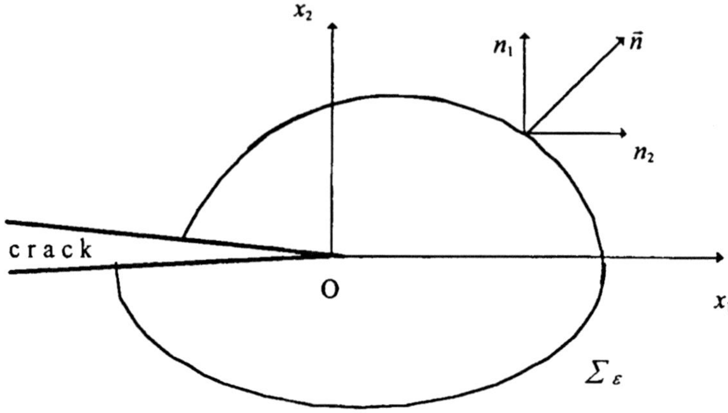

Figure 1.2. A closed path Σ_ε encircling the crack tip O

For brevity, it is convenient to use tensor notation, as shown above. A repeated index means summation on the index, and comma means differentiation on the corresponding Cartesian coordinate, so that $u_{i,k} = \partial u_i/\partial x_k$; $2\varepsilon_{ik} = u_{i,k} + u_{k,i} \equiv \partial u_i/\partial x_k + \partial u_k/\partial x_i$. With tensor notation, Equation (1.10) for surface energy can be written as

$$\gamma = \frac{1}{2} \lim_{\varepsilon \to o} \oint_{\Sigma_\varepsilon} \left[n_1(W + K) - T_i u_{i,1} \right] ds \qquad (i = 1,2,3) \qquad (1.11)$$

Equation (1.10) or (1.11) expresses the law of energy conservation, and 2γ the energy flow rate into the crack tip.

This equation of surface energy is valid for arbitrary solids. It can be simplified for particular constitutive equations of the solid. As a result of these calculations, a more convenient equation was derived in many particular cases[3,6]. Hutchinson, Rice, Atluri, Liebowitz, Nishioka, Morozov, Nikishkov and many other authors have used Equation (1.11) directly in computerized calculations of γ.

Equations (1.10) and (1.11) hold for interface cracking as well. As the direct corollary of the energy conservation law, Σ_ε may be chosen arbitrarily. That is why the integral in Equation (1.10) was called invariant[3,8]. In analytical work, it is most convenient to take Σ_ε in the form of a circle centered at 0, or a narrow symmetrical rectangle with 0 as the center of symmetry. Also, Equations (1.10) and (1.11) hold over an arbitrary contour Σ_ε (and not only shrinking to 0), if the crack tip, the point 0, moves steadily at a constant speed and W, K, T_i and u_i are functions of x_1 and x_2 inside Σ_ε in the moving coordinate frame (steady-state field).

In Section 1.4 below, we provide a summary of some exact analytical calculations based on Equation (1.11) for different solids and configurations often met in practice.

The surface energy characterizes the strength of bonding along the crack path during the crack growth and is usually called the effective surface energy or effective interface surface energy, fracture energy or work, adhesion energy or work, or cohesion energy or work. It is characteristic of material separation or debonding. For the connection or bonding of two pieces of the same material by manufacturing, the value of surface energy may be vastly different.

1.4. Adhesion Energy

Consider debonding of an interface between a thin film and a substrate, a thin plate and substrate, a thin film and plate, two different thin films, two different thin plates, and interface debonding in a multilayer thin shell. The geometry of a material and crack may be arbitrary. We consider the neighborhood of a point 0 at the crack front, the dimension of which is great as compared to the thickness of a plate or film, but small compared to the characteristic in-plane dimension of the plate or film, or to the radius of shell curvature at 0. The sides of the interface crack in the neighborhood of 0 are assumed to be free of tractions.

***A thin film on a substrate** (Figure 1.3)*

Suppose a momentless coating (membrane) was deposited or bonded to a solid substrate. Designate as Γ_c the adhesion energy per unit area. The adhesion energy in this case is equal to[9,6]

$$\Gamma_c = \frac{1}{2} N_1 \left(\frac{\partial u_2}{\partial x_1} \right)^2 = \frac{1}{2} N_1 \alpha^2 \tag{1.12}$$

Here, $N_1 = t\sigma$ is the in-plane force in the membrane at 0 normal to the crack front, t is the thickness, σ is tensile stress in the membrane at 0, and $\partial u_2/\partial x_1 = \alpha$ is the slope angle of the membrane at 0 assumed to be small. The x_3 axis is taken along the crack front at 0, the x_2 axis is perpendicular to the surface at 0, and the x_1 axis is the direction of the crack growth at 0.

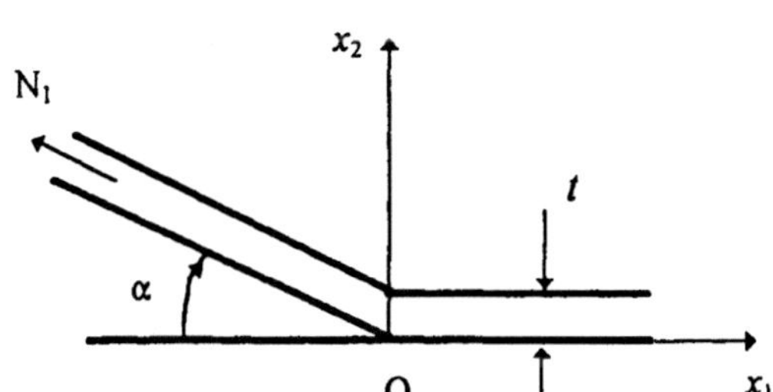

Figure 1.3. Debonding of a membrane bonded to a solid

Measuring α and σ at the growth of debonding at 0 allows one to calculate the adhesion energy at the point by means of (1.12). If the conditions of bonding were identical at all points of the interface during manufacture it is natural to assume that adhesion energy will be constant along the bonding interface, for arbitrary α and σ, for arbitrary shape of a solid, coating, and crack; it characterizes the strength of bonding.

Equation (1.12) was derived from the energy conservation law written in the following form

$$\oint_{\Sigma}(n_1 W - T_i u_{i,1})ds = 0 \tag{1.12a}$$

Here,

$$\Sigma = \Sigma_\varepsilon \cup \Sigma_+ \cup \Sigma_- \cup \Sigma_1 \cup \Sigma_2 \cup \Sigma_3 \cup \Sigma_4 \cup \Sigma_5 \cup \Sigma_6,$$

where Σ_ε shrinks to O as in Equation (1.11) and Figure 1.2;

Σ_+ $(x_2 \to +0,\ +\varepsilon < x_1 < R,\ \varepsilon \to 0, R \to \infty)$;
Σ_- $(x_2 \to -0,\ +\varepsilon < x_1 < R,\ \varepsilon \to 0, R \to \infty)$;
Σ_1 $(x_1 = R,\ R \to \infty,\ 0 < x_2 < t)$;
Σ_2 $(-R < x_1 < +R,\ x_2 = t,\ R \to \infty)$;
Σ_3 $(x_1 = -R,\ 0 < x_2 < t,\ R \to \infty)$;
Σ_4 $(x_2 \to +0,\ -R < x_1 < \varepsilon,\ \varepsilon \to 0,\ R \to \infty)$;
Σ_5 $(x_2 \to -0,\ -R < x_1 < \varepsilon,\ \varepsilon \to 0,\ R \to \infty)$; and
Σ_6 $(x_1{}^2 + x_2^2 = R^2,\ x_2 < 0,\ R \to \infty)$.

The closed path Σ and all its subsections are passed in such a direction that the material is on the left hand side. Only the case when the velocity of front O with respect to the material is much less than the sound speed, is considered (in this case $K<<W$).
By definition, the adhesion energy is

$$\Gamma_c = \lim_{\varepsilon \to 0} \oint_{\Sigma_\varepsilon}(n_1 W - T_i u_{i,1})\, ds \tag{1.12b}$$

Along the interface $x_2 = 0$, $0 < x_1 < N$ we have: $[u_{i,1}] = 0$ because of nonslip condition, and $[T_i] = 0$ because of equilibrium condition where the brackets [...] designate the discontinuity of the quantity in the brackets. Also, $n_1 = 0$ along Σ_+ and Σ_-. From here, it follows that

$$\int_{\Sigma_+ \cup \Sigma_-}(n_i W - T_i u_{i,1})ds = 0 \tag{1.12c}$$

Along Σ_1 and Σ_6 the stresses and strains have the order of $1/R$, so that the integrand function in (1.12a) has the order of $1/R^2$. Therefore, we have

$$\int_{\Sigma_1 \cup \Sigma_6} (n_1 W - T_i u_{i,1}) ds = 0 \tag{1.12d}$$

Along Σ_2, Σ_4 and Σ_5 we have $n_1 = 0$ and $T_i = 0$, so that

$$\int_{\Sigma_2 \cup \Sigma_4 \cup \Sigma_5} (n_1 W - T_i u_{i,1}) ds = 0 \tag{1.12e}$$

Along Σ_3, in the approximation of momentless membrane, we have $ds = -dx_2$, $n_1 = -1$, $T_1 = N_1 / t$, $T_2 = \alpha N_1 / t$, $T_3 = 0$, $u_{1,1} = 0$, $u_{2,1} = -\alpha$, and $W = N_1\, \alpha^2 / (2t)$, so that

$$\int_{\Sigma_3} (n_1 W - T_i u_{i,1}) ds = -\int_0^t \left[-\frac{N_1 \alpha^2}{2t} + \frac{N_1 \alpha^2}{t} \right] dx_2 = -\frac{1}{2} N_1 \alpha^2 \tag{1.12f}$$

Gathering Equations(1.12a) to (1.12f) yields $\Gamma_c = N_1 \alpha^2 / 2$, that is, Equation (1.12).

How close is a real coating to the momentless membrane depends on dimensionless number $\lambda = EtN_1^{-1}$ where E is Young's modulus of the coating. The less λ is, the closer the coating is to the momentless membrane.

Below in the chapter, analogous calculations of contour integrals are omitted.

If a membrane is made of several layers, Equation (1.12) holds provided

$$N_1 = \Sigma t_i \sigma_i \; . \tag{1.13}$$

Here, t_i and σ_i are the thickness and stress for the ith layer.

***A thin plate on a substrate** (Figure 1.4)*

Suppose a thin linearly-elastic coating capable of transferring bending moments was bonded to a solid substrate. Also, assume that normal and shear in-plane forces in the plate are negligibly small (pure bending at 0). In this case, the adhesion energy is equal to[9,6]

$$\Gamma_c = \frac{1}{2} k_b \left(\frac{\partial^2 u_2}{\partial x_1^2} \right)^2 \tag{1.14}$$

Here, k_b is the flexural rigidity or bending stiffness of the beam at 0, and $\partial^2 u_2 / \partial x_1^2$ is the curvature of the neutral line of the beam at 0; the deflections and slopes being considered small. Equation (1.14) is valid for a multilayer plate, if k_b is the bending stiffness of the composite plate.

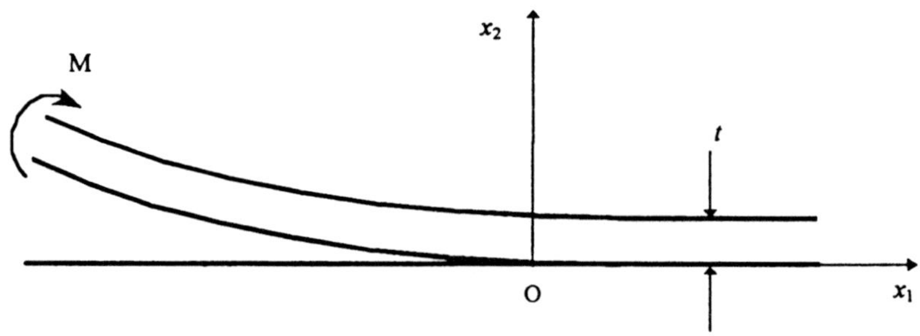

Figure 1.4. Debonding of a thin beam bonded to a solid

If the in-plane stresses are considerable, it is necessary to perform the superposition of bending and plane stress in the plate (Figure 1.5). In this case, the adhesion energy is equal to[9,6]

$$\Gamma_c = \frac{1}{2}\left(M\kappa + N_1\varepsilon_1 + S_{13}\gamma_{13}\right) \tag{1.15}$$

Here:

$$\kappa = \frac{\partial^2 u_2}{\partial x_1^2}, \quad \varepsilon_1 = \frac{\partial u_1}{\partial x_1}, \quad \gamma_{13} = \frac{\partial u_1}{\partial x_3} + \frac{\partial u_3}{\partial x_1},$$

M is the bending moment at 0, u_1, u_2, and u_3 are the displacement components along x_1, x_2, and x_3, ε_1 and γ_{13} are normal and shear strains, N_1 and S_{13} are the normal and shear forces on the plane perpendicular to the x_1 axis (all variables are taken at 0).

If $M = 0$, there are only in-plane peeling forces of plane stress that cause debonding at 0. It is relevant to mention here the names of debonding types in three particular cases (Figure 1.5):

— Opening mode cracking or debonding ($M \neq 0$, $N_1 = 0$, $S_{13} = 0$)

— Transverse shear cracking or debonding ($M = 0$, $N_1 \neq 0$, $S_{13} = 0$)

— Screw shear cracking or debonding ($M = 0$, $N_1 = 0$, $S_{13} \neq 0$).

Equations (1.14) and (1.15) are valid for linearly elastic materials. This condition may be met only within the framework of small scale yielding, that is, when the dimension of the zone of inelastic deformations near the crack end is much less than the plate thickness.

For an arbitrary inelastic material, the adhesion energy at 0 is equal to[9,6]

$$\Gamma_c = M\kappa + N_1\varepsilon_1 + S_{13}\gamma_{13} - W_D \ , \tag{1.16}$$

where

$$W_D = \int M d\kappa + N_1 d\varepsilon_1 + S_{13} d\gamma_{13} \equiv$$
$$\equiv \int_0^\tau \left[M(\tau)\dot{\kappa}(\tau) + N_1(\tau)\dot{\varepsilon}_1(\tau) + S_{13}(\tau)\dot{\gamma}_{13}(\tau) \right] d\tau \quad . \tag{1.17}$$

Here, W_D is the strain work per unit area of the plate in plan view, τ is time, $\dot{\kappa}(\tau), \dot{\varepsilon}_1(\tau)$ and $\dot{\gamma}_{13}(\tau)$ are the rates of the corresponding quantities. The value of W_D depends on the loading path.

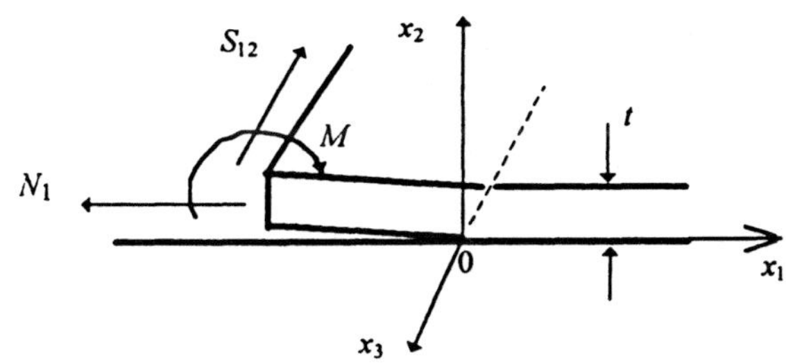

Figure 1.5. Debonding of a thin plate under combined bending and plane stress (The plate is bonded to a substrate)

The physical meaning of Γ_c and W_D for an inelastic material can be clarified by considering pure bending when N_1 and S_{13} vanish at 0. In this case[9,6]

$$\Gamma_c = M\kappa - W_D, \qquad W_D = \int M d\kappa \equiv \int_0^\tau M(\tau)\dot{\kappa}(\tau) d\tau \tag{1.18}$$

Consider the M-κ diagram of pure bending of a beam (Figure 1.6). To obtain the diagram, we bend the beam by a moment, $M = M(\tau)$, and measure the corresponding curvature, $\kappa = \kappa(\tau)$ (or vice versa), then eliminate τ and find the M-κ diagram. The loading path prescribed by the function $M(\tau)$ should reproduce that of the beam near the point 0.

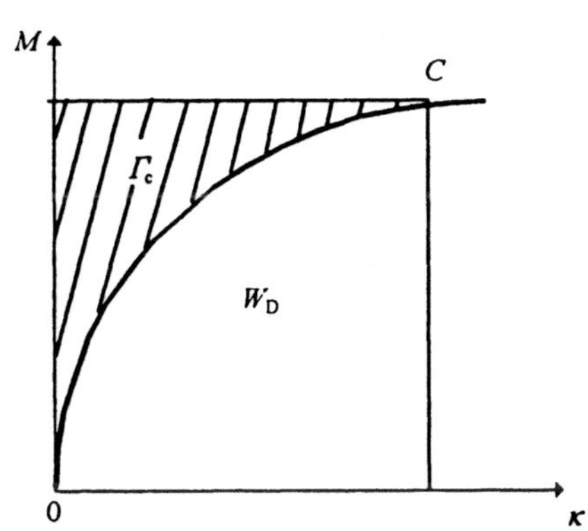

Figure 1.6. The M-κ diagram of pure bending of an inelastic beam

The critical position C of the crack initiation on the M-κ diagram should be found from an independent experiment with a cracked specimen of the type depicted in Figure 1.4. It is clear that W_D is equal to the area under the curve $M(\kappa)$, and Γ_c is the complementary area shaded in Figure 1.6. Here and below we accept that initial stresses and strains equal zero, if there are no special indications.

Equation (1.15) follows from Equations (1.16) and (1.17) in the case of linearly elastic materials, when the diagrams of the type of Figure 1.6 are linear and path-independent.

Equations (1.16) and (1.17) are probably most practical and accurate because debonding of thin films and coatings is, almost always, accompanied by plastic and creep strains occupying a region which is comparable or greater than the film thickness.

Figure 1.6 seems to imply that the adhesion energy depends on the loading path. However, the position of point C can be determined only from independent experiments with cracked specimens, which can provide a final answer to the question. For inelastic materials, the adhesion energy can also depend on the plate thickness. Equations (1.14) - (1.18) hold, as well, for multilayer thin films, beams and plates.

It is unclear whether adhesion energy can depend on the type of loading. For example, will it differ for pure bending and plane stress peeling? Generally speaking, yes, but it is only experiments based on the derived equations that can give a definitive answer.

Two different bonded membranes *(Figure 1.7)*

Consider two thin bonded momentless films made of different materials of thickness, t_1 and t_2. Debonded portions of the films are stretched by forces N_1 and N_2 inclined at angles α_1 and α_2 to the bonded portion.

In this case, the adhesion energy is equal to[9,6]

$$\Gamma_c = N_1\alpha_1^2 + N_2\alpha_2^2 = N_1\alpha_1(\alpha_1 + \alpha_2) \quad , \tag{1.19}$$

because $N_1 \sin\alpha_1 = N_2 \sin\alpha_2$, and α_1 and α_2 are small. The quantities on the right side of Equation (1.19) correspond to the debonding growth. The N_1 and N_2 are taken at 0. Equation (1.19) is also valid for multilayer thin films

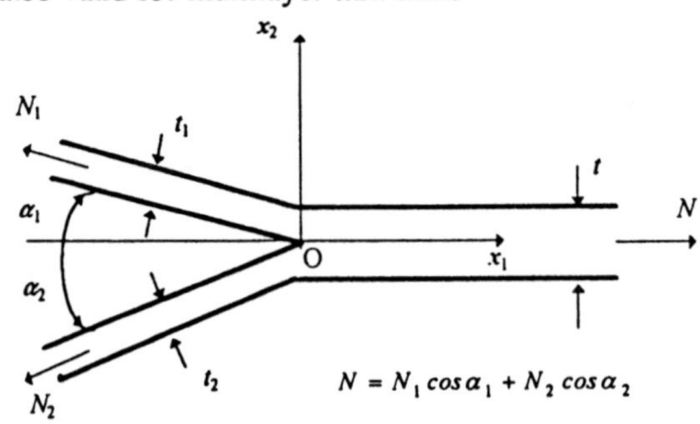

Figure 1.7. Debonding of two bonded thin films

Two different bonded beams (Figure 1.8)

Suppose two linearly elastic beams made of different materials of thickness, t_1 and t_2, are bent by moments, M_1 and M_2. For opening mode debonding, the adhesion energy is equal to[9,6]

$$\Gamma_c = \frac{1}{2}(M_1\kappa_1 + M_2\kappa_2) - M\kappa \tag{1.20}$$

Here, κ_1 and κ_2 are the corresponding curvatures of neutral lines of two debonded beams, κ is the curvature of the composite bonded beam bent by moment, $M = M_2 - M_1$.

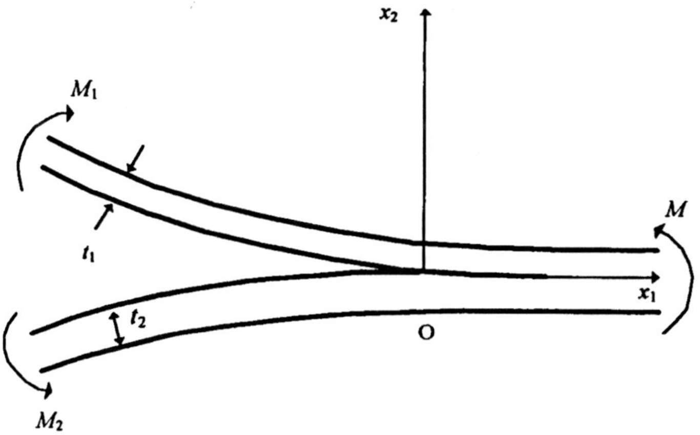

Figure 1.8. Debonding of two bonded beams

In the general case of the debonding of a linearly elastic bimaterial plate under combined bending and plane stress, the adhesion energy is provided by[9,6]

$$\begin{aligned}\Gamma_c = &\frac{1}{2}\left[M_1\kappa_1 + (N_1\varepsilon_1)_1 + (S_{13}\gamma_{13})_1\right] + \\ &+\frac{1}{2}\left[M_2\kappa_2 + (N_1\varepsilon_1)_2 + (S_{13}\gamma_{13})_2\right] - \\ &-(M\kappa + N_1\varepsilon_1 + S_{13}\gamma_{13}),\end{aligned} \tag{1.21}$$

where

$$M = M_2 - M_1\,,\ N_1 = (N_1)_1 - (N_1)_2\,,\ S_{13} = (S_{13})_1 - (S_{13})_2. \tag{1.22}$$

Here, $(N_1)_1$, $(N_1)_2$ and N_1 are the in-plane normal forces in the debonded first and second plates and in the bonded bimaterial plate respectively; $(\varepsilon_1)_1$, $(\varepsilon_1)_2$, and ε_1 are the corresponding normal strains; $(S_{13})_1$, $(S_{13})_2$, and S_{13} are the corresponding in-plane shear forces; and $(\gamma_{13})_1$, $(\gamma_{13})_2$, and γ_{13} are the corresponding shear strains.

In the most general case of an arbitrary inelastic behavior of two bonded dissimilar plates by debonding, the adhesion energy is given by[9,6]

$$\begin{aligned}\Gamma_c &= \left(M\kappa + N_1\varepsilon_1 + S_{13}\gamma_{13} - W_D\right)_1 + \\ &+\left(M\kappa + N_1\varepsilon_1 + S_{13}\gamma_{13} - W_D\right)_2 - \\ &-\left(M\kappa + N_1\varepsilon_1 + S_{13}\gamma_{13} - W_D\right)_0\end{aligned} \tag{1.23}$$

Here, the meaning of quantities in brackets is the same in Equations (1.16) and (1.17) above; the subscripts 1,2, and 0 at a bracket designate that the quantity in the bracket corresponds to the first and second debonded plates and to the composite bonded plate.

All quantities on the right side of Equations (1.20) - (1.23) are taken at 0 (the debonding front) at the moment of debonding growth. These equations are also valid for debonding of two multilayer thin plates bonded to each other.

A thin film deposited on a thin plate *(Figure 1.9)*

Consider debonding of a soft thin coating bonded to a thin plate. Assume that the coating can transfer only tensile in-plane stresses (membrane), and the plate can transfer both in-plane loads and bending moments (plane stress plus bending). Consider first pure bending of a linearly elastic plate.

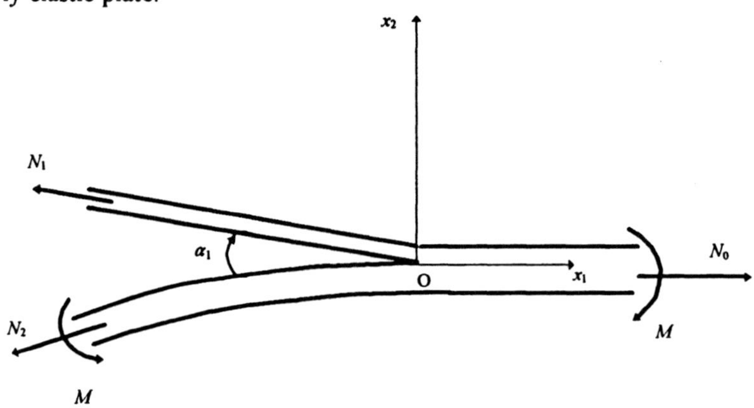

Figure 1.9. Debonding of a membrane bonded to a thin plate

The adhesion energy by debonding is[9,6]

$$\Gamma_c = N_1\alpha_1^2 + \frac{1}{2}M\left(\kappa_2 - \kappa_0\right) \tag{1.24}$$

Here, N_1 is the tension force in the membrane, α_1 is the slope angle of the membrane at 0, M is the bending amount, κ_2 is the curvature of the neutral line of the debonded plate, and κ_0 is the curvature of the neutral line of the bonded composite plate at 0.

Consider now superposition of plane stress and bending in a linearly elastic plate. In this case, we have[9,6]

$$\Gamma_c = N_1\alpha_1^2 + \frac{1}{2}M(\kappa_2 - \kappa_0) + \frac{1}{2}(N_2\varepsilon_2 - N_0\varepsilon_0) + \frac{1}{2}S_{13}(\gamma_2 - \gamma_0) \qquad (1.25)$$

where $N_0 = N_1 + N_2$

Here, N_1, N_2 and N_0 are the normal forces in the x_1 direction in the two debonded plates and the bonded plate respectively, ε_2 and ε_0 are the normal strains in the x_1 direction in the debonded and bonded plates respectively, and S_{13}, γ_2 and γ_0 are the respective shear force and shear strains γ_{13} in the debonded and bonded parts of the plates at 0.

For an arbitrary material behavior of the plate, the equation for adhesion energy is[9,6]

$$\Gamma_c = N_1\alpha_1^2 + (M\kappa_2 + N_2\varepsilon_2 + S_{13}\gamma_2 - W_2) - (M\kappa_0 + N_0\varepsilon_0 + S_{13}\gamma_0 - W_0) \qquad (1.26)$$

Here, W_2 and W_0 are strain energies (per unit area of plate in plan view) of the corresponding debonded and bonded plates at 0, the physical meaning of which was explained above in Equation (1.17) using Figure 1.6.

1.5. Brittle Interface Cracking/Debonding

Consider interface debonding/cracking in a scale which is small compared to the characteristic linear dimensions of the material or structure. For example, in the cases treated in Section 1.4 above, it is the plate or film thickness that plays the role of the characteristic linear dimension. The neighborhood of point 0 on the front of an interface opening mode crack is depicted on Figure 1.10 (two half-spaces bonded along one half of the boundary and debonded along the other half). This case has been well studied in the literature (See, e.g., reviews by Hutchinson and Suo[10] and Deng[11]).

Summarized here are only some basic points. It is also convenient to provide upper bound estimates using the particular limiting case when one of the half-spaces is rigid.

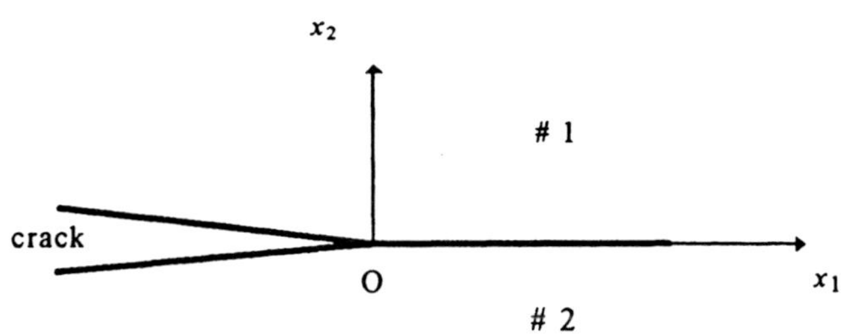

Figure 1.10. Debonding of two bonded solids of different materials

Stresses in the problem of Figure 1.10 differ from the corresponding stresses in a homogeneous material (when the elastic constants of the half-spaces coincide) by a comparatively small amount, less than 10% everywhere, except for a small region near point 0. This conclusion directly follows from comparison of Equations (3.124) in the book[3] for the stresses in the inhomogeneous case with Equations (3.44) - (3.45) in the same book for the stresses in the homogeneous case taking into account the physical condition (3.117).

The dimension d of the region around 0 where the departure from the homogeneous material stresses is significant, is much less than 0.1 percent of the length of the interface crack. This was established first by Abramov[12] in 1937 (see also page 81 in[3]). Later, in 1945, Galin[13] introduced the local zone of closing slip near 0 (to avoid oscillations) and calculated the zone size using the accurate solution of the corresponding mixed boundary value problem of combined slip/bonding with slip on the edge*. Therefore, the d for interface nanocracks has an order of interatomic spacing, and the d for microcracks or macrocracks is usually less or comparable to grain size, crack surface roughness and other random structural irregularities neglected in the theory of the crack size scale. For incompressible solids, d is always equal to zero. Thus, this local effect can be ignored in the calculation of the adhesion energy by using the contour Σ_ε in Equation (1.10) outside this region, to give[14]

$$\Gamma_c = \frac{1}{4}\left(\frac{1}{\mu_1} + \frac{1}{\mu_2}\right)K_{III}^2 + \frac{(\mu_1 + \mu_2)(\mu_2 + \mu_1\kappa_2)}{4\mu_1\mu_2[\mu_2(1+\kappa_1) + \mu_1(1+\kappa_2)]}\left(K_I^2 + K_{II}^2\right) \tag{1.27}$$
$$(\kappa_1 = 3 - 4\nu_1, \kappa_2 = 3 - 4\nu_2)$$

Here, μ_1, μ_2, ν_1 and ν_2 are shear module and Poisson's ratios of the corresponding materials, and K_{I}, K_{II}, and K_{III} are the stress intensity factors to be found from the solution of the corresponding boundary value problems. They are defined as:

$$K_I = \sigma_{22}\sqrt{2\pi x_1}, \quad \text{when } l >> x_1 >> d,\ x_2 = 0 \quad \text{(opening mode crack)} \tag{1.28}$$
$$K_{II} = \tau_{12}\sqrt{2\pi x_1}, \quad \text{when } l >> x_1 >> d,\ x_2 = 0 \quad \text{(transverse shear crack)} \tag{1.29}$$
$$K_{III} = \tau_{32}\sqrt{2\pi x_1}, \quad \text{when } l >> x_1 >> d,\ x_2 = 0 \quad \text{(screw shear crack)} \tag{1.30}$$

where l is the crack length.

Similar equations were derived for the adhesion energy of arbitrary anisotropic bonded solids [3,11]. When $\mu_1 = \mu_2 = \mu$ and $\nu_1 = \nu_2 = \nu$, Equation (1.27) gives

$$\Gamma_c = \frac{1}{2\mu}\left[(1-\nu)\left(K_I^2 + K_{II}^2\right) + K_{III}^2\right] , \tag{1.31}$$

which was derived earlier in Ref.[15] This is the equation of cohesion energy (or adhesion energy, if the solids are bonded by a very thin layer of an adhesive material , the thickness of the layer being neglected).

The corresponding equation of cohesion energy for the dynamic opening mode crack is provided by

* In 1959, V. V. Sokolovsky informed the present author that he came up with the idea even earlier, but L. A. Galin was luckier in calculations.

$$\Gamma_c = \frac{K_I^2(1+\nu)m^2\left[1-m^2(1-2\nu)/(2-2\nu)\right]^{1/2}}{8ER(m,\nu)} , \tag{1.32}$$

where

$$R(m,\nu) = \left[(1-m^2)\left(1-\frac{1-2\nu}{2-2\nu}m^2\right)\right]^{1/2} - \left(1-\frac{1}{2}m^2\right)^2 , \quad m = \frac{V}{c_2}. \tag{1.33}$$

Here, V is the crack speed, and c_2 is the speed of shear waves. When $m \to 0$, Equation (1.32) reduces to Irwin's equation (1.8). Equation (1.32) was first derived in Refs[16,17], and later by many authors.

When $V \to V_R$ ($m \to m_R$) which is the Rayleigh speed, $R(m,\nu) \to 0$ and the surface energy tends to infinity, $\Gamma_c \to \infty$. This implies that the crack speed is always less than the Rayleigh speed and cannot approach it.

Notice that, in contrast, the surface energy by cutting an elastic solid tends to zero when the speed of a thin blade tends to the Rayleigh speed (see Chapter 9 below).

1.6. Elastic-Plastic Solids: Stresses At the Crack Tip

There is a plastic region near the interface crack tip in an elastic-perfectly plastic material. The size of the region has the order K^2/τ_1^2 and K^2/τ_2^2 in the first and second materials, where K is the stress intensity factor computed for ideally elastic material pair with zero region size, and τ_1 and τ_2 are the yield stresses of the corresponding materials.

To estimate stresses at a crack tip inside the plastic region, it is convenient to use the model of a solid as an incompressible elastic - perfectly plastic material obeying the Huber-Mises yielding criterion. The second invariant of stress deviator in the material is a material constant in this model. For a homogeneous material, the stresses at the crack tip were found by Hill and Sokolovsky to be the same as those which Prandtl had earlier discovered at the punch edge.[18,19]

Consider the solution to the problem for an inhomogeneous material when two solids made of different elastic-plastic materials are bonded, with an interface debonding zone being free of tractions. Mathematically, the problem of stresses at the crack front is reduced to the following boundary value problem in plane strain:

$$\frac{\partial \sigma_x}{\partial x} + \frac{\partial \tau_{xy}}{\partial y} = 0 \qquad (|x| < \infty, |y| < \infty), \tag{1.34}$$

$$\frac{\partial \tau_{xy}}{\partial x} + \frac{\partial \sigma_y}{\partial_y} = 0 \qquad (|x| < \infty, |y| < \infty), \tag{1.35}$$

$$(\sigma_y - \sigma_x)^2 + 4\tau_{xy}^2 = \begin{cases} 4\tau_1^2 & \text{when } y > 0 \\ 4\tau_2^2 & \text{when } y < 0 , \end{cases} \tag{1.36}$$

$$\sigma_y = \tau_{xy} = 0 \qquad \text{when } y = 0,\ x < 0 \tag{1.37}$$

$$[\sigma_y] = [\tau_{xy}] = 0 \qquad \text{when } y = 0,\ x > 0\ . \tag{1.38}$$

Here, x and y are Cartesian rectangular coordinates, σ_x, τ_{xy} and σ_y are stresses, and τ_1 and τ_2 are shear stress yield limits of the first and second solids respectively. The brackets [...] denote the discontinuity of the quantity in the brackets along the bonded interface.

The non-linear equation system (1.34) - (1.36) belongs to the hyperbolic type; there are two families of characteristic lines issuing from a free boundary at 45 degrees to the tangent at every point. From here, it follows that, in the isosceles right-angled triangles AOB and AOF having traction-free crack sides as hypotenuses, the only stress field is:

$$\sigma_y = \tau_{xy} = 0, \quad \sigma_x = 2\tau_1 \quad \text{in } \Delta\,\text{ABO}, \quad \sigma_x = 2\tau_2 \quad \text{in } \Delta\,\text{AOF} \tag{1.39}$$

The positive sign of σ_x in both triangles means *a posteriori* that we consider the opening mode tensile crack (this sign in AOB is correct for the model under study, but is incorrect for an elastic-brittle crack!). The crack tip, O, is the singular point of the boundary value problem.

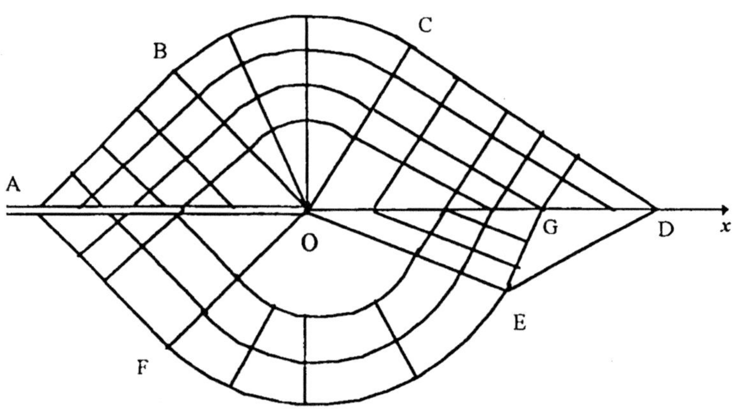

Figure 1.11. The interface crack tip between two perfectly plastic materials (characteristic lines in the crack tip neighborhood correspond to the first possible solution)

The solution will be sought in the following form:

a. In curvilinear triangles OBC and OFE, the stresses σ_r, σ_θ and $\tau_{r\theta}$ are functions of θ only.

b. In rectangular triangles OCD and OGE, the stresses are constants.

Here, r and θ are polar coordinates centered at 0, and σ_r, σ_θ and $\tau_{r\theta}$ are the corresponding stresses.

The equilibrium and yield condition equations in polar coordinates are:

$$\frac{d\tau_{r\theta}}{d\theta} + \sigma_r - \sigma_\theta = 0 \tag{1.40}$$

$$\frac{d\sigma_\theta}{d\theta} + 2\tau_{r\theta} = 0 \tag{1.41}$$

$$(\sigma_r - \sigma_\theta)^2 + 4\tau_{r\theta}^2 = \begin{cases} 4\tau_1^2 & \text{when} \quad y > 0 \\ 4\tau_2^2 & \text{when} \quad y < 0 \end{cases} \tag{1.42}$$

Consider the stress field in the upper region, y >0. Introduce the new function, f (θ):

$$\begin{aligned} \sigma_r - \sigma_\theta &= 2\tau_1 \sin f(\theta) \\ \tau_{r\theta} &= \tau_1 \cos f(\theta) \end{aligned} \tag{1.43}$$

so that Equation (1.42) is satisfied automatically. Substituting (1.43) into (1.40) and (1.41) provides

$$\left(2 - \frac{df}{d\theta}\right) \sin f(\theta) = 0 \tag{1.44}$$

$$\frac{d\sigma_\theta}{d\theta} = -2\tau_1 \cos f(\theta). \tag{1.45}$$

Equation (1.44) has two possible solutions:

either 1. $\sin f(\theta) = 0$, giving $\cos f(\theta) = \pm 1$; (1.46)

or 2. $f = 2\theta + C_1$, $\sigma_\theta = -\tau_1 \sin(2\theta + C_1) + C_2$. (1.47)

The characteristic lines in ΔOBC and ΔOFE are formed by radial and concentric lines (only in the first solution), and in ΔOCD and ΔOGE by two rectangular grids of different orientations.

The corresponding stress fields are:

either 1. $\sigma_r = \sigma_\theta = \mp 2\tau_1\theta + C_3$, $\tau_{r\theta} = \pm\tau_1$; (1.48)

or 2. $\sigma_r = \tau_1 \sin(2\theta + C_1) + C_2$, $\sigma_\theta = -\tau_1 \sin(2\theta + C_1) + C_2$, $\tau_{r\theta} = \tau_1 \cos(2\theta + C_1)$ (1.49)

Here, C_1, C_2, and C_3 are arbitrary constants. In (1.48), either the upper or lower sign should be taken. By replacing τ_1 with τ_2, one can also get the solutions for ΔOFE.

Equations (1.43) and (1.46) imply

$$\sigma_\theta = \tau_1 \, , \, \tau_{r\theta} = \tau_1 \quad \text{along } \theta = \frac{3\pi}{4} \qquad (OB) \tag{1.50}$$

The stresses σ_θ and $\tau_{r\theta}$ should be continuous along OB. Matching the solutions (1.48) and (1.49) with this condition we find that in Δ OBC:

$$\text{either} \quad 1. \; \sigma_r = \sigma_\theta = -2\tau_1\theta + \tau_1\left(1 + \frac{3\pi}{2}\right), \quad \tau_{r\theta} = \tau_1 \; ; \tag{1.51}$$

$$\text{or} \quad 2. \; \sigma_r = \tau_1 \cos 2\theta + \tau_1 \, , \quad \sigma_\theta = -\tau_1 \cos 2\theta + \tau_1 \, , \quad \tau_{r\theta} = -\tau_1 \sin 2\theta \quad . \tag{1.52}$$

Similarly, we find that in Δ OFE:

$$\text{either} \quad 1. \; \sigma_r = \sigma_\theta = +2\tau_2\theta + \tau_2\left(1 + \frac{3\pi}{2}\right), \quad \tau_{r\theta} = -\tau_2; \tag{1.53}$$

$$\text{or} \quad 2. \; \sigma_r = \tau_2(1 + \cos 2\theta), \quad \sigma_\theta = \tau_2(1 - \cos 2\theta), \quad \tau_{r\theta} = -\tau_2 \sin 2\theta \, . \tag{1.54}$$

In the triangles OCD and OGE, the σ_x are different, and the σ_y and τ_{xy} are the same due to boundary condition (1.38). Using (1.36) we have:

$$\sigma_x = \begin{cases} \sigma_y - 2\sqrt{\tau_1^2 - \tau_{xy}^2} & \text{(OCD)} \\ \sigma_y - 2\sqrt{\tau_2^2 - \tau_{xy}^2} & \text{(OGE)} \end{cases} \tag{1.55}$$

It is assumed here that $\sigma_y > \sigma_x$ in these triangles — this is physically evident for opening mode tensile cracks. (Again, it is not valid in the elastic case!)

The stresses, σ_θ and $\tau_{r\theta}$, should be continuous along OC and OE. Designate: $\theta = \theta_1$ along OC and $\theta = -\theta_2$ along OE. From here, it follows for both solutions that:

1. *The first solution*

$$-2\tau_1\theta_1 + \tau_1\left(1 + \frac{3\pi}{2}\right) = \left(\sigma_y - 2\sqrt{\tau_1^2 - \tau_{xy}^2}\right)\sin^2\theta_1 + \sigma_y \cos^2\theta_1 - 2\tau_{xy}\sin\theta_1\cos\theta_1;$$

$$\tau_1 = \sqrt{\tau_1^2 - \tau_{xy}^2}\,\sin 2\theta_1 + \tau_{xy}\cos 2\theta_1;$$

$$-2\tau_2\theta_2 + \tau_2\left(1 + \frac{3\pi}{2r}\right) = \left(\sigma_y - 2\sqrt{\tau_2^2 - \tau_{xy}^2}\right)\sin^2\theta_2 + \sigma_y \cos^2\theta_2 + 2\tau_{xy}\sin\theta_2\cos\theta_2;$$

$$-\tau_2 = -\sqrt{\tau_2^2 - \tau_{xy}^2}\,\sin 2\theta_2 + \tau_{xy}\cos 2\theta_2. \tag{1.56}$$

2. *The second solution*

$$\tau_1(1-\cos 2\theta_1) = \left(\sigma_y - 2\sqrt{\tau_1^2 - \tau_{xy}^2}\right)\sin^2\theta_1 + \sigma_y\cos^2\theta_1 - 2\tau_{xy}\sin\theta_1\cos\theta_1;$$

$$-\tau_1\sin 2\theta_1 = \sqrt{\tau_1^2-\tau_{xy}^2}\,\sin 2\theta_1 + \tau_{xy}\cos 2\theta_1;$$

$$\tau_2(1-\cos 2\theta_2) = \left(\sigma_y - 2\sqrt{\tau_2^2 - \tau_{xy}^2}\right)\sin^2\theta_2 + \sigma_y\cos^2\theta_2 + 2\tau_{xy}\sin\theta_2\cos\theta_2;$$

$$\tau_2\sin 2\theta_2 = -\sqrt{\tau_2^2-\tau_{xy}^2}\,\sin 2\theta_2 + \tau_{xy}\cos 2\theta_2. \qquad (1.57)$$

There are four equations in every solution for determining θ_1, θ_2, σ_y, and τ_{xy}.

Excluding σ_y and τ_{xy} in the equation system (1.56), one can derive the following equations for the first solution:

1. *The first solution*

$$\lambda\cos 2\theta_1 = -\cos 2\theta_2; \qquad \lambda = \tau_1/\tau_2,$$

$$2(\theta_2 - \lambda\theta_1) - (1-\lambda)\left(1+\frac{3\pi}{2}\right) =$$

$$= 2\sin 2\theta_2\sin^2\theta_2 - 2\lambda\sin 2\theta_1\sin^2\theta_1 - \lambda\cos 2\theta_1(\sin 2\theta_1 + \sin 2\theta_2). \qquad (1.58)$$

The stresses, σ_y and τ_{xy}, in the triangles OCD and OGE are equal to:

$$\sigma_y = \tau_1\sin 2\theta_1 + \tau_1\left(1+\frac{3\pi}{2}\right) - 2\tau_1\theta_1, \quad \tau_{xy} = \tau_1\cos 2\theta_1\text{, when } \theta_1 > \theta > -\theta_2 \qquad (1.59)$$

When $\tau_1 = \tau_2$, from (1.58) and (1.59) it follows that $\theta_1 = \theta_2 = \pi/4$, and

$$\sigma_y = \tau_1(2+\pi),\ \sigma_x = \pi\tau_1,\ \tau_{xy} = 0\text{, when } |\theta| < \pi/4. \qquad (1.60)$$

This is the Prandtl solution. The equation system (1.58) can be solved numerically (Problem 1.14 below). Its solution provides θ_1 and θ_2 in terms of λ.

2. *The second solution.* The equation system (1.57) of the second solution can be solved similarly (Problem 1.15 below). For a homogeneous material, with $\tau_1 = \tau_2$, it is easy to find that

$$\theta_1 = \theta_2 = \frac{\pi}{2},\ \sigma_y = 4\tau_1,\ \sigma_x = 2\tau_1,\ \tau_{xy} = 0 \qquad (1.60a)$$

This is the second solution of the classical theory of plasticity unnoticed by Prandtl, Hill, and Sokolovsky. The structure of the field of characteristic lines in this solution is shown in Figure 1.12. In the solution, the stress σ_y has a discontinuity along $x = 0$. It is easy to generalize it and derive the solution to the corresponding honhomogeneous problem when the line, $x = 0$, is an interface between two different elastic-perfectly plastic materials (the crack, along $y = 0$ $x < 0$, is perpendicular to the interface in this case). In contrast to equation (1.60a), the stress σ_y in this case will be equal to $\sigma_y = 2(\tau_1 + \tau_2)$ for $x > 0$ where τ_2 is the yield limit for shear stress in the material $x > 0$. All the other equations are the same as those for the homogeneous case.

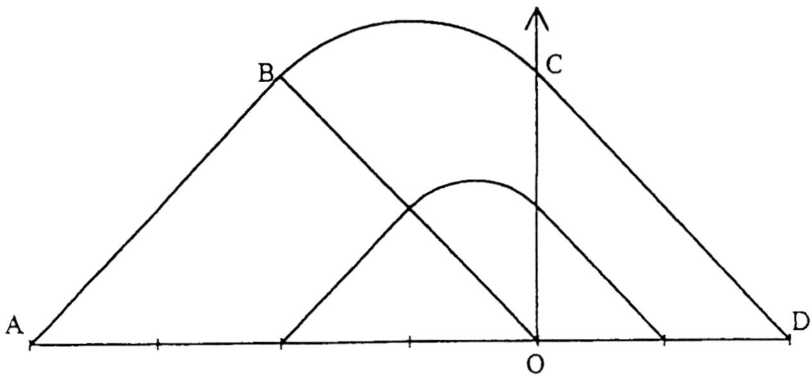

Figure 1.12. Characteristic lines in the second solution of the plasticity theory ($\sigma_y = 4\tau_1$, $\sigma_x = 2\tau_1$, $\tau_{xy} = 0$ when $|\theta| < \pi/2$; the stress σ_y in this solution has a discontinuity along $x = 0$)

Using these solutions, the Reuss equations for strain rate components, and the equation of adhesion energy in the form provided in Problem 1.12, one can show that the adhesion energy is equal to zero in this model (Problem 1.16), as in the homogeneous case[3]. This means physically that the adhesion energy spent to initiate an interface crack in an elastic-perfectly plastic model is equal to zero, as in the homogeneous case[3] . As the loading begins, the crack tip becomes blunt and advances, that is, the crack grows. In the more accurate model of crack initiation based on the consideration of dislocation emission by a crack tip (Chapter 6), it appears that the corresponding adhesion energy is non-zero, but very small (roughly ten times less than that of an ideal brittle crack growth with no dislocations emitted). In the well-developed crack growth range, it is necessary to include unloading and residual stresses and strains (see Problems 1.18 and 1.19 below). The calculation of adhesion energy characterizing the intermediate crack growth range, where the crack growth is comparable with the size of the plastic region near the crack tip, is more difficult. When the well-developed crack growth exceeds many times the size of the plastic region at the crack front, the methods and equations of Sections 1.3 and 1.5 should be used (Problems 1.18 and 1.19).

1.7. Debonding of Two Solids Made of Power - Law Hardening Materials

Suppose two bonded half-spaces are made of different incompressible hardening materials following a power law between octahedral shear stress and strain. Assume that the power in both materials is the same, but the coefficients in the power law are different. Using group theory, one can easily prove the self-similarity of the plane problem for the space debonded along $x_2 = 0$ $x_1<0$, with traction-free sides[3,8] (Figure 1.10).

Let the σ - ε diagram of these materials in the tension test be of the form

$$\varepsilon = A_1\sigma^{k+1} \quad (x_2 > 0)\,, \quad \varepsilon = A_2\sigma^{k+1} \quad (x_2 < 0) \tag{1.61}$$

Here, ε and σ are the strain and stress, k is a certain number, and A_1 and A_2 are some material constants. Equations (1.61) may be convenient approximations of real data. For most structural materials, k is positive, but for some soils, k may be negative.

The stresses have the following form[8]

$$\begin{aligned}
\sigma_r &= -(\lambda+1)^{-1}\, r^{\lambda}\left[f''(\theta)+(\lambda+2)f(\theta)\right],\\
\sigma_\theta &= -(\lambda+2)r^{\lambda} f(\theta),\\
\tau_{r\theta} &= r^{\lambda} f'(\theta);\\
\tau_{z\theta} &= r^{\lambda} g(\theta), \quad \tau_{zr} = -\frac{1}{\lambda} r^{\lambda} g'(\theta);
\end{aligned} \tag{1.62}$$

where

$$\lambda = -\frac{1}{k+2} \tag{1.62a}$$

Here, r, θ, and z are cylindrical coordinates (with z being directed along crack front), and $f(\theta)$ and $g(\theta)$ are some functions of θ.

Equations (1.62) satisfy both equilibrium equations. Substituting the stresses in the compatibility equations provides two coupled ordinary differential equations, one of fourth order and the other one of second order (Problem 1.20). The only material constant in these equations is k. The stresses near the crack tip in a homogeneous incompressible power law hardening material (i.e. $A_1 = A_2 = A$) do not depend on the constant A, because

$$\sigma_\theta = \tau_{r\theta} = 0, \quad \tau_{z\theta} = 0 \quad \text{along} \quad \theta = \pm\pi. \tag{1.63}$$

For the debonding crack between dissimilar materials, six more boundary value conditions must be met:

$$\begin{aligned}
[\sigma_\theta] = [\tau_{r\theta}] = [\tau_{z\theta}] = 0 \quad &\text{along} \quad \theta = 0\\
[u_r] = [u_\theta] = [u_z] = 0 \quad &\text{along} \quad \theta = 0
\end{aligned} \tag{1.64}$$

Now the stresses at the crack tip depend on k and the ratio A_2/A_1. The solution of the homogeneous boundary value problem consists of two different analytical pieces, in $0 < \theta < \pi$ and $0 > \theta > -\pi$, tied by the six condition equations (1.64) along $\theta = 0$. There are 12 homogeneous boundary condition equations and 12 unknowns following the general solutions of the fourth and second order differential equations in these intervals. This is an eigenvalue problem[8], where λ plays the role of eigenvalue. The solution to the eigenvalue problem contains one indefinite factor.

There are three different particular solutions corresponding to opening mode, transverse shear and screw shear debonding. It is convenient to define them using the generalized stress intensity factors[3]:

opening mode debonding

$$K_I = \lim_{r\to 0}\left[\sigma_\theta(\theta)\,(2\pi r)^{-\lambda}\right]\Big|_{\theta=0} \tag{1.65}$$

transverse shear debonding

$$K_{II} = \lim_{r\to 0}\left[\tau_{r\theta}(\theta)\,(2\pi r)^{-\lambda}\right]\Big|_{\theta=0}\ ; \tag{1.66}$$

screw shear debonding

$$K_{III} = \lim_{r\to 0}\left[\tau_{z\theta}(\theta)\,(2\pi r)^{-\lambda}\right]\Big|_{\theta=0}\ . \tag{1.67}$$

Here, K_{I}, K_{II}, and K_{III} are the stress intensity factors. In the particular case of a linear elastic material, when $k = 0$ and $\lambda = -1/2$, we get the known results (Section 1.5).

These types of loading and fracturing are analogous to the types of the field vector polarization in Maxwell's theory of electromagnetism.

The stress-strain law (1.61) makes the problem non-linear, and we cannot use the superposition principle to derive the mixed type stress distribution from three particular types of loading. However, the mode of debonding can be easily guessed for the particular types of external loading, namely:

$$\begin{aligned}
&\tau_{r\theta} = \tau_{z\theta} = 0 \quad \text{along } \theta = 0\text{, for the opening mode;}\\
&\sigma_\theta = \tau_{z\theta} = 0 \quad \text{along } \theta = 0\text{, for the transverse shear mode;}\\
&\sigma_\theta = \tau_{r\theta} = 0 \quad \text{along } \theta = 0\text{, for the screw shear mode.}
\end{aligned} \tag{1.68}$$

Using the general equation of adhesion energy, one can derive the following equation for the adhesion energy for debonding two power law hardening materials:

$$\Gamma_c = \eta_I(\lambda,\delta)A_I K_I^{\,k+2} \qquad \text{(opening mode);} \tag{1.69}$$

$$\Gamma_c = \eta_{II}(\lambda,\delta)A_I K_{II}^{\,k+2} \qquad \text{(transverse shear);} \tag{1.70}$$

$$\Gamma_c = \eta_{III}(\lambda,\delta)A_I K_{III}^{\,k+2} \qquad \text{(screw shear); } (\delta = A_2 / A_1)\ . \tag{1.71}$$

Here, η_I, η_{II} and η_{III} are some functions of λ and δ which must be found numerically (Problem 1.21).

If a loading path is of a mixed type, it is necessary to introduce a stress intensity factor which is specific to the loading path. The equation of the adhesion energy in this case is:

$$\Gamma_c = \eta_m(\lambda,\delta)A_I K_m^{\,k+2} \qquad \text{(mixed loading)} \tag{1.72}$$

Here, η_m is a function of λ, δ, and the loading path; and K_m is the stress intensity factor of this mixed loading path:

$$K_m = \lim_{r \to 0} \left[\sigma (2\pi r)^{-\lambda} \right] \Big|_{\theta=0} , \tag{1.73}$$

where

$$\sigma^2 = \sigma_\theta^2 + \tau_{r\theta}^2 + \tau_{z\theta}^2 \ . \tag{1.74}$$

Note that, opening mode, transverse shear, and screw shear cannot follow from Equation (1.72) as particular cases, because $\eta_m (\lambda, \delta)$ is specific to the loading path.

When $k = 0$, then $\lambda = 1/2$ and Equations (1.68) - (1.71) coincide with those for linearly elastic materials.

The adhesion energy generally depends on the loading path and physical/geometrical parameters of the specimen tested for debonding, for example, the thickness of a bonding layer, the thicknesses of bonded plates, the debonding length, and the elastic and inelastic properties of the materials comprising the bonded plates and bonding layer. In other words, the adhesion energy depends on the method of measurements, specimen sizes and type, and material properties. The latter is the most discouraging fact poorly understood by practitioners, because it makes it impossible to compare the quality of different adhesives for bonding different inelastic material pairs, if the dimensions or loading paths of specimens differ. To take into account these effects, it is convenient to apply the methods of this section based on the general energy conservation equation, Equations (1.10), (1.11) or (1.75) below. Study problems 1.3 - 1.23 to learn now to derive more and more sophisticated and accurate methods for measuring the adhesion energy.

1.8. Debonding of Two Solids of Different Viscoelastic Materials

Consider an interface crack between two dissimilar linearly viscoelastic materials. The neighborhood of the crack front can be studied in the half-infinite scale of Figure 1.10 with the traction-free debonding along $x_2 = 0$ $x_1 < 0$. Suppose the crack tip (and the x_1 x_2 coordinate frame) moves at the speed V in the x_1 direction[3,8,20]. Because of the motion, elastic regions form in the two materials; they have dimensions, VT_1 and VT_2, where T_1 and T_2 are the characteristic relaxation times in the materials. Remember that the relaxation time equals E/η, where E and η are the characteristic Young's modulus and viscosity coefficient in the viscoelastic model. Therefore, shrinking the integration path, Σ_ε, in Equations (1.10) or (1.11) to the crack tip provides the equations of adhesion energy, Equations (1.27) - (1.32), as in an ideally brittle material. This means that a viscoelastic material is ideally brittle if VT_1 and VT_2 are comparable with the size of the specimen. In this case, cracking is reversible for short times, and the adhesion energy equals the interface surface tension[3].

On the other hand, when VT_1 and VT_2 are very small, the elastic "kernel" disappears inside a structural irregularity and the adhesion energy becomes a functional of the first, second or higher time derivatives of the stress intensity factor, depending on the constitutive equations of the materials[8]. In other words, the adhesion energy in this case depends strongly on the loading rate. The adhesion energy can be large for extremely high loading rates and very small crack growth velocity (Problem 1.24). This problem explains the high adhesive properties of polymer materials possessing viscoelastic properties with the small characteristic relaxation time.

1.9. Conclusion

The adhesion and cohesion energy of solid matter corresponding to debonding depends on the physical properties of the constituents, the geometrical parameters of the solid and debonding, and the mode, path and rate of loading. It is very seldom that the adhesion energy can be a physical constant.

An integrated approach to calculating this property of a solid is provided by the equation of adhesion energy based on the invariant contour integral applicable to arbitrary solids. The approach allows one to invent and predict a variety of experimental/theoretical methods for measuring and controlling the adhesion energy.

The invention of practicable methods requires a comprehensive analysis of constitutive equations of the constituents, the geometry of the solids, and the loading conditions; this yields the initial data necessary for applying the general equations of adhesion energy using analytical and numerical methods.

In order to master this approach, the reader is encouraged to solve the following problems.

1.10. Problems

1.1. Find the radius of a thin cylindrical needle of martensite phase embedded in austenite phase as a result of isothermal stress-induced phase transition characteristic for a dynamic self- mainained process of cold detonation or bonding energy release.
Assume that:

a. Latent heat of phase transformation is Q (per unit volume);
b. Interface surface energy is γ_{12} (per unit area);
c. Change of internal energy as a result of the phase transition is ΔU(per unit volume);
d. Thermal expansion mismatch is $\Delta\alpha$, so that the phase expansion is $T\Delta\alpha$ at temperature T.
e. A martensite phase cylinder propagates along its axis, so that the phase transformation occurs at the leading end of the cylinder.

Hint: Consider the invariant integral, Equation (1.10) or (1.11), for the steady-state development of the cylinder along its axis[6].

1.2. An elastic liquid space with a spherical gas bubble is subjected to negative pressure. Find the critical pressure (tensile strength) of the liquid in terms of surface tension, compressibility of the liquid, and the radius of the bubble.

1.3. A soft thin coating is bonded to a flat boundary of a rigid substrate. There is an interfacial blister which initially has an elliptical shape in plan view. Consider the debonding of the blister due to gas pressure in the cavity[6]. Assume that the coating is momentless (membrane) and study the development of the blister under one of the following conditions:

a. The gas pressure is kept constant.

b. The mass of gas in the cavity is constant.

c. The gas pressure is a given function of time.

1.4. Consider Problem 1.3 for a rigid thin coating under the assumption that the coating is an elastic plate under pure bending[6].

1.5. Consider the Problems 1.3 and 1.4 when the initial blister has:

a. A parabolic shape in plan view.

b. A hyperbolic shape in plan view.

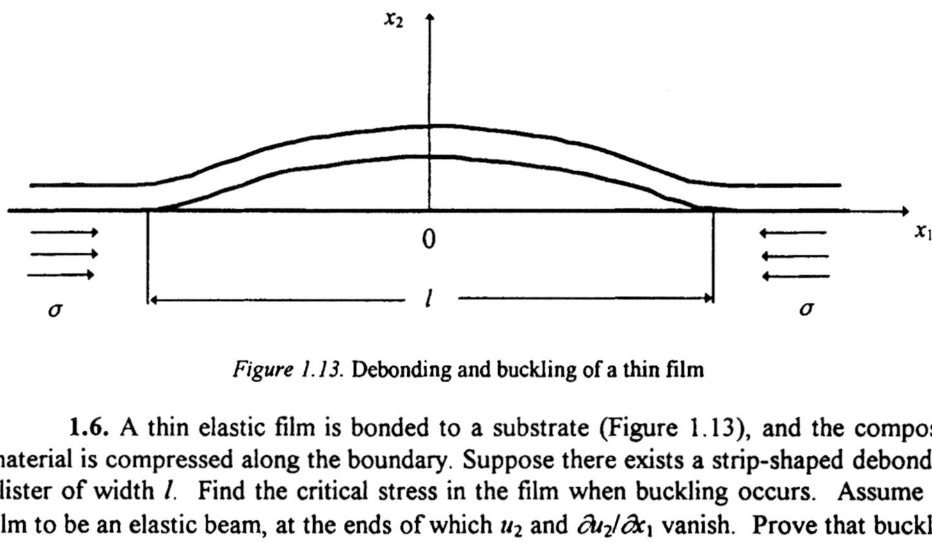

Figure 1.13. Debonding and buckling of a thin film

1.6. A thin elastic film is bonded to a substrate (Figure 1.13), and the composite material is compressed along the boundary. Suppose there exists a strip-shaped debonding blister of width l. Find the critical stress in the film when buckling occurs. Assume the film to be an elastic beam, at the ends of which u_2 and $\partial u_2/\partial x_1$ vanish. Prove that buckling precedes debonding and does not depend on adhesion energy. Prove that debonding starts only after the maximum deflection of the buckled blister achieves a certain critical value depending on adhesion energy.

1.7. A thin elastic film is bonded to a substrate, and the surface is subjected to intensive heating. Prove that, under the assumptions of Problem 1.6, buckling of a pre-existing blister from thermal stresses precedes debonding and does not depend on adhesion energy. Prove that debonding starts only after the maximum deflection of the buckled blister achieves a certain critical value depending on adhesion energy, and debonding proceeds without stopping even under decreasing stresses.

1.8. Study the growth of an interface debonding zone, a blister, under the following conditions:

a. Two bonded membranes of different materials; or

b. Two bonded thin plates of different, linearly elastic materials; or

c. A thin membrane bonded to a thin plate of a linearly elastic material.

The contour of the debonding zone at the initial moment can be:

an ellipse, a parabola or a hyperbola.

The loading conditions may be as follows:

Gas pressure in the blister cavity is constant; or

The gas mass in the blister cavity is constant; or

Gas pressure in the blister cavity is a given function of time; or

Deflection in the middle of the blister is a given function of time while gas pressure equals zero.

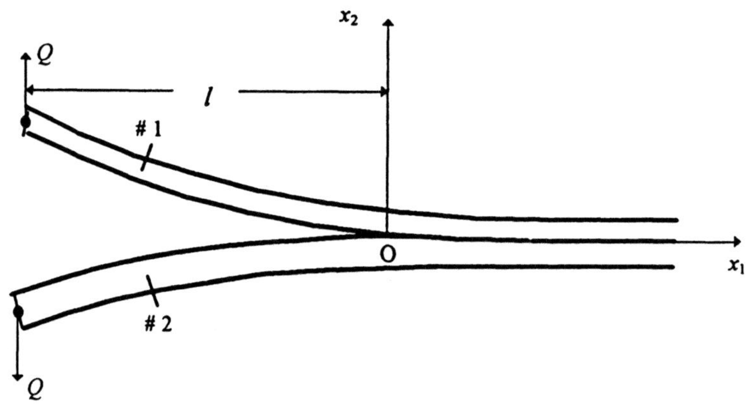

Figure 1.14. Debonding of two elastic-perfectly plastic beams

1.9. Study debonding in a bimaterial beam made of two different linearly elastic-perfectly plastic materials (Figure 1.14). The debonded ends are subject to transverse forces. Consider the following conditions:

a. Debonding starts before a local yielding occurs; or

b. Debonding starts after a local yielding zone develops; or

c. The yielding zone grows until a plastic hinge is formed, while debonding does not start, at all.

Find the bounds of these conditions in terms of adhesion energy, geometrical dimensions, and the elastic and plastic constants of the materials[6].

1.10. A round cylindrical bimaterial linearly-elastic shell has an axisymmetrical delamination zone near an open end of the shell. The length of the debonding zone along the axis is l. Suppose the sides of the debonding zone are subject to constant pressure p. Study the development of debonding as p increases[6].

1.11. Show how to measure the adhesion energy by debonding using the solutions of Problems 1.3 to 1.10.

1.12. Show that the equation of surface or adhesion energy, Equation (1.10), can be written in the form

$$\gamma = \frac{1}{2} \lim_{\varepsilon \to 0} \oint_{\Sigma_\varepsilon} \left(\sigma_{12} \frac{\partial u_1}{\partial x_1} + \sigma_{22} \frac{\partial u_2}{\partial x_1} + \sigma_{23} \frac{\partial u_3}{\partial x_1} \right) ds, \qquad (1.75)$$

if one takes Σ_ε in the shape of a very thin rectangular contour (Figure 1.15) shrinking to 0, that is, the center of symmetry of Σ_ε

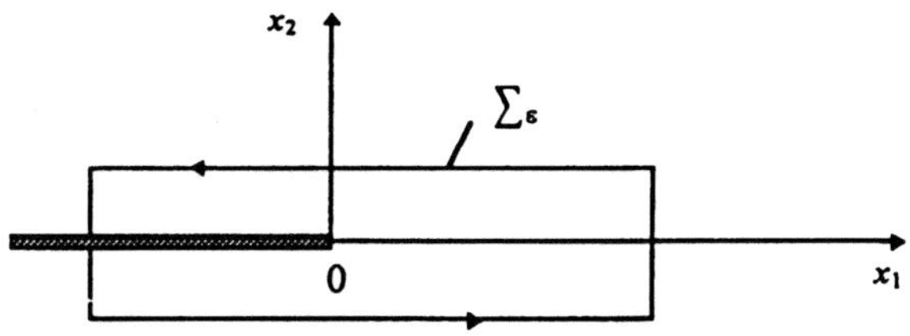

Figure 1.15. Integration path in the form of a thin rectangular contour

1.13. Show that opening mode, transverse shear, and screw shear debonding make sense in any solids, and the terms in Equation (1.75) correspond respectively to transverse shear, opening mode and screw shear contributions to the adhesion energy.

In Equation (1.15) for thin plates, show that the terms correspond, respectively, to opening mode (from bending moment), transverse shear (from in-plane normal force), and screw shear (from in-plane shear force) contributed by a one-sided plate. Notice that, even though the adhesion energy is a non-linear function of stresses, the linear superposition of these three modes in the equation of adhesion energy is still valid.

1.14. Solve the equation system (1.58) numerically and sketch the resulting functions, $\theta_1 = \theta_1(\lambda)$ and $\theta_2(\lambda)$, in the interval $0 < \lambda < 1$. Take into account that $\theta_1 = \theta_2 = \pi/4$, when $\lambda = 1$. Show that θ_1 increases and θ_2 decreases, as λ grows in this interval.

1.15. Exclude σ_y and τ_{xy} in the equation system (1.57) and derive two coupled equations for θ_1 and θ_2. Solve them numerically and find θ_1 and θ_2 in terms of $\lambda = \tau_1/\tau_2$. Take into account that $\theta_1 = \theta_2 = \pi/2$, when $\lambda = 1$.

1.16. Calculate the adhesion energy of interface debonding of two dissimilar solids made of elastic-perfectly plastic materials using the equation of adhesion energy from Problem 1.12, the Prandtl-Reuss equations for strain rates or Hencky equations for strains, and the first/second solutions for stresses at the crack tip treated in Section 1.6. Prove that the adhesion energy is zero.

1.17. Study the influence of the Bauschinger effect on the equation of adhesion energy of two identical bonded beams of elastic-plastic material in loading-unloading by bending moments (See Figure 1.8 for $M_1 = M_2$, $\tau_1 = \tau_2$ and $M = 0$). Show that Bauschinger's effect of hysteresis causes a decrease of the adhesion energy, if the limiting load of crack growth does not change[6].

1.18. Consider an opening mode crack in the middle of a layer of thickness t bonding two identical materials. The in-plane dimensions of the crack are assumed to be much greater than t, so that one can study the neighborhood of the crack front using a half-infinite crack along $x_2 = 0$ $x_1 < 0$ (Figure 1.16).

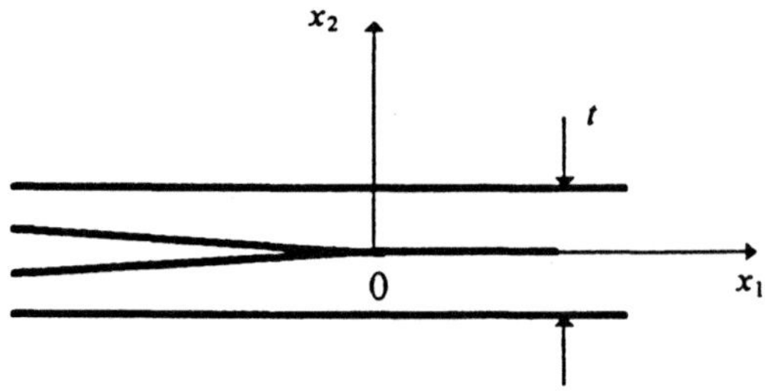

Figure 1.16. Debonding of an adhesive elastic - plastic layer

Suppose that the bonded solids are linearly elastic and the material of the bonding layer is linearly elastic-perfectly plastic. Using Equation (1.10), prove that the equation of adhesion energy has the following form when the crack is large compared to t:

$$\Gamma_c = \frac{1-\nu^2}{E} K_I^2 - W_p - U_L \tag{1.76}$$

$$W_p = t\tau_y \bar{\varepsilon}_{Res} \quad , \tag{1.77}$$

$$U_L = \frac{1}{3} t\tau_y \varepsilon_{Res}(t/2) \quad , \tag{1.78}$$

where

$$\bar{\varepsilon}_{Res} = \frac{1}{t} \int_{-t/2}^{t/2} \varepsilon_{Res}(x_2) dx_2 \tag{1.79}$$

Here, W_p is the energy dissipated as heat by the residual plastic strains far behind the crack front in the bonding layer, U_L is the latent energy of elastic strains in the adjacent yielding - induced zones in the elastic material, K_I is the stress intensity factor computed under the assumption that $t = 0$, E and ν are Young's modulus and Poisson's ratio of the substrates, τ_y is the shear yield stress of the bonding material, and $\varepsilon_{Res}(x_2)$ is the residual plastic strain ε_x far behind the crack front in the bonding layer.

The K_I in (1.76) corresponds to the crack growth and is measured experimentally by using the expression for K_I in terms of loads, crack length and specimen sizes, which follows from the linear theory of elasticity at $t = 0$.

Hint. Assume that the only non-zero stresses and strain are σ_{11}, σ_{33}, and ε_{11}, that they do not depend on x_1 far behind the crack front. Show that the stress is compressive, in the bonding layer, i.e. $\sigma_{11} = -\tau_y$, and that σ_{11} is tensile and is a linear function of x_2 in the induced zone of thickness $t\tau_y(1-\nu^2)/[E\varepsilon_{Res}(t/2)]$ on each side of the crack.

1.19. Consider the crack growth in a homogeneous linearly elastic-perfectly plastic material based on the study of Problem 1.18. Prove that the equation of cohesion energy has the form of Equations (1.76) - (1.79) where $\varepsilon_{Res}(t/2) = \varepsilon_e$ is the elastic limiting strain on the σ-ε diagram and t is the thickness of the layer of residual plastic strains on each side of the crack.

1.20. Consider incompressible power law hardening materials with the Hencky-Ilyushin constitutive equations:

$$\varepsilon_r = -\varepsilon_\theta = \frac{1}{2} a\, I^\kappa (\sigma_r - \sigma_\theta), \tag{1.80a}$$

$$\gamma_{r\theta} = 2a\, I^\kappa\, \tau_{r\theta}, \tag{1.80b}$$

$$\gamma_{rz} = 2a\, I^\kappa\, \tau_{rz}, \gamma_{\theta z} = 2a\, I^\kappa\, \tau_{\theta z}, \tag{1.80c}$$

$$D = 2a\, I^{\kappa+1}, \tag{1.80d}$$

where

$$I = \frac{1}{2}\left[(\sigma_r - \sigma_\theta)^2 + 4\left(\tau_{r\theta}^2 + \tau_{rz}^2 + \tau_{\theta z}^2\right)\right]^{1/2} \tag{1.81a}$$

$$D = \left[4\varepsilon_r^2 + \gamma_{r\theta}^2 + \gamma_{rz}^2 + \gamma_{\theta z}^2\right]^{1/2} \tag{1.81b}$$

Here a and κ are some material constants.

Prove that the stresses in Equations (1.62) satisfy the equilibrium equations:

$$\frac{\partial \sigma_r}{\partial r} + \frac{1}{r}\frac{\partial \tau_{r\theta}}{\partial \theta} + \frac{\sigma_r - \sigma_\theta}{r} = 0 \quad , \tag{1.82a}$$

$$\frac{\partial \tau_{r\theta}}{\partial r} + \frac{1}{r}\frac{\partial \sigma_\theta}{\partial \theta} + 2\frac{\tau_{r\theta}}{r} = 0 \quad , \tag{1.82b}$$

$$\frac{\partial \tau_{zr}}{\partial r} + \frac{1}{r}\frac{\partial \tau_{z\theta}}{\partial \theta} = 0 \quad . \tag{1.82c}$$

Derive the coupled system of two ordinary differential equations for $f(\theta)$ and $g(\theta)$ in Equations (1.62) by using Equations (1.62), (1.80), and (1.81) and the compatibility equations:

$$\frac{\partial^2 \varepsilon_r}{\partial \theta^2} + \frac{\partial}{\partial r}\left(\varepsilon_r + r\frac{\partial^2 \varepsilon_r}{\partial \theta^2}\right) = \frac{\partial^2}{\partial r^2}\left(r\varepsilon_\theta - r^2\frac{\partial \varepsilon_\theta}{\partial r} + r^2\frac{\partial \gamma_{r\theta}}{\partial \theta}\right); \tag{1.83a}$$

$$\frac{\partial \gamma_{rz}}{\partial \theta} = \frac{\partial (r\gamma_{\theta z})}{\partial r} \tag{1.83b}$$

1.21. Prove[8] Equation (1.62a) by using the invariant integral, Equation (1.10).

1.22. Find functions $f(\theta)$ and $g(\theta)$ in Equation (1.62) numerically using the solutions to Problems 1.20 and 1.21. Consider the particular cases of opening mode, transverse and screw shear, and the general case described by Equations (1.72) - (1.74).

1.23. Calculate numerically the $\eta_{I}(\lambda, \delta)$, $\eta_{II}(\lambda, \delta)$, $\eta_{III}(\lambda, \delta)$ and $\eta_{m}(\lambda, \delta)$ in Equations (1.69) - (1.74) by using the results of the solution to Problem (1.22) and the equation of adhesion energy, Equation (1.10).

1.24. Derive the adhesion energy by the debonding of two different viscoelastic solids described by the Maxwell or Kelvin models for extremely low crack growth rate in terms of the stress intensity factor and its time derivative[8]. Assume opening mode and quasistationary approximation[8] of local stresses and strains in the equation of adhesion energy (1.10).

References

1. K. N. Tu, Mayer, J. W. Mayer, and L.C. Feldman (1992), *Electronic Thin Film Science*, Macmillan, New York (Chapter 2).

2. A. W. Adamson (1982), *Physical Chemistry of Surfaces*, Wiley, New York (Chapters 1 and 2).

3. G. P. Cherepanov (1974), Mekhanika Khrupkogo Razrushenia, Moscow, Nauka; English edition (1979), *Mechanics of Brittle Fracture*, R. de Wit and W. C. Cooley (eds.), McGraw Hill, New York (Chapters 2 and 5, and Section 7.8).

4. A. Zangwill (1988), *Physics at Surfaces*, Cambridge University Press, Cambridge.

5. A. A. Griffith (1920), The phenomena of fracture and flow in solids, *Proc. Roy. Soc.* **A 221**, pp. 163-198; republished, as well as papers[7,8] below, in *Fracture: A Topical Encyclopedia of Current Knowledge* (1996), G. Cherepanov (ed.), Krieger Publ., Melbourne, USA.

6. G. P. Cherepanov (1983), *Fracture Mechanics of Composite Materials*, Moscow, Nauka.

7. G. R. Irwin (1957), Analysis of stresses and strains near the end of a crack traversing a plate, *J. Appl. Mech.*, **124**(3), pp. 361-370.

8. G. P. Cherepanov (1967), On crack propagation in continuous media, *App. Math. and Mech. (PMM)*, **31**(3), pp. 476-488.

9. G. P. Cherepanov (1983), Fracture mechanics of multilayer shells. The theory of delamination, *Appl. Math. and Mech. (PMM)*, **47**(5), pp. 968-985.

10. J. W. Hutchinson and Z. Suo (1992), Mixed mode cracking in layered materials, *Advances in Applied Mechanics*, **69**, pp. 63-191.

11. X. Deng (1995), Mechanics of debonding and delamination in composites: asymptotic studies, *Composites Engineering*, **4**(9), pp. 1271-89

12. V. M. Abramov (1937), The contact problem of an elastic half-plane and a rigid punch, *Doklady USSR Academy of Sciences*, **17**(4), pp. 173-178.

13. L. A. Galin (1945), Punch indentation with account of friction and cohesion, *Appl. Math. and Mech. (PMM)*, **9**(5), pp. 413-424.

14. B. M. Malyshev and R. L. Salganik (1965) The strength of adhesive joints using the theory of cracks, *Int. J. Fracture Mech.*, **1**, pp. 114-128.

15. G. I. Barenblatt and G. P. Cherepanov (1961), On the finiteness of stresses at a crack tip, *Appl. Math. and Mech. (PMM)*, **25**(4), pp. 607-610.

16. C. Atkinson and J. D. Eshelby (1968), The flow of energy into the tip of a moving crack, *Int. J. Fracture*, **4**, pp. 3-8.

17. G. P. Cherepanov (1968), Cracks in solids, *J. Solids and Structures*, **4**(4), pp. 811-831.

18. R. Hill (1950), *Mathematical Theory of Plasticity*, Clarendon Press, Oxford.

19. V. V. Sokolovsky (1950), *Theory of Plasticity*, Gostekhizdat, Moscow (in Russian).

20. C. Atkinson and M. L. Williams (1973), A note on the Cherepanov calculation of viscoelastic fracture, *J. Solids and Structures*, **9**, pp. 237-241.

CHAPTER 2

FLUCTUATIONS AND THE KINETIC THEORY OF FRACTURE

In the present chapter, we consider the connection of the classical fracture mechanics based on the crack tip study with the thermal fluctuation theory of fracture. We refine the theory, indicate some limitations and make it more useful for practical applications.

2.1. Thermal Fluctuation Theory of Fracture

The idea of fluctuations was introduced into physics by Ludwig Boltzmann and Albert Einstein when it became clear that the second law of thermodynamics - the law of entropy - was not a universal one. Fluctuations are processes which are not governed by the entropy law, they occur in all scales of space and time. Biological life is a striking example of an anti-entropy phenomenon; physicists consider it as a fluctuation.

In statistical physics, a fluctuation is a deviation from a mean or most probable state, which cannot be described by the macroscopic laws of physics. The probability of a fluctuation is directly proportional to

$$p \sim exp\left(-\frac{U}{kT}\right),$$

where k is Boltzmann's constant, T is the absolute temperature, and U is the activation energy of the fluctuation..

From this point of view, the state with a fracture represents a giant fluctuation from an equilibrium state (without the fracture). This idea can be traced as far back as the early 1930s when the physicists Alexandrov and Zhurkov studied the tensile strength of solids and proposed the empirical equation[1]

$$t = Ae^{-\lambda\sigma}, \tag{2.1}$$

later called by their names. This equation relates the longevity t of a stretched rod to the applied stationary tensile stress σ through the use of two empirical constants A and λ.

Further extensive experimental investigations of the long-term strength of rods made of various materials under stationary tensile loads demonstrated the applicability of this equation over a wide range of loading times ($t = 10^{-5}$ - 10^{7} sec) and temperatures[2-9]. It was established that the experimental points on the $log\ t$-σ diagram lie along a straight line at one and the same value of the absolute temperature T, and with change in the latter, this line shifts to a new position parallel to the original (Figure 2.1a). A treatment of all life vs. stress data on one and the same material in the coordinates $kT\ log\ t$ and σ, showed that they lie within a rather narrow band (dashed section in Figure 2.1b), the breadth of which does not usually exceed 1/10 - 1/5 the width of all of the ranges along the σ axis and the $kT\ log\ t$ axis. The slope of this band is independent of the values of T, σ, and t, but changes significantly depending on the structure of the material and the presence of defects. The point at which the band continuation intersects the ordinate is approximately the value of the chemical activation energy of the given substance, U (in terms of one molecule or gram-molecule). U characterizes the chemical reaction rate υ vs. temperature T of the substance

$$\upsilon = \upsilon_0 exp\left(-\frac{U}{kT}\right). \tag{2.2}$$

This is the well known Arrhennius equation (υ_0 is a temperature - independent quantity). The value of U is equal to 10^2 to 10^3 KJ per gram-mole or gram-atom, or 1 to 10 eV per molecule or atom for condensed matter; this is close to the binding energy of solids. When U is taken per gram-atom or gram-mole, one should replace k by R in Equation (2.2), where R is the universal gas constant.

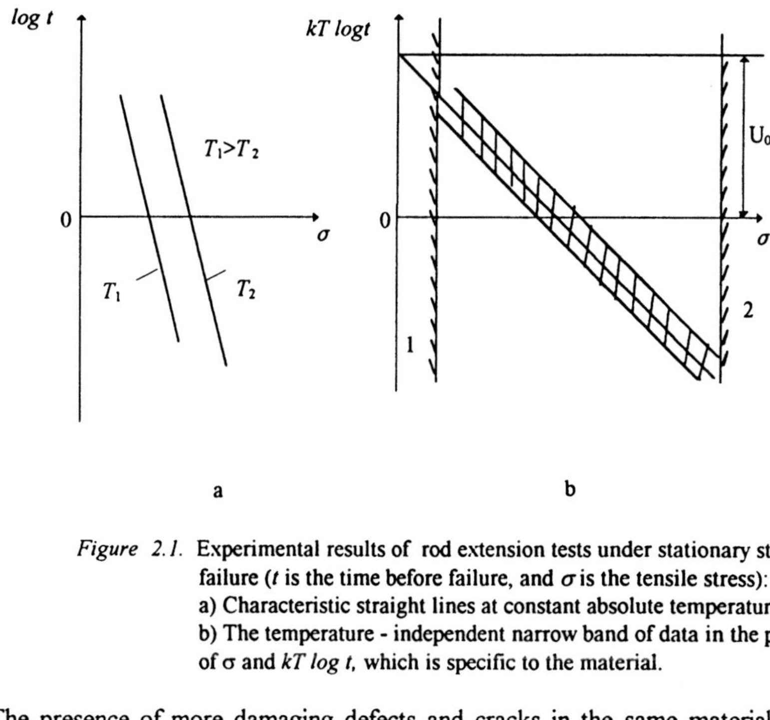

Figure 2.1. Experimental results of rod extension tests under stationary stress up to failure (t is the time before failure, and σ is the tensile stress):
a) Characteristic straight lines at constant absolute temperature T;
b) The temperature - independent narrow band of data in the plane of σ and $kT\ log\ t$, which is specific to the material.

The presence of more damaging defects and cracks in the same material causes the slope of the band in Figure 2.1b to increase. According to the test data of Figure 2.1a, U is a linear function of σ.

These experimental results serve as the basis of the thermal fluctuation theory of strength and fracture of solids, according to which life and failure of a material is described as a stress- affected fluctuation of duration

$$t = \tau_0 \, exp \frac{U_0 - \upsilon_A \sigma}{kT}, \qquad \text{(in atomic units)} \qquad (2.3)$$

where τ_0 is the characteristic time of an elementary thermal fluctuation, namely, the time of propagation of phonons on an interatomic spacing, $\tau_0 = 10^{-13} - 10^{-12}$ sec; and υ_A is the activation or fluctuation volume, in which the active failure occurs.

Equation (2.3) is confirmed by the following considerations from statistical physics. The transition of a loaded sample from a nondamaged state to a damaged one is characterized by some activation energy *U*, which is analogous to the ionization energy during dissociation of atoms into ions. The probability of such a transformation is directly proportional to $exp(-U/kT)$.

On the other hand, failure may occur as the result of a thermal fluctuation during the period τ_0. The probability of this event is equal to τ_0 / t because it is natural to assume that the probability of a failure is unity when the time to failure of the sample equals *t*. From this, it follows that *t* is directly proportional to $exp\ (U / kT)$. Further, assuming that the activation energy decreases linearly with increasing load σ, i.e.

$$U = U_0 - \upsilon_A \sigma, \qquad (2.4)$$

we can find that the constant multiplier in front of the exponential is equal to the time, τ_0, of the failure fluctuation at the stress, $\sigma = U_0 / \upsilon_A$, when the dissociation of the sample into atoms occurs. The latter is obviously equal to the time length of a single elemental thermal fluctuation. In region 1 of Figure 1b at low stresses, the failure process has an athermal character, and the kinetic concept does not hold. In region 2 at high stresses, the Equation (2.3) also does not apply, due to the predominance of the inertial force.

2.2. Engineering Physics Approach to Failure

The phenomena of fracture of solids can be adequately understood only when the role of crack-like defects in the structure of the material is accepted as being of importance, not less than that of the chemical composition of the material. This was proven as far back as 1920 by A. A. Griffith[10] who, through the use of ordinary window glass, showed that the strength of such glass changes by three orders of magnitude (from 1 to 10^3 *KBar*) depending on the size of the crack. Therefore, a crack is a fundamental object of study in the theory of strength and fracture: its role is equivalent to that of the atom in atomic physics.

What exactly is the cause of crack growth? Although a general explanation of this phenomenon, as applied to a crack in glass, was given by Griffith[10], it was Irwin[11] who first understood that it is necessary to look for the cause at the tip of the crack where the process is localized. Irwin introduced the fundamental parameters which characterize the condition of the crack tip before, during and after the moment of localized fracture. For the most important case, opening mode cracks, this parameter is the stress intensity factor, K_I, see Equation (1.3). This began the study of fracture mechanics which currently includes some tens of thousands of papers. Fracture mechanics is a branch of physics, and the term "mechanics" has the same meaning as in the words "quantum mechanics" or "relativistic mechanics". It underscores the precise mathematical character of the discipline. Griffith and Irwin examined brittle and quasi-brittle fractures under quasistatic conditions.

A generalization of the Griffith - Irwin theory to any solid provides the fundamental parameter characterizing the condition at the crack tip[12,13]

$$\Gamma = \lim_{\varepsilon \to \infty} \int_{\Sigma_\varepsilon} [(W + K)n_1 - \sigma_{ij} n_j u_{i,1}] ds \qquad (i, j = 1,2,3). \tag{2.5}$$

Here the designations are the same as in Equation (1.10), but the quantities in the brackets do not necessarily correspond to a crack growth. When $\Gamma = \Gamma_c$, the crack grows. In a physical sense, Γ is the crack driving force, or the energy flow rate at point 0.

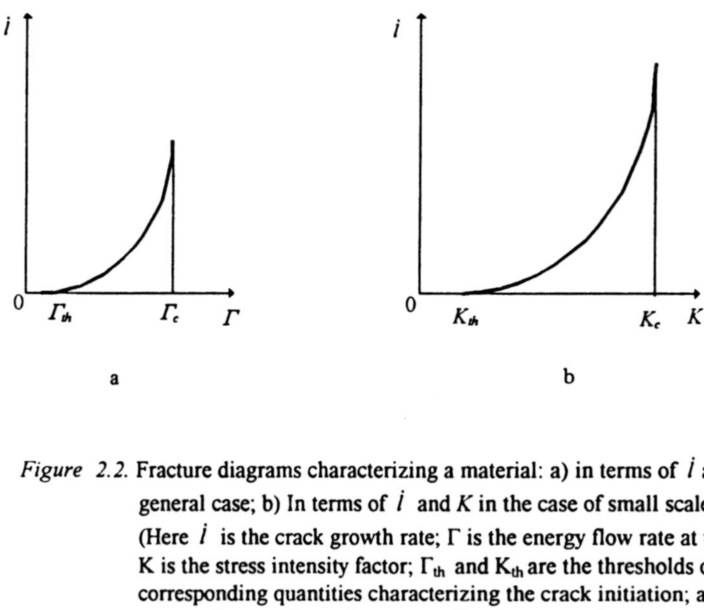

Figure 2.2. Fracture diagrams characterizing a material: a) in terms of $\dot{l}$ and Γ in general case; b) In terms of $\dot{l}$ and K in the case of small scale yielding. (Here $\dot{l}$ is the crack growth rate; Γ is the energy flow rate at the crack tip; K is the stress intensity factor; Γ_{th} and K_{th} are the thresholds of corresponding quantities characterizing the crack initiation; and Γ_c and K_c are maximum values of corresponding parameters).

The crack growth rate has been presented[12,13] in the form of a fracture diagram, $\dot{l} = \dot{l}(\Gamma)$, uniquely characterizing a material (Figure 2.2a). Crack growth begins at a threshold value $\Gamma = \Gamma_{th}$, and $\Gamma = \Gamma_c$ characterizes the maximum value of the crack driving force. The fracture diagram is determined experimentally. The simplest case of this theory when $\Gamma_{th} = \Gamma_c$, corresponds to brittle fracture. For small scale yielding, deviations from linear elasticity occupy a region near a crack tip which is small compared to crack lengths and layer thicknesses or specimen sizes. In this case, $\Gamma = K_I^2(1 - \nu^2)/E$ for opening mode cracks and the fracture diagram can be represented in terms of $\dot{l}$ and K_I as $\dot{l} = \dot{l}(K_I)$ which is the material characterization for small scale yielding (Figure 2.2b). This fracture diagram has been studied [13] for a variety of mechanisms, such as plastic blunting and instability, cyclic loading, hydrogen embrittlement, electrochemical and chemical corrosion, general kinetic mechanisms, adsorption, etc. When yielding is not on a small scale, the Γ can be analytically calculated in many important situations, partly represented in Chapter 1. In the general case, it is necessary to use numerical methods and computer modeling in order to find the Γ-parameter[14-18].

2.3. Amorphous Solids

Consider the crack growth in brittle amorphous polymers and glass studied in Ref.[9,19] using different methods. Suppose an edge crack under plane stress or plane strain conditions issues at right angle from a free boundary[9], and the crack length is small compared to the dimensions of the specimen (Figure 2.3). In this case, the stress intensity factor is equal to

$$K_I = 1.12\sigma\sqrt{pl} \tag{2.6}$$

Here, l is the crack length and σ is the tensile stress on the specimen surface at the point of the crack issue computed under the assumption that $l = 0$.

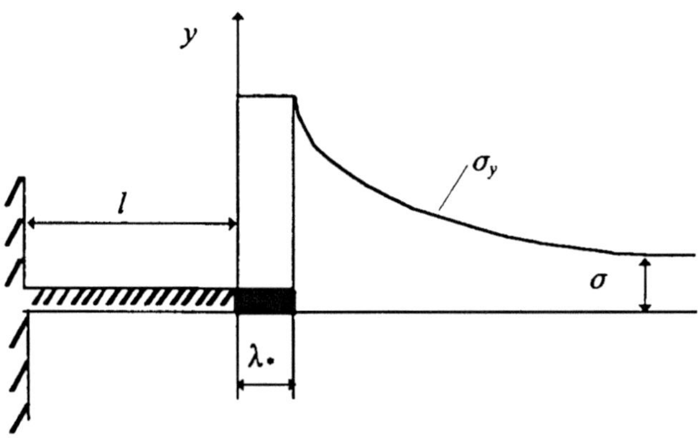

Figure 2.3. The stress distribution along the edge crack continuation in a "tiny plastic particle" by Neuber

In amorphous materials like glass, some polymers, metastable metal glass, and ceramics, there are no conditions for dislocation motion, and these materials are ideally brittle, that is, no dislocations can be emitted by the crack tip. In these materials, all deviations from linear elasticity occur in the process zone of size $\lambda_\bullet$ at the crack tip. The value of $\lambda_\bullet$ does not depend on K_I and is a material constant having the order of the interatomic spacing, a. The value $\lambda_\bullet$ has been accepted by physicists[2-9] to be equal to a for oriented polymers like capron and $3a$ for inorganic glass and unoriented polymers like polymethylmetacrylate (acrylic glass). This concept is very similar to that of a "tiny plastic particle" suggested by Neuber in the 1930's.

The stress in the "plastic particle" is constant and equal to

$$\sigma_p = \frac{K_I}{\sqrt{2p\lambda_\bullet}}, \tag{2.7}$$

because the stress distribution along the crack continuation can be approximated well by the form (Figure 2.3),

$$\sigma_y = \begin{cases} \sigma_p & \text{when} \quad 0 < x < \lambda \\ \dfrac{K_I}{\sqrt{2\pi x}} & \text{when} \quad x > \lambda_*, \quad \text{if} \quad x << l \end{cases} \tag{2.8}$$

and σ_y should be continuous at $x = \lambda_*$.

In experiments, it is easy to measure external load, σ, but it is difficult to measure crack growth accurately.

Increase of σ and/or loading time causes the frontal chemical bond (for example C - C in polymers) at the crack tip to load, straighten and extend, until it achieves its full length, λ_m, and the crack starts (Figure 2.4). At that moment the equilibrium condition implies

$$2\gamma = \lambda_m \sigma_p . \tag{2.9}$$

Here, γ is the surface tension, that is, the surface energy of cohesion.

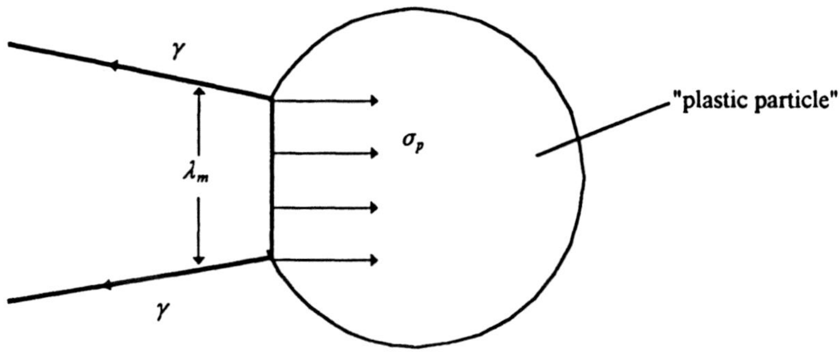

Figure 2.4. Tearing a chemical bond at the crack tip (γ is the surface tension, σ_p is the negative pressure in the "plastic particle", and λ_m is the length of chemical bond, for example C - C in polymers)

Substituting σ_p from (2.9) into (2.7) provides the following value of the threshold stress intensity factor, K_{th}, characterizing the crack initiation:

$$K_{th} = \frac{2\gamma\sqrt{2\pi\lambda_*}}{\lambda_m} \tag{2.10}$$

The value of K_{th} can be measured using Equation (2.6) by observing the crack initiation, and the values of λ_m and λ_* can be estimated by theoretical methods of quantum chemistry, or experimentally using X-ray molecular spectroscopy[3-9]. The value of γ can be found from experiments with liquid substance (Chapter 1) or theoretically by calculating γ using an interpolation of interatomic forces (see, e. g. Chapter 2 in[13]). The value of γ appears to be roughly equal to $E\,\lambda_m/10$. Based on experimental information about λ_m, $\lambda_\bullet$ and γ in[9], the K_{th}, according to Equation (2.10), is equal to 2.35 $N/mm^{3/2}$ for acrylic glass, 1.6 $N/mm^{3/2}$ for capron, and 12 $N/mm^{3/2}$ for silicate inorganic glass. For comparison, the critical value of $K_I = K_G$ calculated on the Griffith - Irwin theory

$$K_G = \sqrt{\frac{2E\gamma}{1-\nu^2}} \tag{2.11}$$

is 0.61 $N/mm^{3/2}$ for acrylic glass, 0.1 $N/mm^{3/2}$ for capron, and 6.22 $N/mm^{3/2}$ for silicate inorganic glass.[9,19]

The further increase of σ and/or loading time, causes most of the chemical bonds in the process zone to load, straighten and extend ("plastic particle"), until the stress there achieves the maximum value, $(\sigma_p)_{max} = U_0/\upsilon_A$, according to Figure 2.1 based on thermal fluctuation theory. At this moment, the whole "plastic particle" and every succeeding "plastic particle" formed at the crack tip will be broken. The critical stress intensity factor characterizing the maximum value, K_c, on the $\dot{l} - K$ diagram of Figure 2.2b can be obtained by substituting $(\sigma_p)_{max}$ for σ_p in (2.7); this yields

$$K_c = \frac{U_0}{\upsilon_A}\sqrt{2\pi\lambda_\bullet} \tag{2.12}$$

Based on experimental data on U_0, υ_A, and $\lambda_\bullet$ in[9], the K_c calculated by Equation (2.12) is equal to 19 $N/mm^{3/2}$ for acrylic glass, 29.2 $N/mm^{3/2}$ for capron and 16 $N/mm^{3/2}$ for silicate glass.

According to Equations (2.3), (2.6), and (2.7), it takes the time,

$$t = \tau_0 exp\left(\frac{U_0}{RT} - \frac{1.12\,\upsilon_A \sigma\sqrt{l}}{\sqrt{2\lambda_\bullet}\,RT}\right), \tag{2.13}$$

to break the "plastic particle" at the crack tip, if the external stress σ is stationary during this time. The $R = kN$ in (2.13) is the universal gas constant, and N is the Avogadro number, so that U_0 and υ_A in (2.13) are taken per gram-mole. From here, it follows that the fracture diagram in the present model is described by the following equation

$$\dot{l} = \frac{\lambda_\bullet}{\tau_0} exp\left(\frac{\upsilon_A K_I}{RT\sqrt{2\pi\lambda_\bullet}} - \frac{U_0}{RT}\right) \quad \left(\dot{l} = \frac{\lambda_\bullet}{\tau_0}\right), \tag{2.14}$$

if the subcritical crack growth, Δl, is great compared to $\lambda_\bullet$.

Empirical data on subcritical crack growth in amorphous materials under stationary loads are well described by the exponential law

$$\dot{l} = V_0 e^{\eta K_I}, \tag{2.15}$$

where V_0 and η are some fitting constants. According to Equations (2.14) and (2.15), we have

$$V_0 = \frac{\lambda_*}{\tau_0} exp\left(-\frac{U_0}{RT}\right), \quad \eta = \frac{\upsilon_A}{RT\sqrt{2\pi\lambda_*}}, \tag{2.16}$$

which relate atomic characteristics to the empirical constants of fracture mechanics.

The predictions (2.10) and (2.12) for K_{th} and K_c are shown to be in satisfactory agreement with experimental measurements of $\dot{l} - K$ diagrams[9,19].

The predictions following from the Griffith - Irwin theory are much less successful for polymers, probably because Equation (2.11) does not take into account the straightening and extension of chemical bonds at the crack tip, which is an important property of fracture of the polymer materials. The value of γ in (2.11) considerably exceeds the surface tension in polymers, due to this effect.

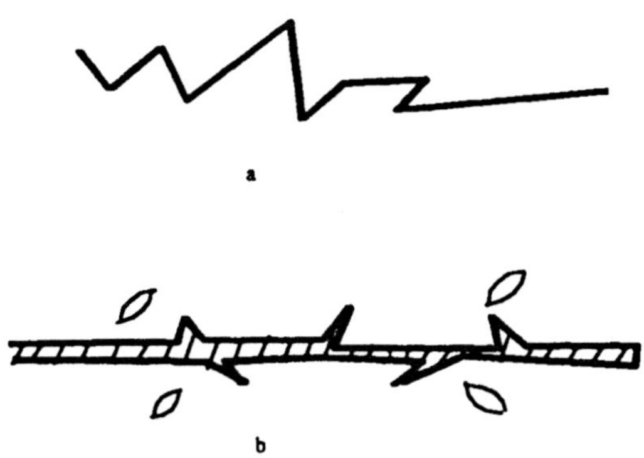

Figure 2.5. Typical mechanisms of energy absorption by a crack in ideally brittle materials like amorphous polymers or glass:
a) The complicated, fractal, nature of the main crack;
b) An additional, induced, cracking in the neighborhood of the main crack

Two other possible mechanisms of energy absorption by fracture in amorphous materials are depicted on Figures 2.5a and 2.5b.

Amorphous materials have several characteristic scales of nonhomogeneity in the structure which are significantly larger than the interatomic spacing. In glass and disordered polymers, this scale has the dimension of a characteristic cluster of atoms (molecules), which usually consists of 10 to 30 atoms (molecules). Organic molecules may be coagulated in lengths up to a coherent meter. In oriented polymers, the scale of nonhomogeneity is associated with the length and thickness of the extended molecule.

Cracks, whose size has the order of the nonhomogeneity scale δ or less, are called microcracks.

The fracture surface of amorphous materials has disordered shape, and the mean characteristic size of the elements of the surface has, obviously, the order of the nonhomogeneity scale. Randomly placed secondary microcracks, which form as the major crack propagates, are essential elements of the surface. Their directions of propagation do not coincide with that of the main crack; and the secondary microcracks "freeze" after the main crack passes them.

If one takes into account the secondary microcracks near the main crack surface, the value of real crack surface area may significantly exceed that of the fictitious surface area which represents the theoretically smooth crack surface. Simple geometrical estimates show that their ratio, Δ can, in principle, have any value (for more detail, see Chapter 8). This means that the crack in an ideally brittle material will, in effect, irreversibly absorb an amount of energy which is Δ times more than that given by the theories based on the smooth crack models.

2.4. Crystals: Dislocation Emission by Thermal Fluctuations

Let us provide some estimates of thermal fluctuations in a crystal. Although dislocations cannot be formed by thermal fluctuations if temperatures and loads are low, temperature will assist the nucleation and movement of dislocations. Suppose the energy of dislocation equals U (per unit thickness of the plane model). The time, t, of the spontaneous nucleation of a dislocation due to thermal fluctuations without loading can be expressed by the Boltzmann equation,

$$t = \tau_o \, exp \frac{U}{kT} \tag{2.17}$$

If t equals say 1 sec for sufficiently high temperature, $T = T_M$, it means that the lattice turns into a state close to liquid; from Equation (2.17), it follows that $U \sim 20\, kT_M$. Taking T_M as the melting temperature gives the value of U to be 3 to 10 eV per atom for many metals, which coincides with the value of the dislocation energy derived from the theory of elasticity.

By means of Equation (2.17), one can conclude also that when, roughly, $T < \frac{1}{2} T_M$, the time t is too small to observe a thermally activated dislocation birth under very low loading. This is the commonly used estimate of creep temperature threshold.

A load reduces the activation energy of dislocation nucleation. The time, t, of the nucleation of a dislocation under a shear stress, τ, due to thermal fluctuations can be expressed by the following equation [2,12]

$$t = \tau_0 \; exp\left[\frac{U}{kT}\left(1 - \frac{\tau}{\tau_s}\right)\right] \tag{2.18}$$

Here, τ_s is the limiting shear load (the Schmid stress of the crystal) when dislocations form very fast. The τ_s is known to be nonzero at the zero absolute temperature.

The value of τ_s essentially depends on the concentration of alloying atoms, impurities, and voids in a lattice. Besides, τ_s depends on temperature. For an estimate, we may take $U = 7eV$ per atom. Then, we have:

when $T = 1°K$ $\qquad \frac{U}{kT} = 10^5$

$$\text{when} \quad T = 1{,}000^\circ\text{K} \qquad \frac{U}{kT} = 100. \tag{2.19}$$

From Equations (2.18) and (2.19), it is seen that when $\tau = 0$, no dislocation can be nucleated for reasonable time due to thermal fluctuations. However, when $\tau \sim \tau_s(T)$, a dislocation will definitely be nucleated in the lattice in this temperature range. Equation (2.18) implies that application of the load τ at temperature T is equivalent to increasing of temperature up to the value of $T=T_1$,

$$T_1 = T\left(1 - \frac{\tau}{\tau_s}\right)^{-1} \tag{2.20}$$

without loading. Hence, we can conclude that if

$$\frac{\tau}{\tau_s} > 1 - \frac{2T}{T_M}, \tag{2.21}$$

thermal fluctuations nucleate a dislocation during a reasonably long span of time (say, one month). If

$$\frac{\tau}{\tau_s} + 2\frac{T}{T_M} < 1, \tag{2.22}$$

thermal fluctuations cannot nucleate a dislocation during that time. The conditions of dislocation nucleation and non-nucleation, Equations (2.21) and (2.22) respectively, are depicted on the plane of τ/τ_s and T/T_M (Figure 2.6).

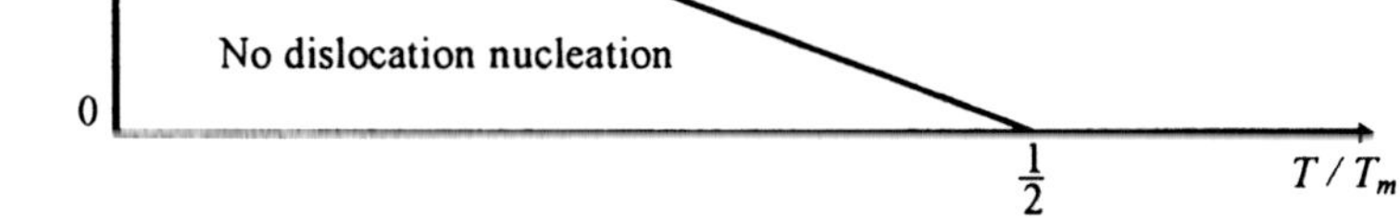

Figure 2.6. Domain of dislocation nucleation

Now, let us use these conditions to determine when a dislocation nucleates at a crack tip. Consider an atomically sharp crack in a crystal for the same specimen and crack geometry of Section 2.3 (Figure 2.3). It has been suggested[13] that there is a certain critical stress intensity factor, K_E, characterizing the crack tip dislocation nucleation. For the corresponding critical value, γ_E, of surface energy, we have, $\gamma_E = K_E^2(1-\nu^2)/2E$.

Estimate γ_E and K_E. When the external load, σ, is very low, so that $K < K_E$, there are no dislocations born near the crack tip, and the crystal response to the load is identical to that of an ideally brittle solid. In this load range, we can apply the "plastic particle" model of Section 2.3 (see Figure 2.4). Consider the crystallographic plane of the most dense atom packing in the "plastic particle" on which the first dislocation nucleation should occur.

The maximum shear stress on this plane, τ_p, as in Equation (2.7), is equal to

$$\tau_p = \frac{\alpha_0 K_I}{\sqrt{2\pi\lambda_\bullet}}, \tag{2.23}$$

where α_0 is a number depending on the orientation of the crack plane with respect to the crystallographic glide-plane. Equations (2.22) and (2.23) show that a dislocation nucleates in the "plastic particle", if

$$\frac{\alpha K_E}{\tau_s\sqrt{2\pi\lambda_\bullet}} + 2\frac{T}{T_M} > 1, \tag{2.24}$$

or

$$K_E > \frac{\tau_s}{\alpha}\sqrt{2\pi\lambda_\bullet}\left(1 - 2\frac{T}{T_M}\right). \tag{2.25}$$

From Equation (2.25), we obtain the threshold of dislocation nucleation,

$$K_E = \frac{\tau_s}{\alpha}\sqrt{2\pi\lambda_\bullet}\left(1 - 2\frac{T}{T_M}\right), \tag{2.26}$$

or the onset of dislocation emission at the crack tip. Put $\lambda_\bullet/\alpha^2 = 10a$ and $T = 0$ for simplicity and estimate K_E and γ_E, for aluminum. We have, according to Honeycomb [20] $\tau_s = 0.75\ MPa$, and $a = 0.285\ nm$, so that $K_E = 100.4\ Pa\cdot m^{1/2} = 3\cdot 10^{-3}\ N/mm^{3/2}$. For aluminum, $\nu = 0.33$ and $E = 70\ GPa$, so that the γ_E corresponding to K_E, is equal to $\gamma_E = 0.1\ \mu J/m^2 = 10^{-4}\ erg/cm^2$. This means that when $K > K_E$ (or $\gamma > \gamma_E$), dislocations are nucleated at a crack-tip by thermal fluctuations. The frequency of dislocation nucleation defined by Equation (2.18) grows fast with the increase of temperature and the stress intensity factor. The values of K_E and γ_E are extremely low. However, the emitted dislocations do not have equilibrium positions and return to the crack tip, because of the tremendous force of crack tip attraction, until K_I increases and the stable equilibrium position appears (for more detail, see Chapter 6).

2.5. Problems

2.1. Consider the fluctuations of the crack tip position, l, and introduce the quadratic fluctuation

$$\Delta^2 = \overline{(\Delta l)^2} = \overline{(l - l_0)^2},$$

where l_0 is the expected value of the random quantity, l. Assume the Gaussian distribution function for l is

$$dp = \frac{1}{\sqrt{2\pi\Delta^2}} exp\left[-\frac{(l-l_0)^2}{2\Delta^2}\right] dl. \tag{2.27}$$

Show that

$$\Delta^2 = \overline{(l-l_0)^2} = \frac{kT}{\eta_l \Gamma}, \tag{2.28}$$

where η_l is a number, and Γ is expressed in atom units. Estimate Δ for glass and other materials.

2.2. Compare two materials with the following properties:

$$\begin{aligned} &\text{1st material:} && U_0 = U_1, && \upsilon_A = \upsilon_1; \\ &\text{2nd material:} && U_0 = U_2, && \upsilon_A = \upsilon_2. \end{aligned} \tag{2.29}$$

Which material has greater longevity, that is, better for making a rod? Consider all possibilities, assume extension and stationary loading.

2.3. Two identical specimens of the same material were subjected to an extension test under stationary stresses, σ_1 and σ_2, at the same temperature. The specimens failed after times t_1 and t_2, respectively. Predict the time before failure if the material is tested under the stationary load σ_3 at the temperature T_3.

2.4. Two identical specimens of the same material were subjected to an extension test under the same stress σ, but at different temperatures, T_1 and T_2. The specimens failed after times t_1 and t_2, respectively. Predict the time before failure if the material is subjected to stress σ_3 at temperature T_3.

2.5. A plate specimen with a small edge crack of initial length, l_0, was subjected to stationary extension by stress σ (see Figure 2.3). Predict the time to failure on the fracture diagrams of the material (Figure 2.2), and consider analytically the case of small scale yielding and the exponential law of Equation (2.15).

2.6. Show that Equation (2.26) predicts a very small threshold of dislocation nucleation, which is caused by a very small value of the Schmid stress, τ_s, in all pure crystals. Show, by using different λ_*'s and materials, that this prediction depends only slightly on the model parameter λ_* and is thus stable with respect to a change of the model.

2.7. Suppose that, by using alloying atoms, by introducing fixed dislocations in the structure, or by diversifying the structure by phase transformation and interphase barriers, one increases the Schmid stress τ_s say 10^2, 10^3 and 10^4 times. How will this influence the prediction of the threshold of dislocation nucleation?

Show that, when the value of τ_s,

$$\frac{\alpha U_0}{\upsilon_A}\left(1-\frac{2T}{T_m}\right)^{-1} \tag{2.30}$$

is achieved, the material becomes ideally brittle, that is, a crack grows with no dislocations emitted. Find τ_s provided that the condition, $K_E = K_{th}$, holds, which is probably the best opportunity for a material designed.

2.8. The basic predictions in this Chapter, Equations (2.10), (2.12) - (2.14), and (2.26), are valid for small scale yielding . Derive the corresponding predictions for any large scale yielding, when the elastic stress intensity factor is meaningless. Use the concept of the Γ defined by Equation (2.5) as the crack driving force or process controlling parameter.

References

1. A. P. Alexandrov and S. N. Zhurkov (1933), *The Phenomena of Brittle Failure* [in Russian], GONTI, Moscow .
2. S. N. Zhurkov (1965), Kinetic concept of the strength of solids (1965), *Int. J. Fracture Mechanics*, **1**(4), pp. 311-323 .
3. V. R. Regel , A. I. Slutsker, and E. E. Tomashevskii (1974), *The Kinetic Nature of the Strength of Solids* [in Russian], Nauka, Moscow .
4. G. M. Bartenev and Yu. S. Zuev (1964), *The Strength and Failure of Highly Elastic Materials* [in Russian], Khimiya, Moscow .
5. G. M. Bartenev (1974), *Superstrong and Highly Strong Inorganic Glasses* [in Russian], Stroiizdat, Moscow .
6. V. E. Gul' (1978), *The Structure and Strength of Polymers* [in Russian], Khimiya, Moscow.
7. G. M. Bartenev (1984), *The Strength and Fracture Mechanics of Polymers* [in Russian], Khimiya, Moscow.
8. S. N. Zhurkov (1986), The kinetic concept of the strength of solids, *Vestnik AN SSSR*, **3**, pp. 46-52.
9. G. M. Bartenev (1989), The thermofluctuation theory of strength and the Griffith criterion, *Problemi Prochnosti [Strength Problems]*, **11**, p. 78.
10. A. A. Griffith (1920), The phenomena of fracture and flow in solids, *Phil. Trans. Roy. Soc.*, **A221**, 163-198; republished in *Fracture: A Topical Encyclopedia of Current Knowledge*, G. Cherepanov (ed.), Krieger Publ., Melburne, USA (1996)
11. G. R. Irwin (1958), Fracture, *Handbook der Physik*, **6**, Springer Verlag, Berlin.
12. G. P. Cherepanov (1994), Nanofracture mechanics approach to dislocation generation and fracturing, *Appl. Mech. Rev.*,**47**(6), part. 2, pp. S326 -330 [Invited paper at the 12th U.S. National Congress of Mechanics].
13. G. P. Cherepanov (1974), Mechanika Khrupkogo Razrushenia, Moscow, Nauka; English edition (1979), *Mechanics of Brittle Fracture*, R. de Wit and W. C. Cooley (eds.), McGraw-Hill, New York.
14. *Computational Methods in the Mechanics of Fracture*, (1986), S. Atluri (ed.), Elsevier Science Publishing, Amsterdam.
15. F. W. Brust, T. Nishioka, S. N. Atluri, and M. Nakagaki (1985), Further studies on elastic-plastic stable fracture utilizing the T*-integral, *Eng. Fract. Mech.*, **22**, pp. 1097-1103 .
16. F. W Brust, J. J. Gowan, and S. N. Atluri (1986), A combined numerical/experimental study of ductile crack growth after large unloading using T*, J, and CTOD criteria, *Ing. Fract. Mech.*, **23**, pp. 537-551.
17. F. W. Brust and S. N. Atluri (1986), Studies on creep crack growth using T*-integral, *Eng. Fract. Mech.*, **23**, pp. 551-575.
18. T. Nishioka (1996), On the dynamic J-integral in dynamic fracture mechanics, in *Fracture: A Topical Encyclopedia of Current Knowledge*, G. Cherepanov (ed.), Krieger Publ., Melbourne, USA.
19. G. P. Cherepanov (1989), Fracture mechanics and kinetic theory of strength, *Problemy Prochnosti [Strength Problems]*, **11**, pp. 3-8.
20. R. W. Honeycomb (1968), *Plastic Deformation of Metals*, Edward Arnold Publ., London.

CHAPTER 3

CRACK NUCLEATION

The present chapter treats the theory of some basic mechanisms of microcrack nucleation in polycrystalline and amorphous materials. The following methods are used to achieve the results:

— Methods of the theory of functions of a complex variable demonstrating the refined work with various singularities of the physical field on the boundary, at infinity, and inside the domain, and the techniques of the Riemann boundary value problem (Section 3.1);

— Mellin transform method and Wiener - Hopf vs. Noble - Jones techniques of solving mixed boundary value problems (Section 3.2);

— The work with complicated singularities and driving forces, and thermodynamics methods (Section 3.3).

Modeling arts, formulation of well-posed boundary value problems of mixed type, and the skills of physical interpretation and discussion of mathematical solutions are also important (Sections 3.1 - 3.3).

3.1 Impinging a Dislocation upon an Interface

The Zener - Stroh mechanism of microcrack nucleation by a dislocation pileup or band coming across an interface has been one of the basic mechanisms of microcrack nucleation in polycrystalline materials. For different elastic properties of the composite constituents, this problem is accurately solved in the present Section. The minimum number of elementary dislocations in the band necessary to form a microcrack, as well as the interface microcrack length are found in terms of physical properties and the number of elementary dislocations in the band. A band tip has a stronger singularity than a pileup, and therefore the solution can serve as an upper bound estimate to the solution for the pileup. The results of this Section may be used for the direct measurement of the true interface surface energy (adhesion energy) by an experimental observation of an interface microcrack nucleation.

Introduction

As far ago as 1948, Zener advanced a conjecture[1] that an opening mode microcrack in a polycrystalline material can be nucleated as a result of an encounter of a pileup or band of dislocations with an interface, due to the stress concentration at the head of the pileup or band on the boundary of different components. Stroh investigated this mechanism theoretically in the particular case when the elastic properties of components were the same[2,3]. This case was studied in more detail in succeeding publications by Bilby and Hewitt[4], Smith and Barnby[5], Francois and Wilshaw[6], Nicolas[7], Morii and Nemat-Nasser[8], and Ashby and Hallam[9]. The more difficult problem when the elastic properties of the components are different, was solved by this author[10].

The accurate solution to the latter problem in the case of a Luders-Chernov band is provided in the succeeding text. The formulation of the boundary value problem uses the representation equations advanced earlier[11]. The problem is reduced to the Riemann problem and the solution is given in explicit form. The equations of the microcrack length and critical number of dislocations necessary to nucleate a microcrack, are derived.

The head of a dislocation band has a stronger singularity than a pileup. Therefore, the solution can serve as an upper bound estimate to that to a dislocation pileup containing the same number of elementary dislocations. Using the solution and an experimental observation of an interface microcrack near a dislocation band provides means of directly determining the true interface surface energy (adhesion energy).

Formulation of the problem

Let us consider the dislocation band, also called the Luders - Chernov band, coming across an interface between two different materials and nucleating a microcrack (Figure 3.1). The *x*-axis is assumed to coincide with an interface, while the origin of the *xy* rectangular coordinate system coincides with the point of intersection of the band and interface. The angle α between the x-axis and band is called the angle of incidence or attack.

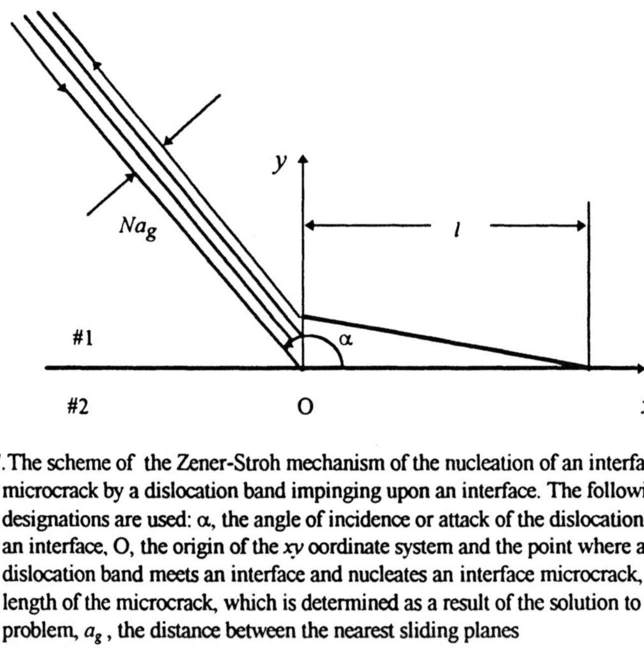

Figure 3.1. The scheme of the Zener-Stroh mechanism of the nucleation of an interface microcrack by a dislocation band impinging upon an interface. The following designations are used: α, the angle of incidence or attack of the dislocation band to an interface, O, the origin of the *xy* oordinate system and the point where a dislocation band meets an interface and nucleates an interface microcrack, *l*, the length of the microcrack, which is determined as a result of the solution to the problem, a_g, the distance between the nearest sliding planes

The plain strain boundary condition equation can be written as follows:

For $y = \pm 0 \quad 0 < x < l$ $$\sigma_y = \tau_{xy} = 0; \tag{3.1}$$

For $y = \pm 0 \quad x < 0$ and $x > l$ $$[\sigma_y - i\tau_{xy}] = 0, \quad [u_x + iu_y] = 0; \tag{3.2}$$

For $y = x \tan\alpha \quad y > 0$ $$[\sigma_x + \sigma_y] = 0, [\sigma_y - \sigma_x + 2i\tau_{xy}] = 0, [u_x + iu_y] = Be^{i\alpha} \tag{3.3}$$

Here, $B = N\,b_e$, where N is the number of elementary edge dislocations in the band, b_e is the absolute value of the displacement discontinuity on an elementary edge dislocation, l is the microcrack length, σ_x, σ_y, and τ_{xy} are the stresses, and u_x and u_y are displacements. The jump $[Q]$ of the value Q is calculated according to the rule of counter-clockwise travel (Figure 3.2).

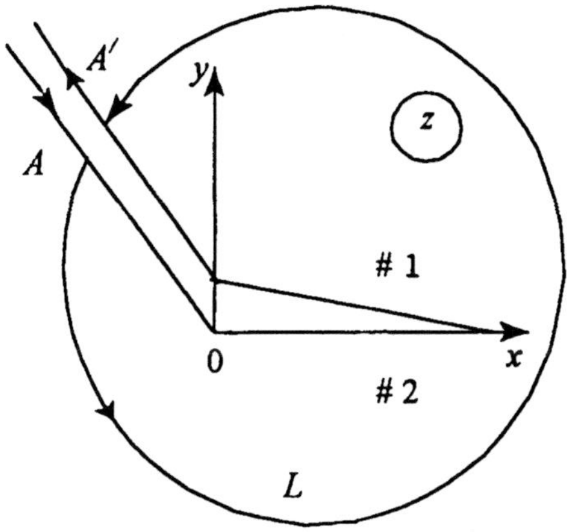

Figure 3.2. The scheme describing the value of a discontinuity, $[Q]$, of a quantity, Q, during travel along the path L from the point A to the point A' located on the opposite bank of a band

For $z = x + iy \to \infty$ all the stresses vanish, i. e.

$$\sigma_x\,,\ \ \sigma_y\,,\ \text{and}\ \ \tau_{xy} \to 0 \tag{3.4}$$

At the coordinate origin, as $z \to 0$, here is a complicated singularity resulting from the crack-tip and dislocation-head singularities. The combined singularity represents a superposition of the logarithmic singularity of a dislocation band and the interface crack-tip singularity which is known in the form:

For $\quad y = 0 \qquad\qquad x < 0$

$$\sigma_y - i\tau_{xy} = \frac{K_I - iK_{II}}{\sqrt{2\pi r}}\exp\left(-i\beta\ln\frac{r}{l}\right) \qquad (r = |x| << l) \tag{3.5}$$

where β is a small dimensionless elastic constant, and K_I and K_{II} are the stress intensity factors.

The stress intensity factors and the crack length are to be sought as a result of the solution. The stresses and strains are represented by the Kolosov - Muskhelishvili formulae:

$$\begin{aligned} &\sigma_x + \sigma_y = 4Re\,\varphi_n'(z) \qquad (z = x + iy) \\ &\sigma_x - \sigma_y + 2i\,\tau_{xy} = 2[\bar{z}\,\varphi_n''(z) + \psi_n'(z)] \\ &2\,\mu_n(u_x + i\,u_y) = \kappa_n\varphi_n'(z) - z\overline{\varphi_n'(z)} - \overline{\psi_n'(z)} \end{aligned} \tag{3.6}$$

Here, $\varphi_n(z)$ and $\psi_n(z)$ are the analytic functions of the complex variable $z = x + iy$, μ_n is the shear modulus, $\mu_n = E_n/(2+2\nu_n)$, and $\kappa_n = 3\text{-}4\nu_n$, where E_n and ν_n are Young's modulus and Poisson's ratio respectively. The subscript $n = 1$ corresponds to the upper half-plane and $n = 2$ to the lower half-plane.

In this problem the sought functions φ_n and ψ_n are multy-valued, in contrast to the usual case of the single-valued potentials studied in Muskhelishvili.[13]

In order to determine the polar components of stresses and displacements the following equations will be used below:

$$u_r + iu_\theta = e^{-i\theta}(u_x + iu_y) \tag{3.7a}$$

$$\sigma_r + \sigma_\theta = \sigma_x + \sigma_y \tag{3.7b}$$

$$\sigma_\theta - \sigma_r + 2i\tau_{r\theta} = e^{2i\theta}(\sigma_y - \sigma_x + 2i\tau_{xy}) \tag{3.7c}$$

where

$$z = x + iy = re^{i\theta}. \tag{3.8}$$

The driving or propelling force of the dislocation band under study may be produced by the stress $\sigma_x = \sigma_1$ in the upper half-plane and matching the analogous $\sigma_x = \sigma_2$ in the lower half-plane, where $\mu_2(\kappa_1 + 1)\sigma_1 = \mu_1(\kappa_2 + 1)\sigma_2$. These stresses do not influence the interface microcrack nucleation, and are omitted in the succeeding study.

The analytical solution

We first study the stress and strain distribution in the composite elastic medium at the moment preceding the nucleation of an interface crack (when $l = 0$ in Figure 3.1). This distribution is defined by the following functions:

For $Im(z) > 0$:

$$\varphi_1(z) = C_1 \ln z\,; \quad \psi_1(z) = C_2 \ln z\,, \quad \chi_1(z) = C_3 \ln z\,, \tag{3.9a}$$

For $Im(z) < 0$:

$$\varphi_2(z) = D_1 \ln z\,; \quad \psi_2(z) = D_2 \ln z\,, \quad \chi_2(z) = D_3 \ln z\,, \tag{3.9b}$$

where $\chi(z) = \int \psi(z)dz$, $\ln z = \ln|z| + i\theta$ is the principal value of the logarithmic function in the z - plane cut along the ray $z = re^{i\alpha}$; and C_1, C_2, C_3, D_1, D_2, and D_3 are some complex constants, whose physical meaning we discover by investigating the singular point O.

Let us consider an arbitrary arc AB in the upper or lower half-plane which runs from A to B. The following equations hold:

$$X + iY = -i\left[\varphi(z) + z\overline{\varphi'_n(z)} + \overline{\psi'_n(z)}\right]_A^B \tag{3.10a}$$

$$2\mu_n[u_x + iu_y]_A^B = [\kappa_n\varphi_n(z) - z\overline{\varphi'_n(z)} - \overline{\psi'_n(z)}]_A^B \tag{3.10b}$$

$$M = Re[\chi_n(z) - z\psi_n(z)z\bar{z}\varphi_n'(z)]_A^B \tag{3.10c}$$

Here, (X, Y) is the resultant vector of forces acting on the arc AB from the right-hand side, M is the moment of these forces with respect to the coordinate origin, the square brackets $[Q]_A^B$ designate the difference, $Q(B)$ - $Q(A)$, of any quantity Q in the brackets. Let AB be a closed contour encompassing the coordinate origin counter-clockwise. Additionally, let the points, $A = re^{i\alpha}$ and $B = re^{i(2p-\alpha)}$ be on opposite sides of the band . Substitute Equations (3.9) into Equations (3.10) and obtain the following equations:

$$X + iY = \pi(C_1 + D_1 - \overline{C_2} - \overline{D_2}) \; . \tag{3.11a}$$

$$2Be^{i\alpha} = \frac{\pi i}{\mu_1}\left(\kappa C_1 + \overline{C_2}\right) + \frac{\pi i}{\mu_2}\left(\kappa_2 D_1 + \overline{D_2}\right), \tag{3.11b}$$

$$M = -\pi\, Im(C_3 + D_3) \; . \tag{3.11c}$$

We also satisfy the boundary condition equations on the interface $z = re^{i0}$ and $z = re^{\pi i}$, Equations (3.2), using Equations (3.6) and (3.9):

$$\overline{C_1} + C_2 = \overline{D_1} + D_2 \; ; \tag{3.12a}$$

$$\mu_2\left(\kappa_1 C_1 - \overline{C_2}\right) = \mu_1\left(\kappa_2 D_1 - \overline{D_2}\right) \tag{3.12b}$$

From Equations (3.11) and (3.12) for M = 0, it follows that:

$$C_1 = \frac{2\mu_1\mu_2 Be^{i\alpha} + \mu_1(X + iY)}{2\pi i(\mu_2 + \mu_1\kappa_2)}, \qquad C_2 = \frac{2\mu_1\mu_2 Be^{-i\alpha} - \mu_1\kappa_2(X - iY)}{-2\pi i(\mu_2 + \mu_1\kappa_2)}$$

$$D_1 = \frac{\mu_2(X + iY)}{2\pi i(\mu_2 + \mu_1\kappa_2)} + \frac{\mu_1\mu_2 Be^{i\alpha}}{\pi i(\mu_2 + \mu_1\kappa_2)}, \qquad D_2 = \frac{\mu_2\kappa_1(X - iY)}{2\pi i(\mu_1 + \mu_2\kappa_1)} - \frac{\mu_1\mu_2 Be^{-i\alpha}}{\pi i(\mu_1 + \mu_2\kappa_1)}$$

The singular solution:

$$\psi_1(z) = \frac{\mu_1(M + iC_0)}{(\mu_1 + \mu_2)\pi i z}; \quad \psi_2(z) = -\frac{\mu_2(M + iC_0)}{(\mu_1 + \mu_2)\pi i z}; \tag{3.13}$$
$$\varphi(z) = 0 \qquad (B = 0, \;\; X = Y = 0)$$

describes the interface point singularity combining the concentrated force moment, M, and the point source of the compression type, when $C_0 < 0$, or extension type, when $C_0 > 0$, where C_0 is a real constant.

Another singular solution:

$$\varphi_1(z) = \frac{\mu_1(X+iY)}{2\pi i(\mu_1+\mu_2\kappa_1)} \ln z; \qquad \psi_1(z) = \frac{\mu_1\kappa_2(X-iY)}{2\pi i(\mu_2+\mu_1\kappa_2)} \ln z \tag{3.14a}$$

$$\varphi_2(z) = \frac{\mu_2(X+iY)}{2\pi i(\mu_2+\mu_1\kappa_2)} \ln z; \qquad \psi_2(z) = \frac{\mu_2\kappa_1(X-iY)}{2\pi i(\mu_1+\mu_2\kappa_1)} \ln z \tag{3.14b}$$

describes the concentrated force (X, Y) applied to the interface.
The singular solution:

$$\varphi_1(z) = \frac{\mu_1\mu_2 B}{\pi i(\mu_1+\mu_2\kappa_1)} e^{i\alpha} \ln z\,; \quad \psi_1(z) = \frac{-\mu_1\mu_2 B}{\pi i(\mu_2+\mu_1\kappa_2)} e^{-i\alpha} \ln z \tag{3.15a}$$

$$\varphi_2(z) = \frac{\mu_1\mu_2 B}{\pi i(\mu_2+\mu_1\kappa_2)} e^{i\alpha} \ln z; \quad \psi_2(z) = \frac{-\mu_1\mu_2 B}{\pi i(\mu_1+\mu_2\kappa_1)} e^{-i\alpha} \ln z \tag{3.15b}$$

describes the interface dislocation band with the Burgers vector, $Be^{i\alpha}$.

We may superpose all the singularities, but we confine ourselves to the particular case $B \neq 0$, $X = Y = 0$, $M = 0$, $C_0 = 0$, which is of immediate interest for our problem. Equations (3.15) and (3.6) provide the stress and strain distribution in the composite body just before the formation of a crack. It is more convenient to introduce the analytic functions $\Phi_n(z)$ and $\Psi_n(z)$, where

$$\Phi_n(z) = \varphi_n'(z), \quad \Psi_n(z) = \psi_n'(z), \quad n = 1 \text{ or } 2. \tag{3.16}$$

The boundary condition Equations (3.3) imply the continuity of $\Phi_n(z)$ and $\Psi_n(z)$ on the slip line $z = re^{i\alpha}$ $(r > 0)$, and so these functions are single-valued and analytic everywhere outside the cut $z = x \pm i0$ $0 < x < l$ in the z - plane. According to Equations (3.7) they have to meet the following condition at infinity:

For $z \to \infty$

$$\Phi_1(z) = \frac{\mu_1\mu_2 B}{\pi i z(\mu_1+\mu_2\kappa_1)} e^{i\alpha}; \qquad \Psi_1(z) = \frac{-\mu_1\mu_2 B}{\pi i z(\mu_2+\mu_1\kappa_2)} e^{-i\alpha} \tag{3.17a}$$

$$\Phi_2(z) = \frac{\mu_1\mu_2 B}{\pi i z(\mu_2+\mu_1\kappa_2)} e^{i\alpha}; \qquad \Psi_2(z) = \frac{-\mu_1\mu_2 B}{\pi i z(\mu_1+\mu_2\kappa_1)} e^{-i\alpha} \tag{3.17b}$$

From Equations (3.6), we get:

$$\sigma_y + i\tau_{xy} = \Phi_n(z) + \overline{\Phi_n}(z) + z\Phi_n'(z) + \Psi_n(z) \tag{3.18}$$

with $n = 1$ for $Im(z) > 0$, and $n = 2$ for $Im(z) < 0$.

Using Equations (3.18) we can write the boundary condition equations (3.1) as follows:

For $z = x + i0$ $\quad 0 < x < l$: $\quad \Phi_1 + \overline{\Phi_1} + z\Phi_1' + \Psi_1 = 0$. (3.19a)

For $z = x - i0$ $\quad 0 < x < l$: $\quad \Phi_2 + \overline{\Phi_2} + z\Phi_2' + \Psi_2 = 0$. (3.19b)

Below we use the following functions introduced by this author[11] :

$$\Phi(z) = \Phi_1(z) \qquad \text{for } Im(z)>0 \qquad (3.20a)$$

$$\Phi(z) = \frac{\mu_1(1+k_2)}{\mu_1 + k_1\mu_2}\Phi_2(z) + \frac{\mu_2 - \mu_1}{\mu_1 + \mu_2\kappa_1}\left[\overline{\Phi_1}(z) + z\overline{\Phi_1'}(z) = \overline{\Psi_1}(z)\right] \qquad \text{for } Im(z)<0 \qquad (3.20b)$$

$$\Omega(z) = \overline{\Phi_1}(z) + z\overline{\Phi_1'} + \overline{\Psi_1}(z) \qquad \text{for } Im(z)<0 \qquad (3.21a)$$

$$\Omega(z) = \frac{\mu_1(1+\kappa_2)}{\mu_2 + \kappa_2\mu_1}\left[\overline{\Phi_2}(z) + z\overline{\Phi_2'}(z) + \overline{\Psi_2}(z)\right] - \frac{\mu_1\kappa_2 - \mu_2\kappa_1}{\mu_2 + \mu_1\kappa_2}\Phi_1(z) \qquad \text{for } Im(z)>0 \qquad (3.21b)$$

Here the short bar above an analytical function, e. g. $f(z)$, designates another analytical function, $\bar{f}(z)$, defined as follows $\bar{f}(z) = \overline{f(\bar{z})}$. Thus the function $\bar{f}(z)$ is analytic in the lower half-plane $Im(z) < 0$, if the function $f(z)$ is analytic in the upper half-plane $Im(z) > 0$. From this, it follows that the functions $\Phi(z)$ and $\Omega(z)$ are analytic in both the upper and lower half-planes z. Moreover, as was shown[11] they are continuous on those parts of the x-axis where the dissimilar elastic media are bonded without sliding. In our problem, the functions $\Phi(z)$ and $\Omega(z)$, are analytic everywhere in the entire complex plane z cut along $z = x \pm 0i$, $0 < x < l$ where they have a discontinuity. Using $\Phi(z)$ and $\Omega(z)$ according to (3.20) and (3.21), we transform the Kolosov - Muskhelishvili representations (3.6) to the following form:

For the upper half-plane $Im(z) > 0$:

$$\sigma_x + \sigma_y = 4Re\,\Phi(z) \qquad (3.22a)$$

$$\sigma_x - i\,\tau_{xy} = \Phi(z) + \Omega(\bar{z}) + (z - \bar{z})\overline{\Phi'(z)} \qquad (3.22b)$$

$$2\mu_l\left(\frac{\partial u}{\partial x} + i\frac{\partial v}{\partial x}\right) = \kappa_l\Phi(z) - \Omega(\bar{z}) - (z - \bar{z})\overline{\Phi'(z)} \qquad (3.22c)$$

For the lower half-plane $Im(z) < 0$:

$$\sigma_x + \sigma_y = 4\frac{\mu_1 + \mu_2\kappa_1}{\mu_1(1+\kappa_2)} Re\left(\Phi(z) + \frac{\mu_1 - \mu_2}{\mu_1 + \mu_2\kappa_1}\Omega(z)\right); \quad (3.23a)$$

$$\sigma_x - i\,\tau_{xy} = \frac{\mu_1 + \mu_2\kappa_1}{\mu_1\,(1+\kappa_2)}\Phi(z) + \frac{\mu_1 - \mu_2}{\mu_1(1+\kappa_2)}\Omega(z) + \frac{\mu_1\kappa_2 + \mu_2}{\mu_1(1+\kappa_2)}\Omega(\bar{z}) + \frac{\mu_1\kappa_2 - \mu_2\kappa_1}{\mu_{1\,l}(1+\kappa_2)}\Phi'(\bar{z})$$

$$+ (z - \bar{z})\left[\frac{\mu_1 + \mu_2\kappa_1}{\mu_1(1+\kappa_2)}\overline{\Phi'(z)} + \frac{\mu_1 - \mu_2}{\mu_1(1+\kappa_2)}\overline{\Omega'(z)}\right]; \quad (3.23b)$$

$$2\mu_2\left(\frac{\partial u}{\partial x} + i\frac{\partial v}{\partial x}\right) = \kappa_2\frac{\mu_1 + \mu_2\kappa_1}{\mu_1(1+\kappa_2)}\Phi(z) + \kappa_2\frac{\mu_1 - \mu_2}{\mu_1(1+\kappa_2)}\Omega(z) - \frac{\mu_1\kappa_2 + \mu_2}{\mu_1(1+\kappa_2)}\Omega(\bar{z})$$

$$- \frac{\mu_1\kappa_2 - \mu_2\kappa_1}{\mu_1(1+\kappa_2)}\Phi(\bar{z}) - (z - \bar{z})\left[\frac{\mu_1 + \mu_2\kappa_1}{\mu_1(1+\kappa_2)}\overline{\Phi(z)} + \frac{\mu_1 - \mu_2}{\mu_1\,(1+\kappa_2)}\overline{\Omega'(z)}\right] \quad (3.23c)$$

Using the new representations for stresses in Equations (3.22) and (3.23) we convert Equations (3.19) into the following form:

For $z = x \pm i0$ $\quad 0 < x < l$

$$\Phi^{+} + \Omega^{-} = 0 \quad (3.24a)$$

$$(\mu_1 + \mu_2\kappa_1)\Phi^{-} + (\mu_1 - \mu_2)\Omega^{-} + (\mu_1\kappa_2 - \mu_2\kappa_1)\Phi^{+} + (\mu_1\kappa_2 + \mu_2)\Omega^{-} = 0 \quad (3.24b)$$

Here: $\quad \Phi^{\pm} = \lim_{y\to\pm 0}\Phi(x+iy)\,; \quad \Omega^{\pm} = \lim_{y\to\pm 0}\Omega(x+iy)$

The representation given by Equations (3.22) and (3.23), which reduces the problem to only two potentials like the corresponding homogeneous problem, makes it possible to derive the solution any number of arbitrary loaded interface cracks along a straight line[11]. Some of these solutions were later derived by England, Sih, Erdogan, Rice, and other researchers using different particular methods.

Equations (3.24) represent the classic Riemann problem for two analytic functions which can be efficiently solved in this case. We divide Equation (3.24b) by $-\mu_1(1+\kappa_2)$ or $\mu_2(1+\kappa_1)$ and add to Equation (3.24a). As a result, we obtain:

$$F_1^{+} = F_1^{-} \quad \text{for} \quad z = x \pm i0 \quad 0 < x < l, \quad (3.25)$$

where

$$F_1(z) = \frac{(\mu_1 + \mu_2\kappa_1)\Phi(z) - (\mu_2 + \mu_1\kappa_2)\Omega(z)}{\mu_1(1+\kappa_2)}; \quad (3.26)$$

and

$$F_2^+ + mF_2^- = 0 \quad \text{for} \quad z = x \pm i0 \qquad 0 < x < l, \tag{3.27}$$

where

$$F_2(z) = \Phi(z) + \Omega(z), \qquad m = \frac{\mu_1 + \mu_2\kappa_1}{\mu_2 + \mu_1\kappa_2}; \tag{3.28}$$

This is a particular case of the Riemann problem for one function, which was studied by Carleman, see in Gakhov[14]. In view of Equations (3.25) the function $F_1(z)$ is analytic in the entire complex plane z except possibly for the coordinate origin, z = 0, where it may have a simple pole. Hence, we obtain:

$$F_1(z) = \frac{A}{z}, \tag{3.29}$$

where A is a certain constant.

For $z \to \infty$

$$F_1(z) = O\left(1/z^2\right) \tag{3.30}$$

Comparing Equations (3.29) and (3.30) we get:

$$A = 0, \text{ i.e. } F_1(z) = 0. \tag{3.31}$$

The solution to the Riemann problem in Equations (3.27) in the class of functions having integrable singularities at the crack-tips, $z = 0$ and $z = l$, is of the form

$$F_2(z) = \frac{D}{z^{\frac{1}{2}-i\beta}(z-1)^{\frac{1}{2}+i\beta}}, \tag{3.32}$$

where D is a certain constant and β equals

$$\beta = \frac{1}{2\pi} \ln m = \frac{1}{2\pi} \ln \frac{\mu_1 + \mu_2\kappa_1}{\mu_2 + \mu_1\kappa_2}. \tag{3.33}$$

For the particular case $\nu_1 = \nu_2 = 1/2$ (incompressible materials) in plane strain we have $m = 1$ and $\beta = 0$. In Equation (3.32) we choose the branch of the function $z^{1/2+i\beta}(z-l)^{1/2-i\beta}$ which is single-valued in z outside the cut along the segment $(0,l)$ of the real axis and tends to $z + O(1/2)$ as z tends to infinity.

According to Equations (3.17), (3.20), (3.21) and (3.28), $F_2(z)$ should be constant at infinity:

For $z \to \infty$

$$F_2(z) = \frac{\mu_1\mu_2 B}{2\pi i} e^{i\alpha} \frac{\mu_1(1+\kappa_2) + \mu_2(1+\kappa_1)}{(\mu_1 + \mu_2\kappa_1)(\mu_2 + \mu_1\kappa_2)} \tag{3.34}$$

Comparing Equations (3.32) and (3.34) we obtain

$$D = \frac{\mu_1 \mu_2 B}{\pi i} e^{i\alpha} \frac{\mu_1(1+\kappa_2)+\mu_2(1+\kappa_1)}{(\mu_1+\mu_2\kappa_1)(\mu_2+\mu_1\kappa_2)} \tag{3.35}$$

Using Equations (3.26), (3.28), (3.29) and (3.32) we arrive at the following solution:

$$\Phi(z) = \frac{\mu_1 \mu_2 B e^{i\alpha}}{\pi i(\mu_1+\mu_2\kappa_1) z^{\frac{1}{2}-i\beta} (z-l)^{\frac{1}{2}+i\beta}} \tag{3.36a}$$

$$\Omega(z) = \frac{\mu_1 \mu_2 B e^{i\alpha}}{\pi i(\mu_2+\mu_1\kappa_2) z^{\frac{1}{2}-i\beta} (z-l)^{\frac{1}{2}+i\beta}} \tag{3.36b}$$

Equations (3.22), (3.23) and (3.36) provide the explicit analytical expression for the stresses and displacements.

Discussion of the solution

Equation (3.36) shows that near the right-hand tip of the microcrack we have the following singular field:

$$\Phi(z) = \frac{\mu_1 \mu_2 N b_e e^{i\alpha}}{\pi i(\mu_1+\mu_2\kappa_1) l^{\frac{1}{2}-i\beta} \varepsilon^{\frac{1}{2}+i\beta}} + O(1) \tag{3.37a}$$

$$\Omega(z) = \frac{\mu_1 \mu_2 N b_e e^{i\alpha}}{\pi i(\mu_2+\mu_1\kappa_2) l^{\frac{1}{2}-i\beta} \varepsilon^{\frac{1}{2}+i\beta}} + O(1), \tag{3.37b}$$

where $\varepsilon = z - l = |\varepsilon|\, e^{i\theta} \qquad (|\varepsilon| \ll l, \ -\pi<\theta<\pi)$.

Using Equations (3.22b) and (3.37) we find the stresses on the continuation of the microcrack (for $\theta = 0$) in the following form

$$\sigma_y - i\,\tau_{xy} = \frac{D}{l^{\frac{1}{2}-i\beta} \varepsilon^{\frac{1}{2}+i\beta}} = \frac{D}{\sqrt{l\varepsilon}} exp\left[-i\beta\, ln\left(\frac{\varepsilon}{l}\right)\right] \tag{3.38}$$

Hence the stress intensity factors at the right-hand end of the microcrack are given by :

$$K_I - i\,K_{II} = \sqrt{\frac{2\pi}{l}} N \frac{e^{i\alpha}}{i} 2\Gamma \frac{l_0}{b_e} \tag{3.39a}$$

i.e.

$$K_I = 2\sqrt{\frac{2\pi}{l}} N\Gamma \frac{l_0}{b_e} \sin\alpha; \qquad K_{II} = 2\sqrt{\frac{2\pi}{l}} N\Gamma \frac{l_0}{b_e} \cos\alpha \tag{3.39b}$$

where Γ is the energy flow rate into the right-hand tip of the interface microcrack

$$\begin{aligned}\Gamma &= \frac{(\mu_1 + \mu_2\kappa_1)(\mu_2 + \mu_1\kappa_2)}{4\mu_1\mu_2[\mu_1(1+\kappa_2) + \mu_2(1+\kappa_1)]}\left(K_I^2 + K_{II}^2\right) \\ &= \frac{2\pi}{l}|D|^2 \frac{(\mu_1 + \mu_2\kappa_1)(\mu_2 + \mu_1\kappa_2)}{4\mu_1\mu_2[\mu_1(1+\kappa_2) + \mu_2(1+\kappa_1)]}\end{aligned} \tag{3.40}$$

and l_0 is a specific length which equals

$$l_0 = \frac{\mu_1\mu_2 b_e^2[\mu_1(1+\kappa_2) + \mu_2(1+\kappa_1)]}{4\pi\gamma_{fm}(\mu_1 + \mu_2\kappa_1)(\mu_2 + \mu_1\kappa_1)} \tag{3.41}$$

During the motion of the right-hand crack tip along the interface, the value of Γ equals $2\gamma_{fm}$ where γ_{fm} is the interface surface energy of adhesion of two different media, $2\gamma_{fm} = \gamma_1 + \gamma_2 - \gamma_{12}$, where γ_1 and γ_2 are the respective surface energies of materials 1and 2, and γ_{12} is the surface energy of bonding between materials 1 and 2 on the continuation of the microcrack. The left-hand end of the microcrack consists of the dislocation band, and cannot propagate.

From Equation (3.40) we find the length of the stable interface microcrack to be

$$l = \frac{|D|^2}{2E_{fm}} \frac{(\mu_1 + \mu_2\kappa_1)(\mu_2 + \mu_1\kappa_2)}{4\mu_1\mu_2[\mu_1(1+\kappa_2) + \mu_2(1+\kappa_1)]} \tag{3.42}$$

or, using also Equation (3.35)

$$l = l_0 N^2 \tag{3.43}$$

where N is the number of elementary dislocation in the band.

The value of l_0 was calculated for various, widely used composite pairs with the help of Equation (3.41). The results of calculations are given in Table 3.1. As compared to a pileup having a singularity $r^{-1/2}$ in stresses, the band head singularity, r^{-1} in stresses, is obviously stronger. The dislocation band with the same number of dislocations is apparently more severethan the corresponding pileup. Therefore, the solution can serve an upper bound for the latter case.

The particular case of a homogeneous material, $\mu_1 = \mu_2 = \mu$ and $\nu_1 = \nu_2 = \nu$, is of independent interest. This case applies to grains of many metals containing second-phase particles which exist as precipitates within the parent crystals and are often introduced deliberately in order to inhibit dislocation motion and thus raise the yield stress. When pileups or bands occur near deposits of these particles segregated usually in the form of layers, either within a grain or at a grain boundary, a microcrack can form within the precipitate if it is sufficiently brittle. It is generally recognized that such microcracks constitute the dominant fracture mechanism in many steels and alloys.

Table 3.1. The specific lengths for various metal-metal or metal-ceramic pairs

[Here γ is the interface surface energy, designated as γ_{fm} in the text ($\gamma \equiv \gamma_{fm}$). The value of γ of the alloys at the melting temperature, T_m, was taken from Miedema[16], Miedema and den Broeder[17], Murr et al.[18], and Rose et al.[19] The value of γ at 300 K was found using the common linear approximation of temperature dependency of γ.]

Composite Pairs	μ_1 GPa	μ_2 GPa	ν_1	ν_2	b_e nm	$\gamma(T_m)$ $\frac{erg}{cm^2}$	$\gamma(300K)$ $\frac{erg}{cm^2}$	$l_0(T_m)$ nm	$l_0(300K)$ nm	$\frac{b_e}{2l_0}$ at T_m	$\frac{b_e}{2l_0}$ at $300K$
Cu - Al_2O_3	48.32	174.2	0..34	0.22	0.256	475	699	0.611	0.415	0.21	0.31
Ni - Al_2O_3	80.15	174.2	0.31	0.22	0.249	645	986.7	0.588	0.384	0.21	0.32
Ag - Al_2O_3	25.6	174.2	0.38	0.22	0.289	735	1.062.1	0.318	0.22	0.45	0.66
Au - Al_2O_3	24.2	174.2	0.43	0.22	0.288	530	840.6	0.447	0.282	0.32	0.51
Fe - Al_2O_3	73	174.2	0.42	0.22	0.258	800	1.248.6	0.533	0.341	0.24	0.38
Cr - Ag	115	25.6	0.19	0.38	0.25	2.900	6.433	0.055	0.025	2.28	5.06
Ag - Cr	25.6	115	0.38	0.19	0.289	2.900	6.433	0.073	0.033	1.97	4.37
Fe - Cu	73	40	0.42	0.37	0.258	3.410	5.965.2	0.066	0.038	1.96	3.43
Cu - Fe	40	73	0.37	0.42	0.256	3.410	5.965.2	0.065	0.037	1.97	3.45
Fe - Ag	73	25.6	0.42	0.38	0.258	2.400	7.655.5	0.069	0.022	1.86	5.93
Ag - Fe	25.6	73	0.38	0.42	0.289	2.400	7.655.5	0.087	0.027	1.66	5.3
Mo - Sn	121	15.6	0.31	0.33	0.273	2.250	13.107.9	0.056	0.01	2.43	14.16
Sn - Mo	15.6	121	0.33	0.31	0.442	2.250	13.107.9	0.114	0.025	1.5	8.74
W - Sn	160	15.6	0.28	0.33	0.274	1.150	14.444.8	0.114	0.009	1.2	15.07
Sn - W	15.6	160	0.33	0.28	0.442	1.150	14.4448	0.298	0.024	0.74	9.33
Cu - Pb	40	6.67	0.37	0.42	0.256	1.831	3.928.8	0.028	0.013	4.57	9.81
Pb - Cu	6.67	40	0.42	0.37	0.35	1.831	3.928.8	0.052	0.024	3.34	7.17
Cu - Bi	40	12	0.37	0.33	0.256	1.840	2.327.3	0.041	0.032	3.16	4
Bi - Cu	12	40	0.33	0.37	0.474	1.840	2.327.3	0.139	0.11	1.7	2.15

In the homogeneous case Equations (3.36) and (3.39) reduce to the following:

$$\varphi(z) = \frac{EiNb_e e^{i\alpha}}{8\pi(1-\nu^2)} ln\left(\frac{z}{l} + \sqrt{\frac{z^2}{l^2} - 1}\right) \qquad \left(z = x + iy - \frac{l}{2}\right) \tag{3.44a}$$

$$\sigma_x + \sigma_y = 4Re\varphi'(z), \quad \sigma_y - i\,\tau_{xy} = \varphi'(z) + \varphi'(\bar{z}) - (z - \bar{z})\overline{\varphi''(z)}, \tag{3.44b}$$

$$2\mu(u + iv) = \kappa\varphi(z) - \varphi(\bar{z}) + (z - \bar{z})\overline{\varphi'(z)}, \tag{3.44c}$$

where $\sqrt{z^2 - l^2} = z + O\left(\frac{1}{z}\right)$ as $z \to \infty$.

The displacements of the crack banks are:

For $z \to x \pm I0$ $\qquad 0 < x < l$

$$u = \pm \frac{1}{2\pi} Nb_e \cos\alpha \left[\frac{\pi}{2} - \arcsin\left(\frac{2x - l}{l} \right) \right], \tag{3.45a}$$

$$\upsilon = \pm \frac{1}{2\pi} Nb_e \sin\alpha \left[\frac{\pi}{2} - \arcsin\left(\frac{2x - l}{l} \right) \right]. \tag{3.45b}$$

The stress intensity factors are:

For $z \to 0$

$$K_I = \frac{-ENb_e \sin\alpha}{2(1-\nu^2)\sqrt{2\pi l}}, \qquad K_{II} = \frac{-ENb_e \cos\alpha}{4(1-\nu^2)\sqrt{2\pi l}}; \tag{3.46a}$$

For $z \to l$

$$K_I = \frac{ENb_e \sin\alpha}{2(1-\nu^2)\sqrt{2\pi l}}, \qquad K_{II} = \frac{ENb_e \cos\alpha}{2(1-\nu^2)\sqrt{2\pi l}}. \tag{3.46b}$$

The length of the microcrack is:

$$l = l_0 N^2, \tag{3.47}$$

where

$$l_0 = \frac{E b_e^2}{16\pi(1-\nu)\gamma_{fm}} \tag{3.48}$$

Equations (3.44) - (3.48) coincide with those of A. G. Cherepanov,[15] who studied the stairway structure of a microcrack in the homogeneous case. Equation (3.48) is the Stroh equation derived by Stroh using the energy conservation law.

Let us introduce the notion of the first critical level of band strength, $N = N_c$, as follows: an interface microcrack does not nucleate if $N < N_c$, and an interface microcrack does form if $N > N_c$.

Equations (3.42) and (3.43) can be used to estimate N_c. Indeed, if the microcrack is embedded inside a dislocation band, these equations imply that the Zener - Stroh mechanism in this case is not effective and an interface microcrack actually does not nucleate. Hence, using (3.42) and (3.43) we obtain

$$a_g = l_0 N_c \sin\alpha \tag{3.49}$$

where a_g is the smallest distance between two nearest sliding planes. For example, in the case of a simple cubic lattice and $\alpha = \pi/4$ we have

$$a_g = a/\sqrt{2}, \qquad b_e = a\sqrt{2}, \tag{3.50}$$

where a is the interatomic spacing.

According to Equations (3.43) and (3.49) we have

$$N_c = \frac{a_g}{l_0 \sin\alpha} = \frac{4\pi a_g \gamma_{fm}(\mu_1 + \mu_2\kappa_1)(\mu_2 + \mu_1\kappa_2)}{b_e^2 \mu_1 \mu_2 [\mu_1(1+\kappa_2) + \mu_2(1+\kappa_1)] \sin\alpha} \tag{3.51}$$

In Table 3.1, the value of $b_e / (2l_0) = b_e N_c \sin\alpha / (2a_g)$ is computed for various composite pairs by means of Equation (3.51). As is seen, in some alloys even a single edge dislocation can produce an interface nanocrack.

Yielding delay (and delay of dislocation nucleation) is an intrinsic property of crystals. This is why we choose the true interface surface energy, γ_{fm}, in order to characterize the nucleation of an interface crack: there are no dislocations to form during the first moments, just after the microcrack starts to propagate. However, in a while dislocations start to be emitted from the right-hand end of the microcrack and act to arrest this crack-tip. From this, if follows that the solution we have found may be combined with experimental observations of interface microcrack to provide a method for the estimation of true interface surface energy.

The other limitation of the theory under study relates to the path of loading. The solution is valid only for increasing loading paths. By unloading, some dislocations of the opposite sign will emanate from the left-hand crack-tip which is clamped by the band, the resultant number of dislocations in the band will decrease, and this will immediately cause contact of the opposite sides of the microcrack, starting from the right-hand crack-tip. The closure effect requires a more sophisticated theory. It is especially important for life-predictive theories of the interface crack growth that occurs by cyclic, or any variable, loadings. The nucleated microcrack, even if not dangerous for simple loading path, grows stably by cyclic loadings and may achieve a dangerous length.

The analysis of the solution shows that the interface surface energy is the most important physical parameter which determines the nucleation of a microcrack. Unfortunately, there are no good theoretical methods for the calculation of this quantity, and direct experiments with the measurement of the growth of man-made interface cracks are of little use because of the effect of many dislocations emanating from a crack-tip. Meanwhile, this theory, namely Equations (3.41) - (3.43) or (3.47) and (3.48), can be used to calculate γ_{fm} using direct experimental observations of the interface microcrack nucleation.

It should be noted that eleven different modes of interaction between a dislocation band and an interface are possible if the number of elementary dislocations in the band is less than the second critical level. The Zener - Stroh mechanism provides only one of these eleven modes. After the second critical level is passed, 121 different interaction modes are possible, before the third critical level is achieved. The dislocation band-interface reactions of the nth order include 11^n different modes. The branching process of dislocations-interface reactions resembles that of chain chemical reactions.

3.2. Encounter of Two Dislocation Pileups

A pileup is a shear crack with closed sides, along which the tangential displacement suffers a discontinuity and the shear stress is equal to the Schmid stress, τ_s ,characterizing the gliding plane. A pileup forms, for example, when a Frank - Read source generates elementary dislocations running one-by-one along the same gliding plane, and the frontal dislocation comes across an obstacle and stops. The stop causes an accumulation of dislocations near the obstacle which is called a pileup in metal physics, or the closed shear crack in fracture mechanics. The pileup has a parabolic distribution of the Burgers vector near the head, and a stress singularity of $r^{-1/2}$ type coinciding, generally, with a superposition of screw and transverse shear described by the two stress intensity factors, K_{II} and K_{III} . (K_I is zero in this case). The pileup breaks through the obstacle and can propagate when $K_{II}^2 - K_{III}^2$ or the corresponding energy flow rate, Γ, achieves the critical value, Γ_c characterizing the obstacle nature and strength. When $\Gamma=0$, a pileup becomes a usual slip line propagating in a perfectly plastic material.

In 1958, Cottrell[20] suggested that an opening mode microcrack can be nucleated as a result of an encounter of two dislocation pileups or bands. The theory of this mechanism of crack nucleation was given by Cherepanov and Kipnis[21]. In the present section, this theory is treated using the solution to the corresponding mixed boundary value problem obtained by the Wiener - Hopf and Noble - Jones methods. The length of the nucleated crack is found in terms of physical and geometric parameters of the problem.

Formulation of the problem

Plastic strains in materials with yielding delay and well-developed post-critical yielding are usually concentrated along slip lines or pileups which initiate at 45 degrees to the free boundary[22] and propagate in the elastic surrounding (Figure 3.3).

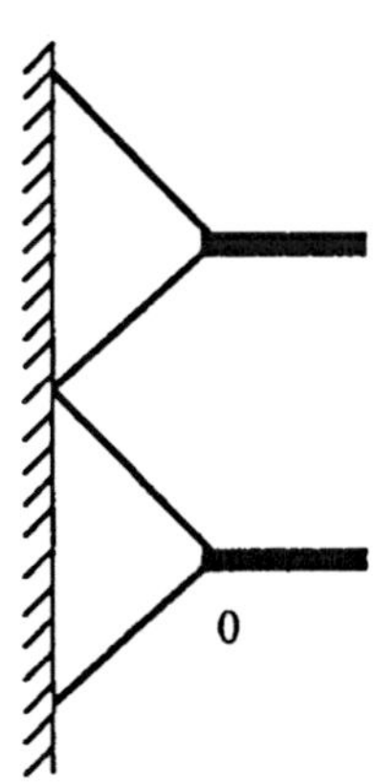

Figure 3.3. Development of slip lines and pileups from the free boundary

We consider plane strain conditions, that is, transverse shear or edge dislocation pileups when $K_{III} = 0$ and $K_{II} \neq 0$. Let us study the neighborhood of the point 0, at which two pileups encounter one another (Figure 3.4). Use polar coordinates r and θ centered at 0. Suppose that the pileups are directed at angles $\pm\,\alpha$ at 0, so that the plane of symmetry, $\theta = 0$, coincides with the plane of the opening mode crack nucleated according to the Cottrell mechanism. We consider the case when the crack length, l, is small compared to the sizes of pileups and specimen. The effect of the encounter of pileups is physically equivalent to the self-wedging of one part of the material by the other part (Figure 3.4a).

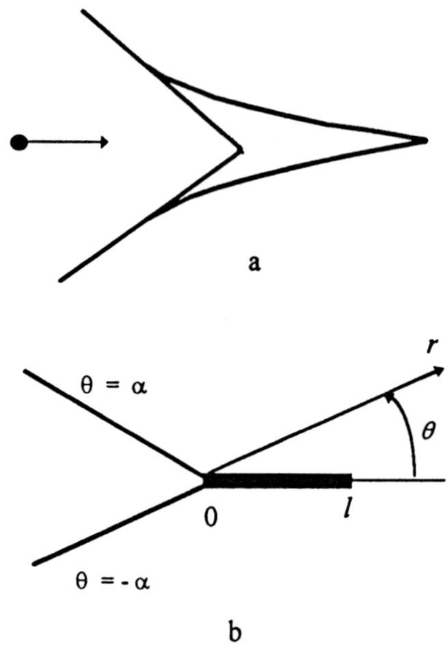

Figure 3.4. Encounter of two dislocation pile- ups and the nucleation of a crack:
a. Physical cheme. b. Mathematical sketch.

The governing equation of the problem are :
Equilibrium equations

$$\frac{\partial \sigma_r}{\partial r} + \frac{1}{r}\frac{\partial \tau_{r\theta}}{\partial \theta} + \frac{\sigma_r - \sigma_\theta}{r} = 0\,, \tag{3.52a}$$

$$\frac{1}{r}\frac{\partial \sigma_\theta}{\partial \theta} + \frac{\partial \tau_{r\theta}}{\partial r} + 2\frac{\tau_{r\theta}}{r} = 0\,; \tag{3.52b}$$

Compatibility equations

$$\Delta\ (\sigma_r + \sigma_\theta) = 0\,, \qquad \text{where} \quad \Delta = \frac{1}{r}\left(r\frac{\partial}{\partial r}\right) + \frac{1}{r^2}\frac{\partial^2}{\partial \theta^2}\ ; \tag{3.53}$$

Hooke's law

$$\frac{\partial u_r}{\partial r} = \frac{1+\nu}{E}[(1-\nu)\sigma_r - \nu\sigma_\theta] \tag{3.54a}$$

$$\frac{u_r}{r}+\frac{1}{r}\frac{\partial u_\theta}{\partial \theta}=\frac{1+\nu}{E}[(1-\nu)\sigma_\theta-\nu\sigma_r] \tag{3.54b}$$

$$\frac{1}{r}\frac{\partial u_r}{\partial \theta}+\frac{\partial u_\theta}{\partial r}-\frac{u_\theta}{r}=2\frac{1+\nu}{E}\tau_{r\theta} \tag{3.54c}$$

Here, σ_r, σ_θ and $\tau_{r\theta}$ are stresses, u_r and u_θ are displacements in polar coordinates r and θ. The boundary conditions of the problem are as follows:

$$[\sigma_\theta]=[\tau_{r\theta}]=0, \quad [u_\theta]=0, \quad \tau_{r\theta}=\tau_s \qquad \text{when } \theta=\alpha \quad (\pi/2<\alpha<\pi) \tag{3.55}$$

$$\tau_{r\theta}=0 \quad u_\theta=0 \qquad \text{when } \theta=0 \text{ and } \theta=\pi \tag{3.56}$$

$$\sigma_\theta=0 \quad \text{when } \theta=0, \quad \text{and } r<l, \quad u_\theta=0 \quad \text{when } \theta=0, \text{ and } r>l \tag{3.57}$$

In the view of symmetry, we consider only the half-plane $0<\theta<\pi$, and therefore Equations (3.55) comprise the boundary conditions on the pileup $\theta=\alpha$. The symmetry conditions are provided by Equations (3.56) and the second equation (3.57). It is assumed that $\pi/2<\alpha<\pi$.

A special asymptotic field is realized far from the opening mode crack, at infinity. The field pre-existed at the moment of the encounter of the pileups, just before the opening mode crack nucleated. Therefore, it corresponds to the solution of the symmetrical problem for the space containing two slip lines that meet at point *0* without cracking (Figure 3.5).

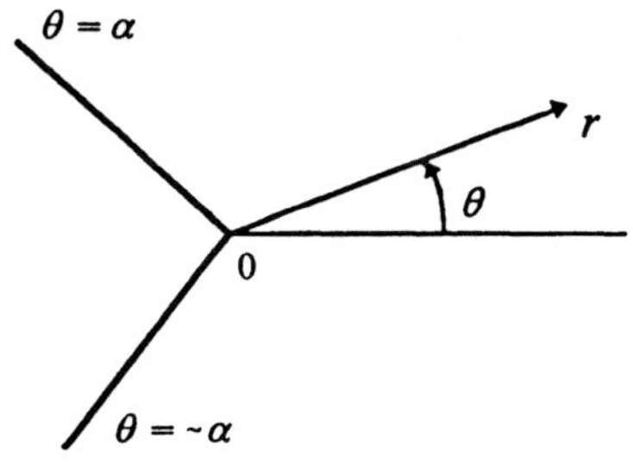

Figure 3.5. The original condition of the encounter of two dislocation pileups with no cracks nucleated

The boundary conditions of the problem can be written as follows:

$$[\sigma_\theta]=[\tau_{r\theta}]=0, \quad [u_\theta]=0, \quad \tau_{r\theta}=\tau_s \qquad \text{when } \theta=\alpha;$$

$$\tau_\theta=0, \quad u_\theta=0 \qquad \text{when } \theta=0 \text{ and } \theta=\pi \tag{3.58}$$

The self-similar solution of the problem given by (3.52) - (3.54) and (3.58) will be as follows:

$$\sigma_\theta = q - \frac{\tau_s}{\sin 2\alpha}\cos 2\theta, \qquad \tau_{r\theta} = -\frac{\tau_s}{\sin 2\alpha}\sin 2\theta,$$

$$\sigma_r = q + \frac{\tau_s}{\sin 2\alpha}\cos 2\theta, \qquad u_\theta = -\frac{1+\nu}{E}\frac{\tau_s r}{\sin 2\alpha}\sin 2\theta, \tag{3.59}$$

$$u_r = \frac{1+\nu}{E}\left[(1-2\nu)q + \frac{\tau_s}{\sin 2\alpha}\cos 2\theta\right]r \qquad (\pi/2 < \alpha < \pi),$$

where q is an arbitrary given constant that characterizes the uniform omnidirectional extension. When $q = 0$ and $\theta = 0$, σ_θ is positive, this corresponds to the self-wedging effect of Figure 3.4a. The field given by Equations (3.59) should be realized as the asymptotic form of the solution of the problem of cracking when $r \to \infty$. In particular, for $\theta = 0$ we must have

$$\sigma_\theta = q - \frac{\tau_s}{\sin 2\alpha} + O\left(\frac{1}{r}\right)$$

According to Equations (3.59), the field existing just before cracking, will be continuous in the stresses and displacements. The stresses in (3.59) satisfy the governing equations of the theory of plasticity because they meet the plasticity condition,

$$(\sigma_r - \sigma_\theta)^2 + 4\tau_{r\theta}^2 = 4\tau_D^2, \tag{3.60a}$$

$$\tau_D = -\frac{\tau_s}{\sin 2\alpha} \qquad (\tau_D \geq \tau_s \text{ and } \pi/2 < \alpha < \pi) \tag{3.60b}$$

Here τ_D is equal to τ_s only for $\alpha = 3\pi/4$, for other α we have $\tau_D > \tau_s$. The value of τ_s characterizes the well-developed plastic deformation along the slip line, while τ_D characterizes the initial condition of the material just before the well-developed yielding occurs.

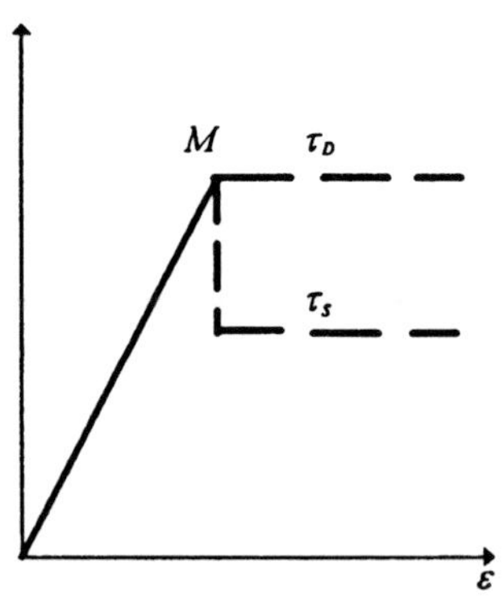

Figure 3.6. The σ - ε diagram of materials with yielding delay

Physically this means that on scales that are large compared to the size of an incipient crack (but small compared to the characteristic dimension of the body and slip lines), we have the onset of plastic deformation, so that the material is at corner point M of the σ -ε diagram (Figure 3.6).

Macroplasticity is realized as a result of formation and development of many slip lines that separate elastic zones[22]. The present analysis shows that cracking and the induced well-developed yielding along the slip lines occur simultaneously.

Let us consider the following boundary conditions, instead of (3.55) - (3.57):

$$[\sigma_\theta] = [\tau_{r\theta}] = 0, \quad [u_\theta] = 0, \quad \tau_{r\theta} = 0 \quad \text{when} \quad \theta = \alpha; \tag{3.61}$$

$$\tau_{r\theta} = 0 \text{ when } \theta = 0 \text{ and } \theta = \pi, \; u_\theta = 0 \quad \text{when } \theta = \pi; \tag{3.62}$$

$$\sigma_\theta - q + \frac{\tau_s}{\sin 2\alpha} \quad \text{when } \theta = 0 \text{ and } r < l; \tag{3.63}$$

$$u_\theta = 0 \quad \text{when } \theta = 0 \text{ and } r > l; \tag{3.64}$$

From the superposition principle, it follows that the solution to the original boundary value problem (3.55) - (3.57) is equal to the sum of solution (3.59) and the solution to the problem formulated by Equations (3.52) - (3.54) and (3.61) - (3.64). The stresses in the latter solution should attenuate as $O(1/r^2)$ at infinity.

The Wiener - Hopf equation and its solution

Apply the Mellin integral transform,

$$\varphi^*(p) = \int_0^\infty \varphi(r) r^p \, dr \tag{3.65}$$

where p is a complex parameter, to the equilibrium equations (3.52) and get

$$\tau^*_{r\theta} = \frac{1}{p-1}\frac{d\sigma^*_\theta}{d\theta}, \quad p\sigma^*_r = \frac{1}{p-1}\frac{d^2\sigma^*_\theta}{d\theta^2} - \sigma^*_\theta \quad . \tag{3.66}$$

From physical considerations, it follows that the stresses should be bounded as $r \to 0$, and should vanish as $O(r^{-2})$ for $r \to \infty$. Hence, the Mellin transforms of the stresses are some analytic functions of a complex variable, p, in the strip, $-1 < Re < 1$.

Substituting σ^*_r from Equation (3.66) into the transformed equation (3.53) provides

$$\frac{d^4\sigma^*_\theta}{d\theta^4} + [(p+1)^2 - (p-1)^2]\frac{d^2\sigma^*_\theta}{d\theta^2} + (p+1)^2(p-1)^2\sigma^*_\theta = 0. \tag{3.67}$$

The solution to this equation can be written as:

$$\sigma^*_\theta = A_1 \sin(p+1)\theta + A_2 \sin(p-1)\theta + A_3 \cos(p+1)\theta + A_4 \cos(p-1)\theta; \quad (\alpha < \theta < \pi) \tag{3.68a}$$

$$\overset{\bullet}{\sigma}_\theta = B_1 sin(p+1)\theta + B_2 sin(p-1)\theta + B_3 cos(p+1)\theta + B_4 cos(p-1)\theta; \quad (\alpha < \theta < \pi) \tag{3.68b}$$

$$\overset{\bullet}{\tau}_{r\theta} = \frac{1}{p-1}\frac{d\overset{\bullet}{\sigma}_\theta}{d\theta}, \quad \overset{\bullet}{\sigma}_r = \frac{1}{p(p-1)}\frac{d^2\overset{\bullet}{\sigma}_\theta}{d\theta^2} - \frac{1}{p}\overset{\bullet}{\sigma}_\theta . \tag{3.68c}$$

Here, A_1, ... , A_4, B_1, ... , B_4 are some functions of p. Seven boundary condition equations (3.61) and (3.62) based on Equations (3.68) provide the following system equations:

$$(A_1 - B_1)(p-\chi)cos(p+1)\alpha + (A_2 - B_2)(p-1)cos(p-1)\alpha - (A_3 - B_3)(p-\chi)sin(p+1)\alpha - (A_4 - B_4)(p-1)sin(p-1)\alpha = 0; \tag{3.69a}$$

$$(A_1 - B_2)sin(p+1)\alpha + (A_2 - B_2)sin(p-1)\alpha + (A_3 - B_3)cos(p+1)\alpha + (A_4 - B_4)cos(p-1)\alpha = 0; \tag{3.69b}$$

$$(A_1 - B_1)(p+1)cos(p+1)\alpha + (A_2 - B_2)(p-1)cos(p-1)\alpha - (A_3 - B_3)(p+1)sin(p+1)\alpha - (A_4 - B_4)(p-1)sin(p-1)\alpha = 0; \tag{3.69c}$$

$$A_1(p+1)cos(p+1)\alpha + A_2(p-1)cos(p-1)\alpha - A_3(p+1)sin(p+1)\alpha - A_4(p-1)sin(p-1)\alpha = 0; \tag{3.69d}$$

$$A_1(p-\chi)cos\, p\pi + A_2(p-1)cos\, p\pi - A_3(p-\chi)sin\, p\pi - A_4(p-1)sin\, p\pi = 0; \tag{3.69e}$$

$$A_1(p+1)cos\, p\pi + A_2(p-1)cos\, p\pi - A_3(p+1)sin\, p\pi - A_4(p-1)sin\, p\pi = 0; \tag{3.69f}$$

$$B_1(p+1) + B_2(p-1) = 0 \quad (\chi = 3 - 4\nu) \tag{3.69g}$$

The solution of the system (3.69) has the form

$$A_2 = \frac{(p+1)sin(\alpha + p\alpha - p\pi)}{(p-1)sin(\alpha - p\alpha + p\pi)} A_1, \qquad A_3 = ctg\, p\pi\, A_1 , \tag{3.70a}$$

$$A_4 = \frac{(p+1)sin(\alpha + p\alpha - p\pi)}{(p-1)sin(\alpha - p\alpha + p\pi)} ctg\, p\pi\, A_1 , \tag{3.70b}$$

$$B_1 = \frac{\delta_1 \Delta_2 + \Delta_1 \delta_2}{4(p\cos p\alpha \sin\alpha + \sin p\alpha \cos\alpha)\sin(\alpha - p\alpha + p\pi)} A_1 , \tag{3.70c}$$

$$B_2 = -(p+1)B_1/(p-1), \tag{3.70d}$$

$$B_3 = \frac{2\delta_2(p\cos p\alpha \sin\alpha + \sin p\alpha \cos\alpha) - (\delta_1\Delta_2 + \Delta_1\delta_2)\cos(p+1)\alpha}{4(p\cos p\alpha \sin\alpha + \sin p\alpha\cos\alpha)\sin p\pi \sin(\alpha - p\alpha + p\pi)\sin(p+1)\alpha} A_1, \quad (3.70e)$$

$$B_4 = -\frac{2\delta_2(p\cos p\alpha \sin\alpha + \sin p\alpha \cos\alpha) - (\delta_1\Delta_2 + \Delta_1\delta_2)\cos(p-1)\alpha}{4(p\cos p\alpha \sin\alpha + \sin p\alpha\cos\alpha)\sin p\pi \sin(\alpha - p\alpha + p\pi)\sin(p-1)\alpha} \frac{p+1}{p-1} A_1, \quad (3.70f)$$

where

$$\Delta_1 = \sin 2p\alpha + p\sin 2\alpha, \qquad \Delta_2 = \sin 2p(\pi - \alpha) - p\sin 2\alpha,$$
$$\delta_1 = \cos 2p\alpha - \cos 2\alpha, \qquad \delta_2 = \cos 2p(\pi - \alpha) - \cos 2\alpha.$$

Using Equations (3.68) and (3.70) we derive:

$$\sigma_\theta^\bullet(p,0) = -\frac{[2\delta_2(\sin^2 p\alpha - p^2\sin^2\alpha) + \Delta_1\Delta_2] A_1}{2(p-1)(p\cos p\alpha\sin\alpha + \sin p\alpha\cos\alpha)\sin p\pi\sin(\alpha - p\alpha + p\pi)}, \quad (3.71a)$$

$$\left(\frac{\partial u_\theta}{\partial r}\right)^\bullet\Bigg|_{\theta=0} = \frac{1-\nu^2}{E}\frac{[\delta_1\Delta_2 + \delta_2\Delta_1] A_1}{(p-1)(p\cos p\alpha\sin\alpha + \sin p\alpha\cos\alpha)\sin p\pi\sin(\alpha - p\alpha + p\pi)}. \quad (3.71b)$$

Eliminating A_1 in (3.71) and taking account of the condition equations (3.63) and (3.64), we arrive at the functional Wiener - Hopf equation

$$\frac{\sigma}{p+1} + \phi^+(p) = -\tan p\pi\, G(p)\phi^-(p) \quad (3.72a)$$

in the strip, $-\varepsilon_1 < Re p < \varepsilon_2, \quad \varepsilon_1 > 0, \ \varepsilon_2 > 0.$

Here,

$$G(p) = \frac{2\delta_2(\sin^2 p\alpha - p^2\sin^2\alpha) + \Delta_1\Delta_2}{(\delta_1\Delta_2 + \Delta_1\delta_2)\tan p\pi}, \quad (3.72b)$$

$$\phi^-(p) = \frac{E}{2(1-\nu^2)}\int_0^1 \frac{\partial u_\theta}{\partial r}\Bigg|_{\substack{r=\rho l \\ q=+0}} \rho^p d\rho \qquad \phi^+(p) = \int_1^\infty \sigma_\theta(\rho l, 0)\rho^p d\rho \quad (3.72c)$$

$$\sigma = \frac{\tau_s}{\sin 2\alpha} - q. \quad (3.72d)$$

Note that $G(it)$ when $-\infty < t < \infty$, is a real even positive function of t that tends to 1 as $t \to \infty$. Therefore the index of $G(p)$ with respect to the imaginary axis is zero, and the following well-known equation by Gakhov[14] is valid:

$$G(p) = \frac{G^+(p)}{G^-(p)} \quad \text{along} \quad Rep = 0\,, \tag{3.73a}$$

where

$$exp\left[\frac{1}{2\pi i}\int_{-i\infty}^{i\infty}\frac{log\,G(\tau)}{\tau - p}d\tau\right] = \begin{cases} G^+(p), & \text{when} \quad Rep < 0, \\ G^-(p), & \text{when} \quad Rep > 0. \end{cases} \tag{3.73b}$$

Using (3.73)and the representation in terms of the Euler Gamma function, $\Gamma(z)$,

$$p\cot p\pi = K^+(p)K^-(p), K^{\pm}(p) = \frac{\Gamma(1 \mp p)}{\Gamma(1/2 \mp p)}\,, \tag{3.74a}$$

$$\frac{\sigma K^+(p)}{p(p+1)G^+(p)} = \frac{\sigma}{p+1}\left[\frac{K^+(p)}{pG^+(p)} + \frac{K^+(-1)}{G^+(-1)}\right] - \frac{\sigma K^+(-1)}{(p+1)G^+(-1)} \quad (Rep = 0,\ p \neq 0)\,, \tag{3.74b}$$

we can rewrite (3.72a) as follows :

$$\frac{K^+(p)\phi^+(p)}{pG^+(p)} + \frac{\sigma}{p+1}\left[\frac{K^+(p)}{pG^+(p)} + \frac{K^+(-1)}{G^+(-1)}\right] = \frac{\sigma K^+(-1)}{(p+1)G^+(-1)} - \frac{\phi^-(p)}{K^-(p)G^-(p)} \quad (Rep = 0). \tag{3.75}$$

The function on the left side of (3.75) is analytic in the half-plane $Rep < 0$, while the function on the right side is analytic in $Rep > 0$. On the basis of the principle of analytic continuation, they are equal to the same function that is analytic in the whole plane p. Using the familiar asymptotic form of the elastic field near the crack,

$$\sigma_\theta \rightarrow K_I(2\pi)^{-1/2} \quad (r - l)^{-1/2} \quad \text{when} \quad \theta = 0 \quad \text{and} \quad r \rightarrow l+\,,$$

we can obtain from (3.72c), the following equation [14]

$$\phi^+(p) = K_I/\sqrt{2pl}, \qquad \phi^-(p) = -K_I/\sqrt{2pl}\,, \quad \text{as} \qquad p \rightarrow \infty. \tag{3.76}$$

From (3.73), (3.74), and (3.76) it follows that the functions on the left and right sides of Equation (3.75) tend to zero as $p \rightarrow \infty$ in the half-planes $Rep < 0$ and $Rep > 0$, respectively. Thus, in accordance with Liouville's theorem, the single analytic function is identically zero in the whole plane p.

From here and Equation (3.75) we derive the solution of functional equation (3.72a),

$$\phi^-(p) = \frac{\sigma K^+(-1)K^-(p)G^-(p)}{(p+1)G^+(-1)} \qquad (Rep > 0) \tag{3.77a}$$

$$\phi^+(p) = -\frac{\sigma\, pG^+(p)}{(p+1)K^+(p)}\left[\frac{K^+(p)}{pG^+(p)} + \frac{K^+(-1)}{G^+(-1)}\right] \qquad (Rep > 0) \tag{3.77b}$$

Here, $K^+(p), K^-(p), G^+(p)$ and $G^-(p)$ are defined by Equations (3.73) and (3.74).

Analysis of the solution

Using Stirling's formula for the Gamma function as $p \to \infty$, we can derive:[14,23]

$$K^+(p) \to \sqrt{-p} \quad \text{as} \quad p \to \infty, \qquad \text{if } Re\,p < 1/2 \tag{3.78a}$$

$$K^-(p) \to \sqrt{p} \quad \text{as} \quad p \to \infty, \qquad \text{if } Re\,p > -1/2 \tag{3.78b}$$

Besides, from (3.73) it follows that

$$\lim\ G^{\pm}(p) = 1 \quad \text{as} \quad p \to \infty. \tag{3.79}$$

Let us determine the stress intensity factor K_I at the crack end. From Equations (3.77a), (3.78b) and (3.79) we obtain:

$$\phi^-(p) = \frac{\sigma K^+(-1)}{G^+(-1)} \frac{1}{\sqrt{p}} \quad \text{as} \quad p \to \infty. \tag{3.80}$$

By means of (3.76)and (3.80) we find:

$$K_I = f(\alpha)\left(-\frac{\tau_s}{\sin 2\alpha} + q\right)\sqrt{l}, \qquad \text{where} \quad f(\alpha) = \frac{2\sqrt{2}}{\sqrt{\pi} G^+(-1)}, \tag{3.81}$$

because $\Gamma(3/2) = +\pi^{1/2}/2$ and $\Gamma(2) = 1$.

Values of $f(\alpha)$ for some α are given below:

α^o	95	105	115	125	135	145	155	165	175
$f(\alpha)$	1.8344	1.7710	1.7117	1.6574	1.6088	1.5633	1.5171	1.4645	1.3853

The equation

$$\int_0^l \left.\frac{\partial u_\theta}{\partial r}\right|_{\theta=+0} r^p\, dr = \frac{2(1-\nu^2)}{E} l^{p+1} \phi^-(p), \tag{3.82a}$$

at $p = 0$ yields

$$u_\theta(0, +\theta) = -2(1-\nu^2) l \Phi^-(0) / E. \tag{3.82b}$$

Taking into account that $u_\theta(r, \theta)$ is an odd function in θ, we get

$$[u_\theta]\Big|_{\substack{r=0\\ \theta=0}} = -4(1-\nu^2)l\phi^-(0)/E \tag{3.82c}$$

Using (3.72d), (3.77a) and (3.82), we obtain the crack tip opening displacement, *U*, at point *0* as the function of α, τ_s, q and l

$$U = \frac{1-\nu^2}{E}\frac{f(\alpha)}{\lambda(\alpha)}\left(-\frac{\tau_s}{\sin 2\alpha}+q\right)l, \tag{3.83a}$$

where

$$\lambda(\alpha) = \frac{[8\sin^2\alpha(\alpha^2-\sin^2\alpha)+2(2\alpha+\sin 2\alpha)(2\pi-2\alpha-\sin 2\alpha)]^{1/2}}{8\sqrt{\pi}\sin\alpha}. \tag{3.83b}$$

In accordance with (3.81) and (3.83),

$$K_I = \frac{\lambda(\alpha)EU}{(1-\nu^2)\sqrt{l}} \tag{3.84}$$

The quantity *U* plays the role of the internal loading process parameter controlling the self-wedging effect. Physically, it characterizes the combined Burgers vector of slip lines at 0.

Equations (3.81), (3.83) and (3.84) constitute the principal findings of the present theory. They enable us to predict nucleation and initial development of a microcrack in metals and alloys obeying the Cottrell mechanism.

In the simplest case of a quasi-brittle Griffith - Irwin crack, $K_I = K_{Ic}$ (K_{Ic} is the fracture toughness), and from (3.81) and (3.84) it follows that

$$\sqrt{l} = \frac{K_{Ic}}{f(\alpha)}\left(q-\frac{\tau_s}{\sin 2\alpha}\right)^{-1}; \tag{3.85a}$$

$$\sqrt{l} = \frac{\lambda(\alpha)EU}{(1-\nu^2)K_{Ic}}. \tag{3.85b}$$

In this problem, there are two loading parameters, q and U, characterizing the microcrack nucleation. Due to (3.85), they are connected by the equation,

$$q - \frac{\tau_s}{\sin 2\alpha} = \frac{\Gamma_c}{Uf(\alpha)\lambda(\alpha)}, \qquad \left(\frac{\pi}{2}<\alpha<\pi,\quad \Gamma_c = K_{Ic}^2(1-\nu^2)/E\right) \tag{3.86}$$

while the crack nucleates. Equations (3.85) predict the length of the nucleated microcrack in terms of q or *U*.

Consider some particular cases. If the external stress is zero, $q = 0$,

$$l = \frac{K_{Ic}^2 (\sin 2\alpha)^2}{\tau_s^2 [f(\alpha)]^2}, \qquad U = -\frac{\Gamma_c \sin 2\alpha}{\tau_s f(\alpha)\lambda(\alpha)}; \tag{3.87}$$

If the external stress is high compressive, so that $q < \tau_s / \sin 2\alpha$, the crack nucleation by the present mechanism is impossible for any pileups or slip lines. An increase of external omnidirectional tensile stress leads to a decrease of the length of a nucleated crack and the crack opening displacement.

If we assume that the slip lines or pileups begin to develop from a free rectilinear boundary of the body (Figure 3.3), then, as shown in [22], the angle α is equal to $3\pi/4$, and the pileups meet at a distance from the free boundary inside the material. In this case, according to (3.86) and (3.87)

$$\alpha = \frac{3p}{4}, \qquad q = 0, \qquad l = 0.386\frac{K_{Ic}^2}{\tau_s^2}, \qquad U = 0.964\frac{\Gamma_c}{\tau_s} \tag{3.88}$$

and the length of the nucleated microcrack appears to be approximately five times greater than the maximum size of the plastic zone at the tip of a macrocrack. The plane of a microcrack nucleated underneath a surface is perpendicular to the free surface, and such a microcrack can turn into an initial fatigue crack by succeeding cyclic loading with the routine scenario of failure from subcritical fatigue crack growth. According to (3.88), this mechanism of crack nucleation requires intensive slip lines or pileups, with the number of elementary edge dislocations being of the order of $\Gamma_c/(\tau_s\ b_e)$, where b_e is the value of the Burgers vector of an elementary edge dislocation in the material.

3.3. Hole Coalescence in Amorphous Metals

Cracks and dislocations are not the only objects examined in fracture mechanics. Small defects in the material such as foreign inclusions and pores are also important. They are important because they are present in almost all materials (in contrast to cracks and dislocations) and because at relatively small sizes they are capable of moving inside the material. The migration and diffusion of these defects influence the yielding and creep strains, and have a strong effect on the movement of the cracks and dislocations. A small pore with a the traction-free spherical surface is referred to as a hole[24], by analogy with the similar object in semiconductor theory.

Examination of point defects like holes is of considerable interest for one of the most promising groups of materials for application in the near future, namely amorphous metals. Amorphous metals do not contain initial cracks and dislocations and therefore have exceptionally high strength which is close to the theoretical strength (of the order of $E/10$, where E is Young's modulus). For example[25], the strength of $Fe_{60}Cr_6Mo_6B_{28}$ amorphous alloy reaches 4.9 *GPa* (at Young's modulus of 150 *GPa*), i.e., it is of the order of E/30. Amorphous metals also have many other excellent properties, especially exceptionally high fatigue endurance and corrosion resistance because there are no initial cracks and dislocations in these metals [25,26]. Amorphization of the structure of materials and alloys in the surface layer takes place during laser treatment of materials — a promising future technology [26,27].

The main structural defect of amorphous metallic alloys is their density fluctuation leading to the formation of a cluster consisting of several dozens of atoms[25,27]. This density fluctuation is referred to as a structural defect of n-type [25,27]. This type of defect is characterized mainly by the induced elastic field determined by external loading (its inherent elastic field is negligible). Consequently, in examining the problems of strength and failure of amorphous metals, their structural defects can be simulated by holes with the characteristic size of the structural defect of n-type (i. e., of the order of several dozens of interatomic spacings).

We need models which can explain and predict the strength and failure of amorphous metals (and, possibly, of all materials whose structure does not contain crack- or dislocation-type defects). In this section such a theoretical model is based on Reference[28].

A hole-type point defect

We simulate the hole by an infinitely small spherical cavity free from external loading[24]. In contrast to voids and inclusions, the inherent elastic field of the hole is induced by the external elastic field. The hole can also be used to simulate efficiently any inclusions and voids in the elastic lattice if their inherent elastic field, formed as a result of the thermal and processing prior history, is small in comparison with the induced external field. Initially, we shall discuss the main concepts for the well-known spherical symmetrical field of the inclusion, void, or hole:

$$u_R = \frac{pR(1-2\nu)}{E} - \frac{(1+\nu)a^3(q-p)}{(2ER)^2}, \qquad u_{Ri} = \frac{qR(1-2\nu_i)}{E_i},$$

$$\sigma_R = p + \frac{(q-p)a^3}{R^3}, \qquad \sigma_\theta = \sigma_\psi = p - \frac{(q-p)a^3}{2R^3} \qquad (R>a) \qquad (3.89)$$

Here, R, ψ and θ are spherical polar coordinates, u_R, σ_R, σ_θ, σ_ψ are the displacement and non-zero stresses at $R > a$, q and p are the limits of σ_R at $R = a$ and $R \to \infty$ respectively, the subscript i denotes a foreign inclusion with radius a, ν is Poisson's ratio and E is Young's modulus. All the quantities without the subcript i relate to the parent material.

For $a << R$, Equations (3.89) yield an inherent field for a general point defect. For q satisfying the condition $q >> p$, it is a void (for $q > 0$) or inclusion (for $q < 0$). At $p>>q$ it is a hole.

Cooling the material with an inclusion (from the melt) by ΔT leads to the formation of an inherent elastic field of the inclusion which is determined by Equations (3.89) at $p=0$ and

$$q = \frac{2(\alpha_i - \alpha)\Delta T}{\dfrac{1-2\nu_i}{E_i} + \dfrac{1+\nu}{2E}} \tag{3.90}$$

(α is the coefficient of thermal expansion).

Since $\alpha \sim 10^{-5}$ to $10^{-6}\ K^{-1}$, $\Delta T \sim 10^3\ K$, $E \sim 100\ GPa$, Equation (3.90) shows that the stress q can reach the values 10^2-$10^3\ MPa$. When $q << p$, the inclusion can be regarded as a hole.

The second reason for the formation of the inherent field of the inclusion (regardless of external loading) is the difference between the radius of the inclusion and the radius of the cavity. For example, if the ion radius of an interstitial atom in the solid solution is greater than the ion radius of the atom of the solvent, the inherent field of the defect will be large and may be described using the theory of inclusions. We assume that Δu is the difference of the radii of the inclusion and the cavity at some temperature, consequently, the inherent field of the inclusion at the same temperature is characterized by Equations (3.89) with $p = 0$ and

$$q = -\frac{\Delta u}{a\left(\dfrac{1+\nu}{2E} + \dfrac{1-2\nu_i}{E_i}\right)}. \tag{3.91}$$

For example, when a manganese ion with radius 0.13 *nm* penetrates to the lattice of iron with ion radius 0.126 *nm*, the value of q is approximately 10 *GPa*.

The internal energy of the point defect of the inclusion type is

$$U_i = -q\Delta u \cdot 2\pi a^2 = 2\pi a^3 q^2\left(\frac{1+\nu}{2E} + \frac{1-2\nu_i}{E_i}\right) \tag{3.92}$$

At $E_i \to 0$ for any given E and Δu, Equations (3.91) and (3.92) indicate that the stress q and the energy of the defect tend to zero. This corresponds to a hollow spherical cavity or pore. In this case, the inherent field of the defect which is independent of p forms as a result of interatomic cohesion forces creating the surface tension γ at the surface of the pore which is equivalent to the stress $\sigma_R = 2\gamma/a$ at $R = a$ according to Laplace's equation. The latter point defect is called a vacancy in metal physics. The inherent field of the vacancy is defined by Equations (3.89) at $p = 0$, $q = 2\gamma/a$, the vacancy energy is $\pi a^3 q^2 E^{-1}(1+\nu)$. Since $\gamma \sim 10\ N/m$, at $a \sim 0.1\ nm$ we have $q \sim 10\ GPa$, and at $a \sim 10\ nm$ $q \sim 100\ MPa$. At $p >> q$ the vacancy can also be regarded as a hole.

As a result of the combined effects of thermal expansion, the difference of the ion radii, and the cohesion force, we have:

$$q = 2a(\alpha_i - \alpha)\Delta T - \frac{\Delta u}{a\left(\dfrac{1+\nu}{2E} + \dfrac{1-2\nu_i}{E_i}\right)} + \frac{2\gamma}{a}. \tag{3.93}$$

Thus, there is the temperature range ΔT in which the inclusion-type defect for $q<<p$ becomes a hole.

The elastic energy of the body with a hole is determined using Clayperon's theorem and Equations (3.89) in the form

$$\frac{1}{2}\sigma_R u_R = 2U_{\infty}(1-\alpha_c\varepsilon_p), \text{ where } \alpha_c = \frac{2.5(1-\nu)}{1-2\nu}, \quad \varepsilon_p = \frac{a^3}{R^3}. \tag{3.94}$$

Here ε_p is the porosity (the ratio of the volume of the holes to the body volume), and U_{00} is the energy of the specimen without the hole.

Equation (3.94) also holds when the number of holes in the specimen is arbitrary and the specimen is subjected to uniform compression or extension (the distribution of the holes in the material is assumed to be uniform). Consequently, the self-energy of the holes in the specimen is

$$U_h = \alpha_c \varepsilon_p U_{00} \, . \tag{3.95}$$

The physical meaning of U_h can be described as follows. Examine an elastic cylindrical specimen made of a material in two structural states: a) with uniformly distributed holes, and b) without holes. Assume that the specimen is loaded by the same uniaxial stress p (Figure 3.7).

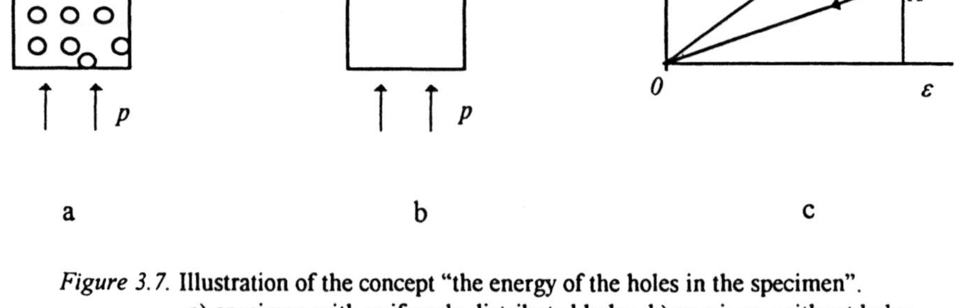

Figure 3.7. Illustration of the concept "the energy of the holes in the specimen". a) specimen with uniformly distributed holes, b) specimen without holes, c) "loading-unloading" diagram of the specimen from which holes have been removed

Specimen b is more rigid than specimen a. Fix the displacements at the points of load application and assume that the holes are capable of escaping from the specimen under the stress, which is traced by the AB section of the p-ε diagram (Figure 3.7c). Unload the specimen from B to 0, it is evident that the area of the $0AB$ triangle is equal to U_h .

A cloud of holes

Consider the general case of an arbitrary nonuniform cluster (cloud) of holes in an elastic specimen and examine the process of condensation of the holes into a crack. It is important to note that the inclusion-type defects do not interact with each other. Therefore, in the specimens saturated with these defects and subjected to uniform external loading, the inclusion-type defects are stationary since the force moving the defect is zero. (This force is directly proportional to the gradient of the first invariant of the stress tensor, see Chapter 5.) For an isotropic elastic material and a spherical hole, the force moving the hole is[24]

$$\vec{\Gamma} = -grad\,U, \quad U = a^3(\alpha\sigma^2 + \beta I)/E, \qquad (\alpha < 0, \ \beta > 0). \tag{3.96}$$

Here α and β are some constants, σ and I are the first and second invariant of the stress tensor at the place where the hole is situated, but without the hole. The contribution of uniform stresses to $\vec{\Gamma}$ is evidently zero, so that $\vec{\Gamma}$ is completely determined by the interaction forces with other holes and boundaries.

The force of interaction between two identical holes is directly proportional to $p^2a^6E^{-1}R^{-4}$, where R is the distance between the holes (See Chapter 5, for more detail). When there is a cluster or cloud of holes, the work required to separate one hole from the cloud is

$$U_1 = \eta p^2 E^{-1}\varepsilon_p a^3 \qquad (\varepsilon_p = a^3 / R_0^3). \tag{3.97}$$

Here η is a dimensionless coefficient, R_0 is the mean distance between the holes in the cloud, and ε_p is porosity. The quantity U_1 is the interaction energy of a single hole in the cloud. In the theory of crystal lattices, the corresponding energy is equal to the energy of the crystal lattice per single atom. The whole energy of interaction of the cloud of the holes is equal to the product of the number of holes by U_1 .

For ideal periodic distribution of the holes in the space, the total force acting on the hole is zero because of symmetry

Spontaneous condensation

Let us assume that the condensation of n holes from a cloud leads to the formation of a disk-shaped cavity with radius r (a crack). The condensation process can take place from the thermodynamic viewpoint, if the total energy of these n holes is equal to the energy of the resultant crack, i.e., the following condition must be satisfied

$$n(-U_1 - U_0 + 4\pi\gamma a^2) = -U_c + 2\pi r^2\gamma \qquad (3.98$$

Here U_0 is the elastic self-energy of the hole (independent of p and r), U_c is the elastic energy of the specimen released during crack formation in the uniform field of constant tensile stress, p, namely,

$$U_c = 2\pi p \int_0^r r_0 w(r_0) dr_0 = \frac{8(1-\nu^2)}{3E} p^2 r^3, \tag{3.99a}$$

where

$$w(r_0) = \frac{4(1-\nu^2)p}{\pi E} \sqrt{r^2 - r_0^2}. \tag{3.99b}$$

The elastic self-energy of the hole U_0 is the energy of the local elastic stresses in the neighborhood of the hole caused by surface tension, this energy coincides with the vacancy energy at $q = 0$. Using Equations. (3.89), (3.92), (3.93), we find:

$$U_0 = \frac{4\pi(1+\nu)a\gamma^2}{E}. \tag{3.100}$$

The total energy of the hole (that is, the energy of the cloud of holes, per single hole) is equal to the sum of its surface energy, the elastic self-energy, and the mean energy of interaction with other holes.

Equation (3.98) represents the law of energy conservation in the condensation process. The number n can be expressed as follows: $n = f_1 S_c/S_h$. Here S_c and S_h are the area of the crack and the cross-sectional area of the hole, respectively, f_1 is a certain coefficient. Since $S_c = \pi r^2$, $S_h = \pi a^2$, it follows that

$$n = f_1 \frac{r^2}{a^2}. \tag{3.101}$$

Discussion

In the particular case of $r = 0$ and $U_c = 0$, when there is no crack, Equation (3.98) expresses the necessary thermodynamic condition for the spontaneous formation of the hole in the material, namely

$$U_0 + U_1 = 4\pi\gamma a^2, \tag{3.102}$$

which, according to Equations (3.97) and (3.100) makes it possible to determine the stress p_* in terms of the radius of the hole a (Figure 3.8):

$$p_* = \left[\frac{4\pi E\gamma}{a\eta\varepsilon_p}\left(1 - \frac{1+\nu}{Ea}\gamma\right)\right]^{1/2} \tag{3.103}$$

The parameter $a_* = (1+\nu)\gamma E^{-1}$ is very small (for metals its value equals approximately 0.01 nm) and, therefore, only the descending portion of the curve on Figure

3.8 has a physical meaning. To estimate p_* in the metals at $U_0 = 0$, put $(\eta\ \varepsilon_p)^{1/2} = 1$ for simplicity. Since in metals $\sqrt{4\pi E\gamma} \sim 1\ Mpa \cdot m^{1/2}$, Equation (3.103) provides: $p_* = 100\ GPa$ for $a = 0.1\ nm$, and $p_* = 1 GPa$ for $a = 1\ \mu m$.

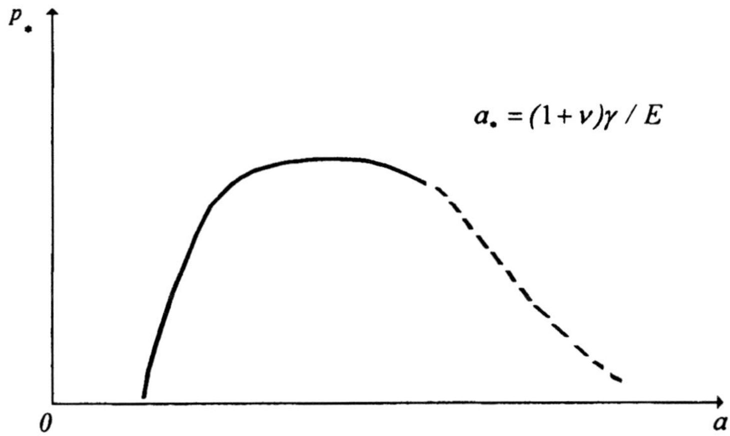

Figure 3.8. The p_* vs. a curve in spontaneous nucleation of a hole

(a is the radius of hole, and p_* is the uniform tensile stress).

In liquids, holes can probably form spontaneously. Equation (3.102) describes the nucleation of spherical voids in liquid matter, e.g. by cavitation.

In the second particular case in which $n = 0$ in Equation (3.98) the material does not contain any holes. In this case Equation (3.98) provides the necessary condition for the spontaneous formation of a crack, namely

$$U_c = 2\pi r^2 \gamma \quad \text{and} \quad p_{**} = \left[\frac{3\pi E\gamma}{4r(1-\nu^2)}\right]^{1/2} \tag{3.104}$$

This equation determines the stress p_{**} necessary for creation of a disk-shaped crack of radius r in the brittle material without holes. At $\nu = 1/3$ and $r = a$, the stress p_{**} is approximately half that of p_*. The spontaneous nucleation of nanocracks of radius 1 to 100 nm is probably possible under high tensile stress in amorphous metals even with no holes involved. The holes, and especially mobile holes, make easier this process of crack nucleation.

Substituting n, U_1, and U_c in Equation (3.98) by Equations. (3.97), (3.99) and (3.101) yields the stress vs. radius condition equation necessary for the holes-enhanced formation of a crack,

$$\bar{p}^2 = \frac{2f_1 - 1 - \dfrac{2f_1\gamma(1+\nu)}{aE}}{f_1\eta\varepsilon_p - 32\bar{r}(1-\nu^2)3^{-1}}, \tag{3.105a}$$

where

$$\bar{p} = p\left(\frac{a}{2\pi\gamma E}\right)^{1/2}, \qquad \bar{r} = \frac{r}{a} \tag{3.105b}$$

Equation(3.105a) was used to construct Figure 3.9 where

$$\bar{r}_* = \frac{3\eta f_1 \varepsilon_p}{32(1-\nu^2)}, \qquad \bar{p}_* = \frac{2f_1 - 1 - \dfrac{2f_1\gamma(1+\nu)}{aE}}{\eta f_1 \varepsilon_p}. \tag{3.105c}$$

$$(\bar{r} < \bar{r}_*, \quad \bar{p}_* > \bar{p}_*)$$

The larger radius corresponds to the higher stress, if the crack is nucleated. However, the crack can nucleate only if $\bar{p} > \bar{p}_*$, and the radius of the nucleated crack cannot be greater than $\bar{r}_*$ for any stress, that is $\bar{r} < \bar{r}_*$. For comparison, the curve,

$$4(1-\nu^2)\bar{p}^2\bar{r} = 1, \tag{3.106}$$

is depicted in Figure 3.9 by a broken line. This equation determines the tensile stress, p, which is necessary to propagate a pre-existing crack of radius r.

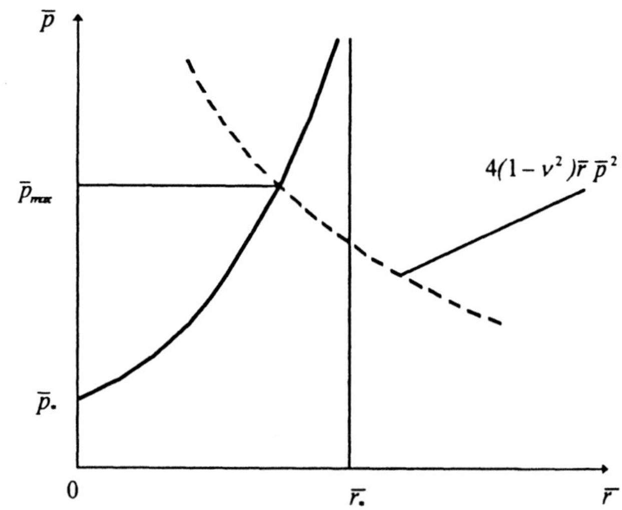

Figure 3.9. The dimensionless crack radius $\bar{r}$ versus the dimensionless stress $\bar{p}$ in the material with a cloud of holes which condense into a crack

The nucleation of a crack as a result of condensation of holes can take place only at very high stresses. This mechanism probably works in amorphous metals which do not contain initial cracks and dislocations in their structure. The tensile stress of the amorphous alloys predicted by Equations. (3.103) and (3.106) of this theory is in satisfactory agreement with the test estimates given in References[25-27]. The curve of Figure 3.9 resembles the λ-shaped diagrams that are typical not only of the physics of phase transitions[29] but also of the technological evolution of social systems[30] experiencing similar transformations.

3.4 Problems

3.1. Consider an edge dislocation at $z = 0$ in a homogeneous plane as the particular case of Equations (3.7). Take $\mu_1 = \mu_2 = \mu$, $\kappa_1 = \kappa_2 = \kappa$, $\nu_1 = \nu_2 = \nu$.
Derive the dislocation field in the coordinate frame Ox_0y_0 (Figure 3.10):

$$
\begin{aligned}
u_0 &= -\frac{B}{4\pi(1-\nu)}[-2(1-\nu)\theta_0 - \sin\theta_0\cos\theta_0],\\
\upsilon_0 &= -\frac{B}{4\pi(1-\nu)}[(1-2\nu)\ln r + \cos^2\theta_0] \qquad (3.107)\\
\sigma_{x_0x_0} &= -\frac{\mu B\sin\theta_0(2+\cos 2\theta_0)}{2\pi(1-\nu)r}, \quad \sigma_{y_0y_0} = \frac{\mu B\sin\theta_0\cos 2\theta_0}{2\pi(1-\nu)r},\\
\sigma_{zz} &= \nu(\sigma_{x_0x_0} + \sigma_{y_0y_0}), \quad \tau_{x_0y_0} = \frac{\mu B\cos\theta_0\cos 2\theta_0}{2\pi(1-\nu)r},
\end{aligned}
$$

Here $u_0, \upsilon_0, \sigma_{x_0x_0}, \sigma_{y_0y_0}, \tau_{x_0y_0}$ are displacement and stress components in the Ox_0y_0 coordinate frame.

$$\theta_0 - \theta = \pi - \alpha$$
$$r^2 = x_0^2 - y_0^2 = x^2 - y^2$$

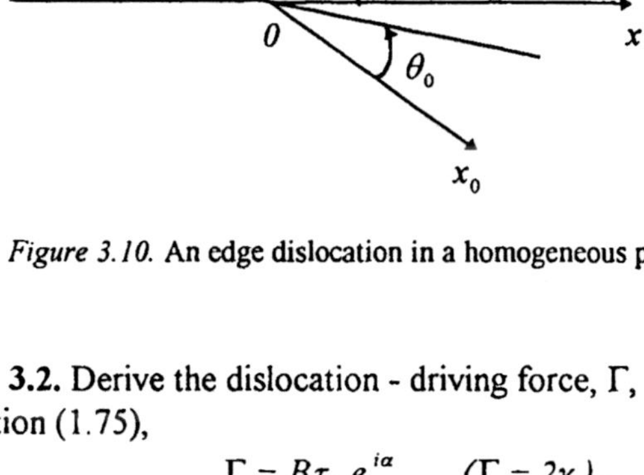

Figure 3.10. An edge dislocation in a homogeneous plane

3.2. Derive the dislocation - driving force, Γ, using the solution to Problem 3.1. and Equation (1.75),

$$\Gamma = B\tau_0 e^{i\alpha} \qquad (\Gamma = 2\gamma). \qquad (3.108)$$

Here τ_0 is the stress $\tau_{x_0y_0}$ of the external stress field (assumed to be zero in Problem 3.1.). This is the Peach-Koehler equation.
Hint. Find the solution for the dislocation in terms of $\Phi(z)$ and $\Omega(z)$, see Equations (3.13) and (3.14).

3. 3. Consider an edge dislocation or dislocation band in the upper half-plane of the bimaterial space (Figure 3.11). Find the dislocation driving force induced by the interface. Show that the force is proportional to $B^2 L^{-1}$ where L is the distance between the dislocation and interface, and B is the magnitude of the Burgers vector. Find the conditions under which the dislocation is attracted to or repelled from the interface.

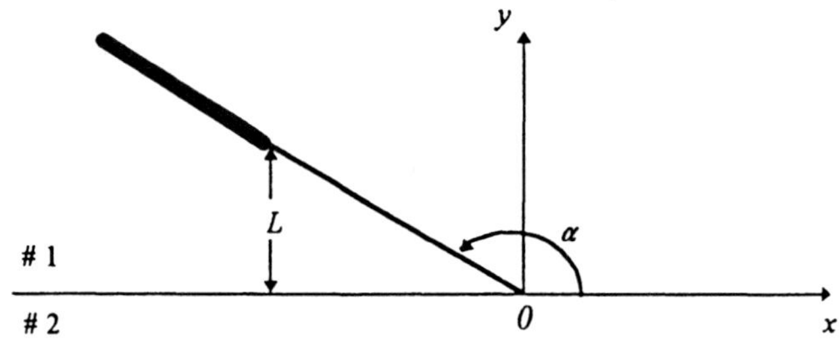

Figure 3.11. A dislocation band approaching an interface

3.4. Consider the mechanism of crack nucleation between two parallel dislocation bands (Figure 3.12). Find stress intensity factors in terms of the Burgers vectors of the dislocation bands. Predict possible scenarios for the crack growth in terms of the Burgers vectors.

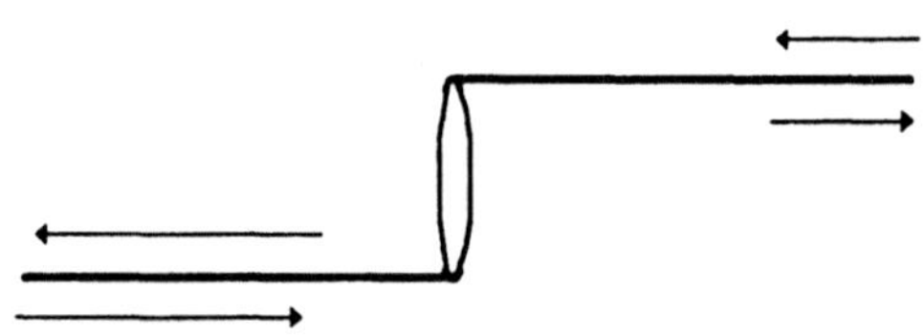

Figure 3.12. A crack nucleated between two dislocation bands

3.5. Find the interaction force between two arbitrary edge dislocations or dislocation bands using Equations (3.107) and (3.108).

3.6. Using the method of Section 3.1, study the case when the nucleated interface crack on Figure 3.1 is that of the transverse shear (mode II) with closed sides.

3.7. Using the method of Section 3.1, study the problem of nucleation of the opening mode interface crack due to the induced stresses of a dislocation band approaching the interface (Figure 3.11), but not impinging on it.

3.8. Study the problem of refraction of an edge dislocation or dislocation band, D, through an interface (Figure 3.13a).

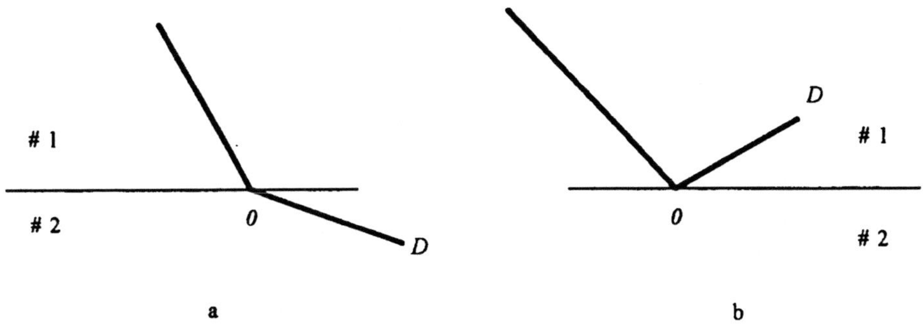

Figure 3.13. Refraction (a) and reflection (b) of a dislocation band at an interfase

3.9. Study the problem of reflection of an edge dislocation or dislocation band, D, from an interface (Figure 3.13b).

3.10. Using the method of Section 3.2, solve the problem of opening mode crack nucleation in one material due to an edge dislocation or dislocation band approaching or impinging on the bimaterial interface Figure 3.14).

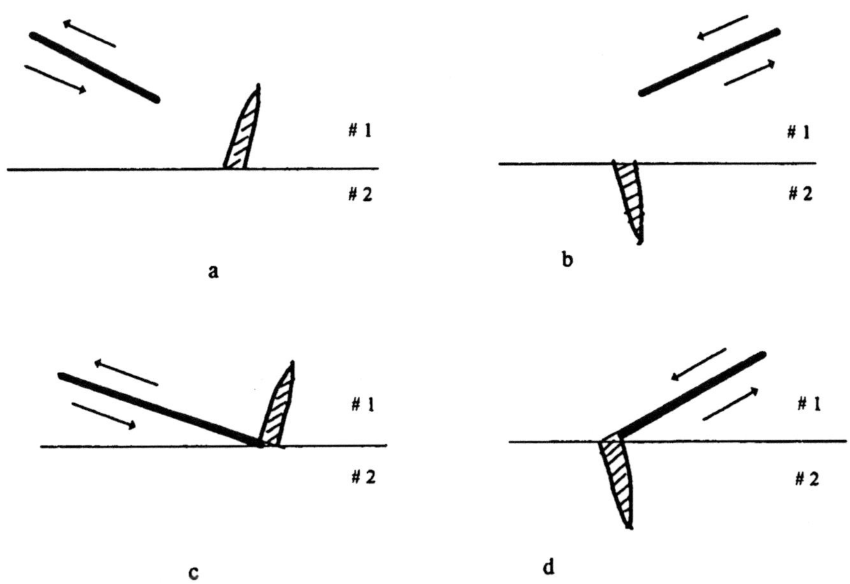

Figure 3.14. Opening mode crack nucleation by a dislocation band approaching (a and b) or impinging on (c and d) a bimaterial interface

3.11. Study Problem 3.7 and 3.10 in the case when the nucleated crack is a mode II crack with closed sides (pileup).

3.12. Study the problem of refraction of a dislocation band on an interface, with an opening mode interface crack being nucleated. In this case, the end B of the crack on Figure 3.1 emits a stable dislocation or dislocation band into the second material.

3.13. A dislocation pileup develops between the barriers at points (- l, 0) and (+l, 0) along the bimaterial interface $y = 0$ in the infinite plane. Find the stress and displacement field for a very large number of dislocations. Derive the pileup-driving force in this case. Assume that the external, unperturbed stresses, $(\sigma_{xx}^{\infty})_1$, $(\sigma_{xx}^{\infty})_2$, σ_{yy}^{∞} and τ_{xy}^{∞}, are constant. Find the relation between $(\sigma_{xx}^{\infty})_1$ and $(\sigma_{xx}^{\infty})_2$ that is, the corresponding stresses in the upper and lower half-planes.

3.14. Show that the tensile stress which is necessary to initiate a pre-existing disk-shaped crack of radius r in a brittle material, is $(2/3)^{1/2}$ times the tensile stress which is necessary to produce a crack of the same radius in the material without holes, in a termodynamically - reversible way.

3.15. Consider a material with holes, and find the ratio of tensile stresses necessary to produce a crack and initiate the same crack. Analyse different possibilities using Figure 3.9 and Equation (3.105a).

3.16. Consider the crack growth due to the flow of mobile holes into a pre-existing crack tip. Assume the following condition of the crack growth: the released energy of the crack per unit area is equal to the surface energy of the material minus the flow of released energy of the holes,

$$\frac{\partial U_c}{\partial r} = 4\pi r\gamma + \frac{\partial n}{\partial r}(U_1 + U_0 - 4\pi\gamma\, a^2)$$

Compare this equation with Equation (3.98).

3.17. For a liquid, find the ratio of tensile stresses necessary to form a spherical void and initiate the extension of a pre-existing void of the same radius.

References

1. C. Zener (1948), The micro-mechanism of fracture, In: *Fracture of Metals*, F. Jonson, W. P. Rop and R. T. Bayles (eds.), *ASM*, Cleveland, Ohio, pp. 3-31.

2. A. N. Stroh (1954), The formation of cracks as a result of plastic flow, *Proc. Roy. Soc.*, London, **A 223**, pp. 404-414.

3. A. N. Stroh (1957), A theory of the fracture of metals, *Advanced Physics*, **6**, pp. 418-465.

4. B. A. Bilby and J. Hewitt (1962), Hydrogen in steel - the stability of microcracks, *Acta Metal.*, **10**, pp. 587-600.

5. E. Smith and J. T. Barnby (1967), Crack nucleation in crystalline solids, *Metal Science*, **1**, pp. 56-64.

6. D. Francois and T. R. Wilshaw (1968), The effect of hydrostatic pressure on the cleavage fracture of polycrycralline materials, *J. Appl. Phys.*, **39**, pp. 4170-4177.

7. M. Nicholas (1968), The strength of metal-alumina interfaces, *J. Materials Science*, **3**, p. 571.

8. M. Morii and S. Nemat-Nasser (1986), Brittle failure in compression: splitting, faulting and brittle-ductile transition, *Philos. Trans. Roy. Soc.*, London, **319**, pp. 337-374.

9. M. F. Ashby and S. D. Hallam (1968), The failure of brittle solids containing small cracks under compressive stress, *Acta Metal.*, **34**, pp. 497-510.

10. G. P. Cherepanov (1994), Interface microcrack nucleation, *J. Mech. Phys. Solids*, **42**(4), pp. 665-680.

11. G. P. Cherepanov (1962), The stress state in a heterogeneous plate with slits, *Izvestia AN SSSR, OTN, Mekhan. i Mashin*, **1**, p. 131 (in Russian).

12. G. V. Kolosov (1909), *Application of the Theory of Functions of One Complex Variable to the Plane Problem of the Mathematical Theory of Elasticity*,Yuriev University Press, (in Russian).
13. N. I. Muskhelishvili (1963), *Some Basic Problems in the Theory of Elasticity* (Trans. by J. R. M. Radock), Noordhoff, Amsterdam, 840 p.
14. F. D. Gakhov (1966). *Boundary Value Problems*, Pergamon Press, London.
15. A. G. Cherepanov (1987), Slip-line as a source of microcracks. In: *Interaction of Bodies in a Fluid Having Free Boundaries*, Cheboxary: State University Press, pp. 141-153.
16. A. R. Miedema (1978), Surface energies of solid metals, *Z. Metallkunde*, **69**, pp. 1280-89.
17. A. R. Miedema and F. J. A. den Broeder (1979), On the interfacial energy in solid-solid and solid-liquid combinations. *Z. Metallkunde*, **70**, pp. 367-379.
18. J. Murr, S. Daw, and M. I. Baskes (1984), Embedded-atom method: derivation and application to impurities, surfaces, and other defects in metals, *Physical Review*, **B 29**(12), pp. 6443 -6453.
19. J. H. Rose, J. Ferrante, and J.R. Smith (1981), Universal binding energy curves for metals and bimetallic interfaces, *Physical Review Letters*,**47**(9), pp. 675-678.
20. A. H. Cottrell (1958), *Trans. AIME*, **212**, p. 192.
21. G. P. Cherepanov and L. A. Kipnis (1984), Some problems of meeting, deviation and branching of slip lines, In *Advances in Fracture Research*, Rama Rao (ed.), Oxford, Pergamon Press.
22. G. P. Cherepanov (1974), *Mekhanika Khrupkogo Razrushenia*, Nauka, Moscow; English edition, *Mechanics of Brittle Fracture*, R. de Wit and W. C. Cooley (eds.), Mc Graw Hill, New York (1979).
23. J. Noble (1962), *The Techiques of the Wiener-Hopf Method*, Pergamon Press, New York.
24. G. P. Cherepanov (1985), Point defects in solids, pp. 605 - 623, In *Fundamentals of Deformation and Fracture (Eshelby Memorial Volume)*, B. Bilby, K. Miller and J. Willis (eds.), Cambridge University Press, Cambridge.
25. Yu. K. Kovneristyi, E. K. Osipov, and E. A. Trofimova (1983), *Physicochemical Fundamentals of Development of Amorphous Metallic Alloys* (in Russian), Nauka, Moscow.
26. G. Bourke and F. Weiss (eds.) (1985), *Surface Treatment and Reliability of Metals* (in Russian translation), Mir, Moscow.
27. R. Ray and L. E. Tanner (1980), Density defects of n-type in amorphous alloys, *Mater. Sci. Eng.*, **45**(2), pp. 824-830.
28. A.G. Cherepanov and G. P. Cherepanov (1990), On one mechanism of the crack nucleation in amorphous metals, *Physics of Metals and Physical Metallurgy*, **2**, pp. 345.
29. S. Ma (1980), *Modern Theory of Crystal Phenomenona* (Russian translation), Mir, Moscow.
30. E. W. Montroll (1987), On the dynamics and evolution of some sociotechnical systems, *Bull. Am. Math. Soc.* **16**(1), pp. 1-46.

CHAPTER 4

PHYSICS OF SINTERING

The rich literature on physical sintering seems to be short of an interdisciplinary approach substantiated both mathematically and experimentally. The present Chapter represents an attempt to create such an approach to physical sintering based on logical analysis of fundamental laws. As to the methods, the Chapter is of primary interest for those who are interested in mathematical modeling of complicated phenomena realized through an interplay of several different physical mechanisms. Constructing self-similar solutions may also be instructive (Sections 4.3 - 4.5). The physical sintering is, in particular, important in its relation to the intriguing problem of crack reversibility and healing.

4.1. Introduction

Sintering is the physical process in which a well-mixed powder aggregate of one or several materials is subjected to high temperature and pressure for a while, and turns into a solid material. Porcelain, pottery and bricks were the first man-made materials manufactured by this method. Today, ceramics are considered to be one of the most promising materials of the future, particularly because of their high strengths at high temperatures.

The process proceeds in two successive stages. The first stage is the powder formation, and the second is the physical sintering itself, in which any two adjacent particles of the powder cohere and lose a relative mobility in the course of time, due to cohesion and diffusive mass exchange between the particles.

Packing stage

The first stage involves packing solid particles of different sizes and shapes into an array with the higest density. Even in the simplest case of the packing of identical spheres, geometry provides a good variety of packing arrays, as presented in Table 4.1.

Table 4.1. Packing geometries for identical spheres

Geometry	Relative density, $\bar{\rho}$	Coordination number, n_c
Simple cubic packing	0.524	6
Body-centered cubic (bcc) packing	0.681	8
Face-centered cubic (fcc) packing (most dense packings)	0.742	12
Hexagonal close-packed (hcp) packing	0.742	12
Least dense packing	0.124	4
Tetragonal packing	0.34	4
Loose random packing	0.54	~8
Dense random packing	0.64	~11

Typical powder packings of different particles before sintering have a relative density in the range 0.5 to 0.8, depending on the mechanical treatment of the powder (pressure, vibrations, grinding and so on). By pressurizing plastic particles, the relative density of the powder can achieve 1, that is, full density. Natural particulated soils like sand or sandy loam consist of particles of different shape and size; the relative density of the soils is usually equal to 0.5 - 0.7, and the coordination number to 6 - 12.

In the geometrical theory of packing, the coordination number refers to the number of neighboring spheres in contact with the sphere under consideration. Note the evident equations,

$$\bar{\rho} + \varepsilon = 1 \qquad \rho = \rho_m \, \bar{\rho}. \tag{4.1}$$

Here, ε is the volume of pores in the unit volume of space (porosity), ρ_m is the material density of particles, ρ is the effective macroscopic density of powder, and $\bar{\rho}$ is the relative density.

Packing is a problem of geometry and mechanical treatment. Shrinkage and densification of the powder while sintering, depend substantially on the packing stage.

In the succeeding text, only the physical problem of sintering is addressed.

Sintering of two particles

The sintering of any two particles in contact is generally determined by a combined action of the following coupled mechanisms / phenomena:

1. Cohesion and adhesion described by the specific surface and interface surface energies;
2. Boundary diffusion of atoms and voids induced by surface concentration gradients;
3. Bulk diffusion of atoms and voids induced by volume concentration gradients;
4. Surface diffusion induced by surface curvature gradients;
5. Condensation and evaporation;
6. Elastic and creep deformations and stresses induced by loading and temperature/concentration gradients;
7. Surface and volume material drift induced by gradients of electric field, temperature, stresses, chemical field and other drift-driving forces.

Sintering is in a sense the opposite to fracturing – it causes the vanishing of the cavity between two materials. Understanding this process could also help us to heal cracks in materials and restore old structures.

During the last fifty years of the development of the theory of physical sintering, numerous attempts have been undertaken to single out the prevalent mechanism of sintering (see, for example, References[1-14]). However, the role of these mechanisms in physical sintering is still not clear. The theories of sintering are based on the concept of the dihedral angle of sintering (Figure 4.1).

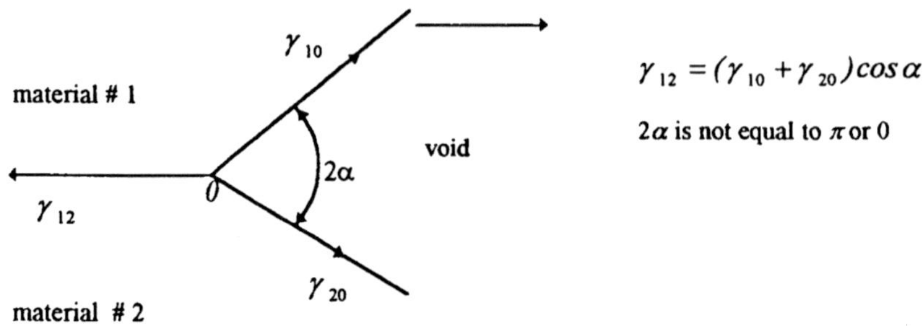

Figure 4.1. Illustration of the concept of dihedral angle of sintering

According to this concept, there is an angle 2α between the sintering materials at point 0, the moving sintering front. This is strictly valid only for liquid matters, when the equilibrium position of the surface membrane created by cohesion forces is not influenced by shear stresses. For a solid matter, this concept may be valid only if point 0 is fixed in space, and the time of observation is sufficiently large to get shear stresses annihilated by creep in the neighborhood of 0.

Cohesion forces in a surface layer of a solid typically lead to a "beak" at point 0, whose length is equal to one to three interatomic spacings, that is, the thickness of the surface layer of the cohesion effect (Figure 4.2). The jaws of the "beak" forming a cusp are subjected to cohesion forces by separation or closure. The surface outside the "beak" (the only one observed experimentally) is not affected by cohesion forces, so that always $2\alpha = \pi$, according to the theories of elasticity and plasticity. This controversy will be explained below.

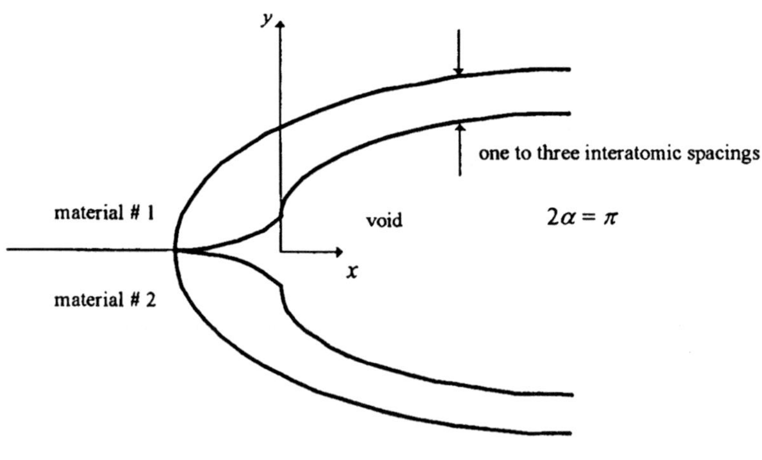

Figure 4. 2. Cohesion forces effect in a solid

Based on experimental observations of the sintering of two identical spheres (Figure 4.3) or a sphere on a plate, the radius of the sintering neck, a is taken to be power function of time[1-14],

$$a = R(t/t_0)^m \qquad (0.05 < a/R < 0.45;\ 0.14 < m < 0.5) \qquad (4.2)$$

Here, R is the sphere radius, t is time, m is the exponent, and t_0 is a specific time of observation (from some minutes to a month). In the experiments, a/R varied typically from 0.15 to 0.35 (generally from 0.05 to 0.45) and m from 0.14 to 0.5 depending on the materials tested and experiment conditions.

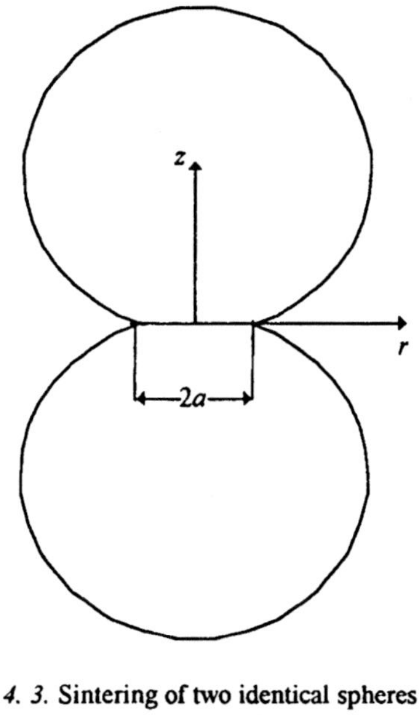

Figure 4. 3. Sintering of two identical spheres

Some experimental data are presented in Table 4.2

The empirical equation (4.2) cannot describe the initial stage of sintering because it provides a non-physical infinite speed for $t \to 0$. Moreover, Equation (4.2) appears to be too simple to describe all possible conditions and materials. As an empirical correlation it cannot be used for prediction of sintering of new materials or under new conditions.

Physical sintering is also concerned with the properties of the final product, as distinct from its chemical composition and its shape. This concern will be partially addressed in this Chapter, as much as it relates to sintering of two particles.

TABLE 4. 2. Some experimental data on sintering of spheres or spheres on plates

Material	R, *mm*	t_0 hours	a/R	T°C	m	major mechanism suspected	Source
Copper	57	0.02-5	0.12-0.26	950- 1050 (0.9-0.97T_M)	~ 1/4	surface diffusion	[3,6]
Copper	50	0.5-41	0.1-0.33	700-900 (0.72-0.86T_M)	~ 1/5	surface diffusion	[1,6]
Silver	180	1-90	0.033-.162	500-800 (0.63-0.87T_M)	~ 1/5	surface, boundary and bulk diffusion	[1,6]
Sodium chloride	60	0.025-0.25	0.07-0.22	700-750 (0.91-0.96T_M)	~ 1/3	vapor transport	[3,6]
Copper	75	1-240	0.11-0.32	750-950 (0.75-0.9T_M)	~ 1/4	surface diffusion	[6]

4.2. Governing Equations of Mass Transport

Consider the sintering of two particles of generally different materials caused by cohesion forces, diffusion, creep and condensation/evaporation. Other important mechanisms of mass transport including compound formation, self-propagating high-temperature synthesis, metastable structural bond energy release, and cold detonation synthesis, are due to chemical reactions and phase transitions. They are ignored in the present treatment and require a special study because of their importance in some systems (for example, sintering of compressed powder of $Ti - C$, $Ni - Al$, $Ti - 2B$, $2Al - F_2O_3$, and $2Al - 3CuO$ mixtures).

The macroscopic density of a material can be represented as

$$\rho = \rho_0 + c^1 + c^2 + \dots + c^N - \varepsilon \rho_m . \tag{4.3}$$

Here, ρ_0 is the immobile part of the material, c^i is the mass concentration of the ith mobile component ($i = 1, 2, \dots, N$), ε is the porosity (volume concentration of voids and pores), and ρ_m is the material density. For example, in sintering of two particles of the same pure material, $\rho_0 = 0$, $N = 1$ and $c^1 = \rho_m$. In sintering of two particles of different pure materials, $\rho_0 = 0$ and $N = 2$.

The transport of the αth mobile component in the bulk material is described by the bulk diffusion equation,

$$\frac{\partial c^{\alpha}}{\partial t} = (D_{\alpha} c^{\alpha}{}_{,k})_{,k} + (m_{\alpha\beta} c^{\alpha} \varphi^{\beta}_{,k})_{,k} \qquad \text{(no sum over } \alpha\text{ !)} \tag{4.4}$$

$$(k = 1, 2, 3;\ b = 0, 1, 2, \ldots, N_0\ ;\ \ \alpha = 1, 2, \ldots, N)$$

Here, D_a is the self-diffusivity of the αth mobile component, and m_{ab} is the mobility coefficient of atoms of the αth component in the drift transport induced by the driving force of the βth external field with the potential φ^{β}.

The main drift potential fields may be:

– Electric field with potential φ_E, so that $\varphi^{\alpha} = q_{\alpha}\varphi_E$, where q_{α} is the electric charge of an ion of the αth component ($\alpha = 1, 2, \ldots, N \leq N_0$);

– Temperature T, so that $\varphi^{0} = T$;

– Stress field with the first invariant of stress tensor, σ, acting as the potential of the stress field upon inclusion atoms and voids (the mobile components),

$$\varphi^{N+1} = \sigma$$

– Chemical field created by concentrations of other components

$$\varphi^{n} = \Omega_n c^{n} \quad (n \neq \alpha,\quad n = 1,2,3,\ldots, N)$$

where Ω_n is the atom volume of the nth component. The other forces can be gravitation, centrifugal force, magnetic field and so on.

From Onsager's principle it follows that

$$m_{\alpha\beta} = m_{\beta\alpha} \tag{4.5}$$

The mass transport along grain boundaries obeys the same equations simplified by the condition that the thickness of the grain boundary layer is very small, but the diffusivity in the layer is very great compared to the bulk diffusion. (See Section 4.5, where the competition between bulk and boundary diffusion in sintering is studied in more detail).

In Sections 4.3 and 4.4 we consider, in particular, the initial stage of sintering governed mostly by cohesion forces (surface tension), surface curvature gradients (surface diffusion), vapor transport, and elastic deformation.

The diffusion of voids and holes often creates the major constituent of the mass transport describing the creep of polycrystalline materials. Voids travel along dislocation lines from the stretched zone to the compressed one, thus forming a material flux in the opposite direction.

The diffusion of voids satisfies the equation,

$$\frac{\partial \varepsilon}{\partial t} = (D\varepsilon_{,k})_{,k} + e_v(1-2\nu)exp\left(-\frac{U - v_p\sigma}{kT}\right) \quad (k = 1,2,3) \tag{4.6}$$

Here, ε is porosity, D_v is the diffusity of voids, U is the activation energy, v_p is the activation volume, σ is the first invariant of the stress tensor, and ν is Poisson's ratio. The quantuty e_v is an empirical constant equal to the strain rate at the theoretical strength, $\sigma = U / v_p$, that is, $e_v = c/a$, where c is the sound velocity and a is the interatomic spacing.

The volume dilatation is equal to

$$\varepsilon_{ii} = \varepsilon, \qquad (i = 1,2,3) \tag{4.7}$$

if one neglects contribution from atom diffusion (ε_{ij} is the strain tensor).

Equation (4.6) is well confirmed by experimental data on steady creep of polycrystalline materials in a uniform stress field of extension-compression, when $\varepsilon_{,k} = 0$. In this case, the first term on the right side of Equation (4.6) is zero, and this equation provides the well-known empirical equation of strain rate by creep in term of stress (generation rate of voids and holes). Note that the Nabarro-Herring equation of creep predicts that $d\varepsilon/dt$ is directly proportional to σ, in disagreement with experimental data. However, we will retain and use the Nabarro-Herring prediction for shear strain rates described by the strain rate deviator,

$$d_{ij} = e_{ij} - \frac{1}{3}e_{nn}\delta_{ij}, \tag{4.8}$$

where

$$e_{ij} = \frac{1}{2}(v_{i,j} + v_{j,i})$$

is the strain rate tensor (v_i are material velocities). Data on drift properties of voids and holes acted upon by other fields are unknown except for the stress field (for more detail, see Chapter 5).

It should be emphasised that the simultaneous consideration of the diffusion of both atoms and voids makes sense if there are some mobile atoms not included in the list of N components and accounted for by the diffusion of corresponding voids. (The transfer of a void is equivalent to that of an atom in the opposite direction). Otherwise, we have double description of the same mass transport. As a matter of fact, we may neglect the consideration of ε, if all mobile atoms are taken into account in the diffusion equations (4.4). Or, on the contrary, we may consider only transport of voids, if we take into account in ε the negative volume of all mobile atoms. It is probably reasonable to retain both c^{α} and ε keeping in mind that they describe the transport of different components.

The issue of stresses induced by diffusion mass transport is most controversial in the sintering problem. The following constitutive equations are reasonable in this case:

$$d_{ij} = \frac{\dot{\tau}_{ij}}{2G} + \frac{\tau_{ij}}{\mu} + \alpha \dot{T} \delta_{ij} + \eta_n \dot{c}^n \delta_{ij} \,, \tag{4.9}$$

$$e_{ij} = \frac{\dot{\sigma}_{ij}}{3E} + \frac{\sigma_{ij}}{3\mu_v} + 3\alpha \dot{T} + 3\eta_n \dot{c}^n + \dot{\varepsilon}, \tag{4.10}$$

where

$$\tau_{ij} = \sigma_{ij} - \frac{1}{3}\sigma_{kk}\delta_{ij}, \quad \delta_{ij} = \begin{cases} 1 & \text{if} \quad i = j \\ 0 & \text{if} \quad i \neq j \end{cases}; \qquad n = 1,2,\ldots,N; \quad i,j,k = 1,2,3$$

Here G is the shear modulus, μ is the shear viscosity and μ_v is the volume viscosity (of linear creep), α is the coefficient of thermal expansion, η_n is the expansion coefficient induced by concentration of the nth component, and τ_{ij} is the stress deviator. The upper dot denotes the time derivative. (The traditional meaning of α as the coefficient of thermal expansion should not be confused with the earlier use of α as an index).

On the right side of Equations (4.9) and (4.10), the first term describes the local Hookean response, the second the Nabarro-Herring prediction, the third the thermal expansion, the fourth the strain rate induced by the concentration of mobile components, and the fifth represents the concentration rate of voids.

The closed system of equations (4.3) - (4.10) provides a full description of mass transport processes inside a material.

4.3. Cohesion Contact of Two Spheres

In the initial stage of sintering, external pressure and cohesion forces cause the contact of spheres. This is the part of the process which may be very important for the whole sintering. We consider it in the simplest case of two identical spheres of one and the same material (Figure 4.3). In this case the shear stress on the area of contact, which is in the plane $z = 0$, is zero, and the resultant force, F, on the contact area is equal to

$$F = -2\pi \int_0^a r\,\sigma_z dr \tag{4.11}$$

Here, r and z are cylindrical coordinates, z is the axis of symmetry, $z = 0$ is the plane of symmetry, and a is the radius of the contact area of the sintering zone. For uniform particulated medium (powder), F can be expressed in terms of porosity, ε, and the overall pressure, p, in the powder as

$$F = \pi p(1-\varepsilon)R^2 \,. \tag{4.12}$$

Let σ_r, σ_z, σ_θ, and τ_{rz} denote the axisymmetric field stresses, and u and w the radial and axial components of the displacement.

Elastic spheres

A simple solution to the problem of contact of two arbitrary elastic ellipsoids was found by Hertz who guessed and proved that the contact area forms the surface of an ellipsoid, the contour of the contact area is an ellipse, and the contact pressure distribution in terms of coordinates represents an ellipsoid with principal axes coinciding with those of the ellipse. Hertz's solution is valid under the following assumptions: (i) $a<<R$, (ii) Cohesion forces can be ignored, and (iii) Shear stresses on the contact area are zero. In the case of two identical spheres, the latter assumption is met because of symmetry condition.

For a circular contact area we study the contact pressure distribution satisfying the balance condition (4.11) is given by

$$\sigma_z = -\frac{3F}{2\pi a^2}\left(1-\frac{r^2}{a^2}\right)^{1/2}, \quad r \le a \tag{4.13}$$

where

$$F = \frac{4a^3 E}{3R(1-\nu^2)} . \tag{4.14}$$

(E is Young's modulus, and ν is Poisson's ratio).
From (4.12) and (4.14) it follows that:

$$p = \frac{4Ea^3}{3\pi(1-\varepsilon)(1-\nu^2)R^3} . \tag{4.15}$$

For example, if $a/R = 1/2$, $\varepsilon = 1/3$, and $\nu = 1/3$, then $p \sim E/10$. For metals, typically $E\sim 1MBar$, so that $p \sim 100\ KBar$. Such a tremendous pressure would be necessary to compress the powder to the densification state roughly corresponding to the latest stage of sintering, when $a/R \sim 1/2$. Due to plastic yielding of metals, this bound appears to be considerably lower. Let us estimate it based on experimental and theoretical data on hardness. According to these data the yield normal stress on the area of contact of two spheres by indentation is roughly three times greater than the yield stress, σ_Y, by uniaxial compression of a bar. From here, it follows that $p = 3\sigma_Y A/(\pi R^2)$ where the area of the contact is taken to be $A \approx \pi(R/2)^2$. For metals, σ_Y equals 1 to 10 $KBar$ so that plasticity decreases the limiting pressure, at least ten times. One should keep in mind that the densification does not, generally, provide for the integrity and strength of the powder mixture – this is what sintering is making.

The shape of a deformed sphere outside the contact area near the front is :

$$z = \frac{F(1-\nu^2)}{\pi a E}\left(\frac{r^2}{a^2}-1\right)^{3/2} \qquad (r \geq a) \tag{4.16}$$

The surfaces of the elastic spheres thus form a cusp near the contact area, so that the dihedral angle equals zero, according to Hertz. It can be shown that this result remains valid for any deformable solid if cohesion forces are ignored.

Another important observation is quite elementary. At the initial moment of contact, the surface of the upper sphere is given by

$$z = R - \sqrt{R^2 - r^2} \tag{4.17}$$

At point, $r = R/2$, we have $z = 0.134R$ and the slope angle of the tangent plane is equal to 30°. Therefore, in the range of $r < R/2$, the cavity outside the spheres can be considered as a narrow crack. This conclusion is still more true if one takes into account deformation of the surface because of diffusion processes and creep. According to Hertz, the relative displacement of the centers of the spheres ("shrinkage") is equal to

$$w_0 = \left[\frac{9F^2(1-\nu^2)^2}{2RE^2}\right]^{1/3}. \tag{4.18}$$

Viscous spheres

Consider slow creeping deformation of two identical spheres of a linearly viscous material in the contact problem (Figure 4.3). Inertial forces are negligibly small, so that the governing equations in this case coincide with the static equations of the theory of elasticity, if the displacements are replaced by the corresponding velocities, and Young's modulus, E, by $2(1+\nu)\mu$, where μ is the dynamic viscosity of the fluid material. (Also, Poisson's ratio ν is usually taken 1/2 for fluids).

The counterpart of Hertz's solution for viscous spheres is:

$$\sigma_z = -\frac{3F}{2\pi a^2}\left(1-\frac{r^2}{a^2}\right)^{1/2}, \tag{4.19}$$

$$F = \frac{16a^3\mu}{3Rt}, \tag{4.20}$$

$$\upsilon_0 = \left[\frac{9F^2t^2(1-\nu^2)^2}{2R\mu^2}\right]^{1/3}. \tag{4.21}$$

Here, t is the time since the beginning of the contact of the spheres, and υ_0 is the centre velocity of the sphere (shrinkage velocity). The dihedral angle is again equal to zero.

According to Equation (4.20), a grows proportional to $t^{1/3}$ if the mechanism of linear creep prevails. It is noteworthy that from the physical point of view, creep is a macroscopic manifestation of the stress-enhanced bulk diffusion of atoms and voids.

Non-linear creep

Linear creep is well confirmed only in some polymer materials. Metals and other polycrystalline materials are strongly non-linear. For steady creep, experimentalist often use the power law with the exponent in the range from 3 to 8.

The governing equations of the steady power law creep for axisymmetric problems are: Equilibrium equations,

$$\frac{\partial \sigma_r}{\partial r}+\frac{\partial \tau_{rz}}{\partial z}+\frac{\sigma_r-\sigma_\theta}{r}=0, \qquad \frac{\partial \tau_{rz}}{r}+\frac{\partial \sigma_z}{\partial z}+\frac{\tau_{rz}}{r}=0 \tag{4.22}$$

Strain rate vs velocity equations,

$$e_r=\frac{\partial u}{\partial r}, \quad e_z=\frac{\partial w}{\partial z}, \quad e_\theta=\frac{u}{r}, \quad e_{rz}=\frac{1}{2}\left(\frac{\partial u}{\partial z}+\frac{\partial w}{\partial r}\right), \tag{4.23}$$

(u and w are the radial and axial components of the material velocity).
Constitutive equations,

$$e_r+e_z+e_\theta=0, \tag{4.24}$$

$$e_r=\mu_c^{-1}I^\chi(\sigma_r-\frac{1}{3}\sigma_{ii}), \tag{4.25}$$

$$e_z=\mu_c^{-1}I^\chi(\sigma_z-\frac{1}{3}\sigma_{ii}), \tag{4.26}$$

$$e_\theta=\mu_c^{-1}I^\chi(\sigma_\theta-\frac{1}{3}\sigma_{ii}), \tag{4.27}$$

$$e_{rz} = \mu_c^{-1}I^\chi\tau_{rz}, \quad \mu_c\Gamma_v = 2I^{\chi+1}, \tag{4.28}$$

where

$$4I^2=(\sigma_r-\sigma_z)^2+(\sigma_r-\sigma_\theta)^2+(\sigma_\theta-\sigma_z)^2+4\tau_{rz}^2, \tag{4.29}$$

$$\Gamma_v^2 = (e_r - e_z)^2 + (e_r - e_\theta)^2 + (e_\theta - e_z)^2 + e_{rz}^2 ,$$
$$\sigma_{ij} = \sigma_r + \sigma_z + \sigma_\theta . \tag{4.30}$$

Here, μ_c is the temperature-dependent dynamic viscosity of power law creeping material, χ is a temperature-dependent constant (for a majority of metals, $2 \le \chi \le 5$). When $\chi = 0$, equations (4.22) - (4.30) are the classic Stokes' equations for viscous fluids. Equations (4.19) - (4.30) well describe steady creep of many materials. According to (4.25), the materials are assumed to be incompressible. This means that, based on this theory, the densification of a powder by sintering is assumed to be due to the closure of voids.

It can be rigorously proven that the solution to the equations (4.22) - (4.30) near the sintering front (crack tip) has the following form:

If the surface tension is zero:
stresses σ_z, σ_r, σ_θ, and τ_{rz} are proportional to $(a - r)^{(\chi+1)/(\chi+2)}$,
velocities u and w are proportional to $(a - r)^{(\chi+2)/\chi+20}$ where $|a - r| << a$; (4.31)

If the cohesion forces (surface tension) are non-zero:
stresses σ_z, σ_r, σ_θ, and τ_{rz} are proportional to $(a - r)^{-1/(\chi+2)}$,
velocities u and w are proportional to $(a - r)^{1/(\chi+2)}$ where $|a - r| << a$. (4.32)

From (4.31), it follows that the spheres form a cusp on the sintering front with zero dihedral angle if the surface tension is zero, and the dihedral angle equals π when the surface tension is non-zero.

The counterpart of Hertz's solution to the contact problem of two creeping spheres of the power law material is given by :

$$\sigma_z = \frac{(2\chi + 3)F}{\pi(\chi + 2)a^2}\left(1 - \frac{r^2}{a^2}\right)^{\frac{\chi+1}{\chi+2}} , \tag{4.33}$$

$$F = \eta_F(\chi)\frac{16a^3}{3R}\left(\frac{\mu_c}{t}\right)^{\frac{1}{\chi+1}} , \tag{4.34}$$

$$\upsilon_0 = \eta_v(\chi)\frac{2^{1/3}F^{2/3}}{(9R)^{1/3}}\left(\frac{t}{\mu_c}\right)^{\frac{2}{3(\chi+1)}} , \tag{4.35}$$

Here $\eta_F(\chi)$ and $\eta_v(\chi)$ are some coefficients which are equal to 1 when $\chi = 0$, and υ_0 is the shrinkage rate or velocity, compare with Equation (4.21).
According to Equation (4.34), the neck growth rate by the creep mechanism is equal to

$$\frac{da}{dt}=\frac{1}{6(\chi+1)}\left[\frac{3FR}{2\eta_f(\mu_e)^{1/(\chi+1)}}\right]^{1/3} t^{-(3\chi+2)/(3\chi+3)}, \tag{4.36}$$

even if surface tension is equal to zero. The time exponent varies from -2/3 (at $\chi=0$) to -17/18 (at $\chi=5$). Correspondingly, for the neck radius a, the time exponent varies from 1/3 (at $\chi=0$) to 1/18 (at $\chi=5$).

Consequently, in a powder the neck grows under pressure p even if surface tension is zero. Remember that p enters into (4.33) - (4.36) through F expressed by Equation (4.12).

Cohesion effect in the contact of two smooth elastic spheres

We take into account the cohesion forces (surface tension), while considering the contact of two smooth elastic spheres. Cohesion forces at the sintering front, $z = 0$, $r = a$ (the crack end), cause an advancement of the contact area edge, so that a toroidal zone of tensile stresses appears in the neighborhood of the front. As a result, the tensile stress singularity, proportional to $(a-r)^{-1/2}$ and typical of cracks, forms at the contact area edge, the cusp at the front disappears, and the dihedral angle becomes equal to π.

In this case the elastic solution to the contact problem can easily be constructed by using the superposition of Hertz's and Boussinesq's solutions:

$$\sigma_z = C_1\left(1-\frac{r^2}{a^2}\right)^{1/2}+C_2\left(1-\frac{r^2}{a^2}\right)^{-1/2} \quad (r\le a), \tag{4.37}$$

where C_1 and C_2 are some constants. The first term corresponds to Hertz's solution and the second to Boussinesq's solution derived originally for the flat round punch pressure on the half-space boundary (Boussinesq's term provides a constant displacement w on the entire area of contact). From the energy conservation equation on the front, $z = 0$ $r = a$, it follows (Chapter 1) that

$$\sigma_z = \frac{K_{Ic}}{\sqrt{2\pi(a-r)}}, \quad \text{when} \quad z=0, \quad r\to a-, \tag{4.38}$$

$$w=\pm\frac{4(1-\nu^2)}{E}K_{Ic}\sqrt{\frac{r-a}{2\pi}}, \quad \text{when} \quad z=0, \quad r\to a+, \tag{4.39}$$

where

$$K_{Ic}=\sqrt{\frac{2E\gamma}{1-\nu^2}}, \tag{4.40}$$

$$2\gamma = 2\gamma_s-\gamma_b . \tag{4.41}$$

Here, γ_s is the surface energy of elastic spheres (in the cavity), and γ_b is the interface surface energy in the contact area.

If the surfaces of the spheres in contact are rough, so that the actual area of contact on asperities is Δ times the apparent area of contact, γ_b is equal to

$$\gamma_b = 2\gamma_s(1-\Delta), \quad \Delta \leq 1 \tag{4.42}$$

and

$$\gamma = \gamma_s \Delta . \tag{4.43}$$

The value of Δ may vary from 0.001 to 1 depending on surface treatment, ratio of pressure to Young's modulus, environment, and other factors.

Using Equations (4.38), (4. 11), and (4.14) one can derive C_1 and C_2 in (4.37) and find

$$\sigma_z = -\frac{2aE}{\pi R(1-\nu^2)}\left(1-\frac{r^2}{a^2}\right)^{1/2} + \frac{K_{Ic}}{\sqrt{\pi a}}\left(1-\frac{r^2}{a^2}\right)^{-1/2}, \quad r \leq a \tag{4.44}$$

where

$$F = \frac{4a^3E}{3R(1-\nu^2)} - 2a\left(\frac{2\pi\gamma aE}{1-\nu^2}\right)^{1/2} . \tag{4.45}$$

Figure 4.4 shows the picture of pressure distribution on the contact area, and radius of the contact area in terms of F, based on Equations (4.44) and (4.45). As seen, for $r_* < r < a$, where

$$r_* = a(1-\bar{k})^{1/2}, \quad \bar{k} = \frac{\pi R(1-\nu^2)K_{Ic}}{aE\sqrt{\pi a}} = \frac{R}{a}\left[\frac{2\pi\gamma(1-\nu^2)}{aE}\right]^{1/2}, \tag{4.46}$$

the contact pressure is negative (tensile stress). Cohesion creates the resultant force of attraction which appears at the beginning of the process of contact (Figure 4.4b). The maximum value of attraction force given by

$$F = F_{ma} = \frac{3}{2}\pi\gamma R , \tag{4.47}$$

is achieved at $a = a_*$, where

$$a_* = \left[\frac{9\pi\gamma R^2(1-\nu^2)}{2E}\right]^{1/3}. \tag{4.48}$$

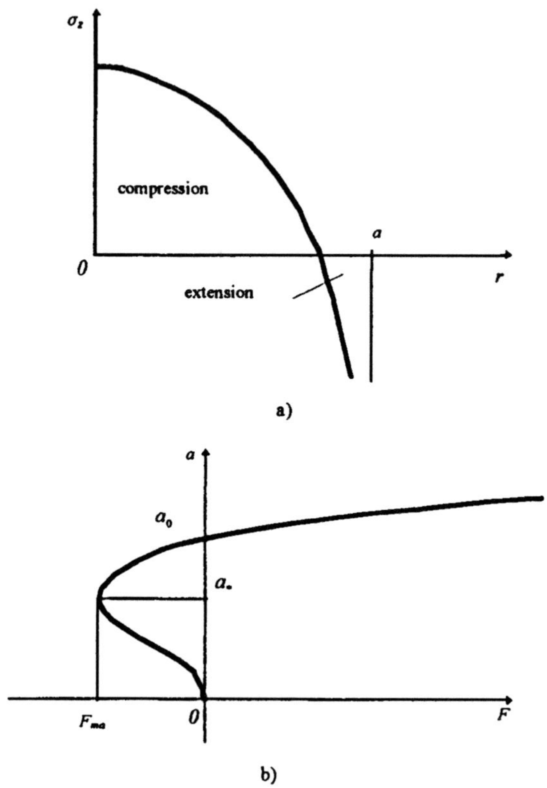

Figure 4.4. The graphs showing: a) pressure distribution on the contact area, σ_z vs r, and b) radius of contact are in terms of equivalent force, a vs F.

The value of F_{ma} characterizes the ultimate strength of coherence of two spheres.
If the pressure in powder is negligibly small, that is, $F = 0$ by contact of two adjacent spheres, the radius of cohesion area of contact given by Equation (4.45) is

$$a = a_0 = R\left[\frac{9\pi\gamma(1-\nu^2)}{2RE}\right]^{1/3}. \tag{4.49}$$

This a_0 is the initial radius of the neck at the beginning of sintering. For example, we calculate a_0 for the conditions of the experiments by Kuczynski and Kingery and Berg (see Table 4.2):

Two copper spheres
$E = 1.39\,MBar$, $\gamma = 1670\ ergs/cm^2$ at 1000 °C, $\nu = 0.33$, $R = 57\ \mu m$ and $a_0 = 0.014R$;
Two silver spheres
$E = 1.1\,MBar$, $\gamma = 1140\ ergs/cm^2$ at 900 °C, $\nu = 0.34$, $R = 180\ \mu m$ and $a_0 = 0.009R$;
Two sodium chloride spheres
$E = 0.3\,MBar$, $\gamma = 225\ ergs/cm^2$ at 25 °C, $\nu = 0.33$, $R = 60\ \mu m$ and $a_0 = 0.011R$.
So, the cohesion-induced radius of sintering neck at the beginning of the process makes up about one percent of the sphere radius, if the pressure is negligibly small.
If cohesion is important, the relative displacement of the centers of the spheres is determined by[15]

$$\frac{w_0}{w_c} = \frac{3(F/F_{ma}) + 2 + 2[1 + (F/F_{ma})]^{1/2}}{3^{2/3}\left\{(F/F_{ma}) + 2 + 2[1 + (F/F_{ma})]^{1/2}\right\}^{1/3}} \tag{4.50}$$

where

$$w_c = \frac{3}{2}\left[\frac{\pi^2 R\gamma^2(1-\nu^2)^2}{E^2}\right]^{1/3}. \tag{4.51}$$

Here w_c is the maximum displacement of spheres by extension when $F < 0$, and F_{ma} is defined by Equation (4.47). From (4. 50) it follows that, under zero pressure, the initial shrinkage by sintering caused by cohesion forces is equal to

$$w_0 = w_{0c} = (4/3)^{2/3} w_c = 6^{1/3}\left[\frac{\pi^2 R\gamma^2(1-\nu^2)^2}{E^2}\right]^{1/3}. \tag{4.52}$$

When pressure is applied and grows, the initial shrinkage is found by Equation (4.50) until yielding begins which is characterized by $F = F_Y$, where

$$F_Y = 2.7\pi\sigma_Y a^2 \tag{4.53}$$

(σ_Y, uniaxial compression yield stress).

4.4. Surface Diffusion and Vapor Transport

From Equations (4.16), (4.39), and (4.44) it follows that the shape of the void cavity outside the contact area near the front at the initial moment of sintering is as follows:

$$z = \frac{2(1-\nu^2)K_{Ic}a^{1/2}}{E\pi^{1/2}}\left(\frac{r^2}{a^2} - 1\right)^{1/2} \tag{4.54}$$

So, the dihedral angle equals π at the parabolic meniscus of the front. Far from the front, the void shape approaches that of Hertz's solution, Equations (4.16).

Now suppose that a toroidal zone of tensile stresses embracing the circular front of sintering at $z = 0$ $r = a$ is very small. Due to (4.44), the small radius, r_*, of the toroidal zone centered at the sintering front is equal to

$$r_* = 4a\left[\frac{16Ea^{3/2}}{\pi^{1/2}RK_{Ic}(1-\nu^2)} - 1\right]^{-1} \tag{4.55}$$

The first term in the brackets is usually much greater than the second one, so that we have:

$$r_* = \frac{\pi^{1/2}RK_{Ic}(1-\nu^2)}{4Ea^{1/2}} = \frac{1}{4}R\left[\frac{2\pi\gamma(1-\nu^2)}{aE}\right]^{1/2} \tag{4.56}$$

The cohesion-induced tensile zone makes it especially favorable for mass transport towards the sintering front. Typically, for $a \sim R \sim 10\ \mu m$ and $\gamma/E \sim 0.1\ nm$, we have $r_* \sim 20 nm$.

Surface diffusion

Surface diffusion has been most often suspected to be the major mechanism of sintering[1-14]. Following Herring and Mullins[2,4] we derive the equation of surface diffusion.

Denote by $\mu(K)$ the increase in chemical potential per atom that is transferred from a point of zero curvature to a point of curvature K on the surface. The μ equals[2]

$$\mu(K) = K\gamma\Omega\ , \tag{4.57}$$

where Ω is the atom volume, and γ is the surface free energy per unit area.

Gradients of chemical potential along the surface via gradients of curvature will produce a drift of surface atoms with an average velocity given by the Nernst - Einstein equation[17]

$$V = -\frac{D_s}{kT}\frac{\partial\mu}{\partial s} = -\frac{D_s\gamma\Omega}{kT}\frac{\partial K}{\partial s}\ , \tag{4.58}$$

where D_s is the coefficient of surface diffusion, T is absolute temperature, and s the arc length along the profile in the radial cross section of the sintering void (k, Boltzmann's constant). The surface current J of atoms is the product of V by the number of atoms per unit area n,

$$J = -\frac{D_s\gamma\Omega n}{kT}\frac{\partial K}{\partial s} \tag{4.59}$$

The surface divergence of J, which is the decrease in the number of atoms per unit area per unit time, is balanced by the speed of movement v_n of the surface element along its normal. From here and Equation (4.59), one then obtains

$$v_n = -\frac{D_s \gamma \Omega^2 n}{kT}\frac{\partial^2 K}{\partial s^2} \tag{4.60}$$

For a general surface, $\partial^2 K/\partial s^2$ is replaced by $\nabla_s^2 K$, the surface Laplacian of K.

We express Equation (4.60) in terms of void shape $z(r, t)$ where t is time. From geometrical considerations, it follows that

$$v_n = -\frac{\partial z}{\partial t}\left[1+\left(\frac{\partial z}{\partial r}\right)^2\right]^{-1/2}, \tag{4.61}$$

$$K = -\frac{\partial^2 z}{\partial r^2}\left[1+\left(\frac{\partial z}{\partial r}\right)^2\right]^{-3/2}, \tag{4.62}$$

$$\frac{\partial}{\partial s} = \frac{\partial x}{\partial s}\frac{\partial}{\partial x}, \quad \frac{\partial x}{\partial s} = \left[1+\left(\frac{\partial z}{\partial r}\right)^2\right]^{-1/2}. \tag{4.63}$$

Substituting v_n and K in (4.60) by (4.61) and (4.62) yields

$$\frac{\partial z}{\partial t} = -B_s\frac{\partial}{\partial r}\left[\left(1+\left(\frac{\partial z}{\partial r}\right)^2\right)^{-1/2}\frac{\partial}{\partial r}\frac{\dfrac{\partial^2 z}{\partial r^2}}{\left(1+\left(\dfrac{\partial z}{\partial r}\right)^2\right)^{3/2}}\right] \tag{4.64}$$

$$\left(B_s = \frac{D_s \gamma \Omega^2 n}{kT}\right)$$

This is the governing equation of surface diffusion. Let us use it to find the sintering rate da/dt at the very beginning just after the contact occurred. According to (4.54), the maximum of K and $\partial^2 K/\partial s^2$ is at the nose of the parabolic profile of the sintering void and is equal to

$$K_{max} = \frac{2}{l}, \quad \left(\frac{\partial^2 K}{\partial s^2}\right)_{max} = -\frac{24}{l^3} \tag{4.65}$$

where

$$l = \frac{16\gamma(1-\nu^2)}{\pi E} \ .$$

Here, l is a specific length parameter having an order of nanometer, because γ/E has an order of 0.1 *nm* in solid materials[16,18,19]. The value of $v_n = da/dt$ at the sintering front corresponding to $(\partial^2 K/\partial s^2)_{max}$ is very large:

$$\frac{da}{dt} = v_n = \frac{24\,B_s}{l^3} = \frac{3\pi^3 n\gamma\, D_s \Omega^2 E^3}{2^9 kT\gamma^3(1-\nu^2)^3} \ . \tag{4.66}$$

It means that the near zone of the sintering front can grow very fast if sufficient flux of diffusion atoms is available to create a material support relaxing the tensile stress concentration. As a result, the void tip seldom has the shape defined by Equation (4.54) except for the very beginning of the process, and the tensile zone specific for cracks rapidly vanishes.

The rate of sintering is determined by the slower transport of atoms along the surface of negative curvature. The small surface of positive curvature near the sintering front can be considered as a specific sink, and its size may be ignored. Outside this small zone, $\partial z/\partial r \ll 1$ almost everywhere along the surface of diffusion (See the discussion of Equation 4.17). In this area, the governing equation (4.64) can be simplified to

$$\frac{dz}{dt} = -B_s \frac{\partial^4 z}{\partial r^4} \qquad \text{for} \qquad t \ge 0, \quad r \ge a(t) . \tag{4.67}$$

Equation (4.67) must be solved for the function $z(r,t)$ subject to the following initial and boundary conditions:

$$z = \frac{r^2}{2R} \quad \text{at} \quad t = 0 ; \tag{4.68}$$

$$z = 0 \ \text{ and } \ \frac{\partial z}{\partial R} = 0 \quad \text{at} \quad r = a(t) . \tag{4.69}$$

In the initial condition (4.68) we neglect the small cohesion zone at the beginning of sintering. The second condition (4.69) expresses that cohesion forces and surface tension are ignored at the sintering front except for the very beginning of the process.

The sintering proceeds due to the surface transport of atoms moving along the void surface into the sintering front, where the atoms are deposited in the narrow boundary layer of thickness δ along $z = 0$, $r \le a(t)$. This layer similar to a grain boundary layer has some specific properties of high diffusive conductivity and will be considered in detail in Section 4.5.

The quantity $2Jdt$ at $z = 0$, $r = a$ is the number of atoms deposited at the moving sintering front per unit length for time dt, which is equal to $\delta da/\Omega$. Thus following Equation (4.59) we find

$$\delta \frac{da}{dt} = -\frac{2\, D_s \gamma\, \Omega^2 n}{kT} \frac{\partial K}{\partial s} \quad \text{at} \quad z = 0 \;\; r = a\,, \tag{4.70}$$

or, in the linear approximation,

$$\delta \frac{da}{dt} = 2\, B_s \frac{\partial^3 z}{\partial r^3} \quad \text{at} \quad r = a(t) \tag{4.71}$$

This is the equation characterizing the driving force of sintering by surface diffusion, that is, the specific material sink at the sintering front.

The moving boundary $F(r,z,t) = 0$ should also satisfy the kinematic condition

$$\frac{\partial F}{\partial t} + \frac{\partial F}{\partial r} V_r + \frac{\partial F}{\partial z} V_z = 0\ , \tag{4.72}$$

where V_r and V_z are the corresponding components of the velocity of the boundary point under consideration. In our case, $F = z - z(r,t) = 0$ so that we have:

$$\frac{\partial z}{\partial t} + \frac{\partial z}{\partial r}\frac{da}{dt} + 0 \quad \text{at} \quad z = 0 \;\; r = a \tag{4.73}$$

because $V_z = 0$ and $V_r = da/dt$ at the sintering front. Hence, $\partial z/\partial t = 0$ at the sintering front (as $\partial z/\partial r = 0$ at $r = a$).

The quantity $a(t)$ should be found from the solution of the boundary value problem, Equations (4.67) - (4.69), (4.71) and (4.73). This is a problem with unknown boundary which is strongly non-linear.

In dimensionless variables,

$$\tilde{r} = \frac{r}{R}, \quad \tilde{z} = \frac{z}{R}, \quad \tilde{t} = \frac{t\, B_s}{R^4}, \quad \tilde{a} = \frac{2\delta a(t)}{R^2} \tag{4.74}$$

the boundary problem (4.67) - (4.69), (4.71) and (4.73) is written as

$$\frac{\partial \tilde{z}}{\partial \tilde{t}} = -\frac{\partial^4 \tilde{z}}{\partial \tilde{r}^4} \quad (\tilde{t} \geq 0\, , \tilde{r} \geq \tilde{a}) \tag{4.75}$$

$$\tilde{z} = \partial \tilde{z} / \partial \tilde{r} = \partial \tilde{z} / \partial \tilde{t} = 0\,, \quad d\tilde{a} / d\tilde{t} = \partial^3 \tilde{z} / \partial \tilde{r}^3 \quad (\tilde{r} = \tilde{a}) \tag{4.76}$$

$$2\tilde{z} = \tilde{r}^2 \quad (\tilde{t} = 0)\,. \tag{4.77}$$

This problem can be solved numerically. We provide the asymptotic solution in the far zone based on the self-similar solution of the form

$$\tilde{z}(\tilde{r},\tilde{t}) = \tilde{t}^{1/2} Z(\xi) \qquad \text{where} \quad (\xi) = \tilde{r} / \tilde{t}^{1/4}. \tag{4.78}$$

Here $Z(\xi)$ should satisfy the following conditions:

$$Z^{(4)}(\xi) + \frac{1}{2} Z(\xi) = \frac{1}{4} \xi Z'(\xi) \quad \text{for} \quad \xi_0 < \xi < \infty \tag{4.79}$$

$$Z(\xi_0) = Z'(\xi_0) = 0; \quad Z(\xi) = \frac{1}{2}\xi \quad \text{for} \quad \xi \to \infty \tag{4.80}$$

$$\tilde{a} = \xi_0 \tilde{t}^{1/4} \tag{4.81}$$

The solution to Equation (4.79) can be represented in the form of an expansion on small number 1/4 as follows:

$$Z(\xi) = Z_0(\xi) + \frac{1}{4} Z_1(\xi) + \frac{1}{4^2} Z_2(\xi) + \ldots \tag{4.82}$$

The series is obviously convergent if functions $Z_0(\xi)$, $Z_1(\xi)$, $Z_2(\xi)$, ... are bounded. Here, $Z_0(\xi)$, $Z_1(\xi)$, $Z_2(\xi)$, ... satisfy the equations following from (4.79) - (4.81):

0)
$$\begin{gathered} Z_0^{(4)}(\xi) + \frac{1}{2} Z_0(\xi) = 0 \quad (\xi_0 < \xi < \infty) \\ Z_0 = Z_0' = 0 \ \text{ at } \xi = \xi_0; \quad Z_0(\xi) = \frac{1}{2}\xi^2 \quad \text{when } \xi \to \infty \end{gathered} \tag{4.83}$$

1)
$$\begin{gathered} Z_1^{(4)}(\xi) + \frac{1}{2} Z_1(\xi) = \xi Z_{0'} \quad (\xi_0 < \xi < \infty) \\ Z_1 = Z_1' = 0 \ \text{ at } \xi = \xi_1; \quad Z_1(\xi) = 0 \quad \text{when } \xi \to \infty \end{gathered} \tag{4.84}$$

2)
$$\begin{gathered} Z_2^{(4)}(\xi) + \frac{1}{2} Z_2(\xi) = \xi Z_{1'}(\xi) \quad (\xi_0 < \xi < \infty) \\ Z_2 = Z_2' = 0 \ \text{ at } \xi = \xi_0; \quad Z_2(\xi) = 0 \quad \text{when } \xi \to \infty \end{gathered} \tag{4.85}$$

and so on.

In zero approximation, the solution has the following form:

$$Z_0 = e^{-\alpha\xi}(A_1 \cos\alpha\xi + A_2 \sin\alpha\xi) + \frac{1}{2}\xi^2 \qquad (\alpha = 2^{-3/4}) \tag{4.86}$$

where A_1 and A_2 are given by

$$A_1 = e^{\alpha\xi_0}\left[\left(\frac{\xi_0}{\alpha} + \frac{\xi_0^2}{2}\right)\sin\alpha\xi_0 - \frac{1}{2}\xi_0^2\cos\xi_0\right],$$
$$A_2 = -e^{\alpha\xi_0}\left[\frac{1}{2}\xi_0^2 \sin\alpha\xi_0 + \cos\xi_0\left(\frac{\xi_0}{\alpha} + \frac{\xi_0^2}{2}\right)\right]. \tag{4.87}$$

This solution describes surface diffusion far from the sintering front: it satisfies all conditions except for Equation (4.71).

The zone near the sintering front is described by the other self-similar solution meeting condition (4.71):

$$\tilde{z}(\tilde{r},\tilde{t}) = Z_N(\xi) \quad (\xi = \tilde{r}/\tilde{t}^{1/4}), \tag{4.88}$$

where $Z_N(\xi)$ should satisfy the following conditions

$$Z_N^{(4)}(\xi) = \frac{1}{4}\xi Z_N'(\xi) \qquad (\xi_0' < \xi < \infty) \tag{4.89}$$

$$Z_N(\xi_0) = Z_N'(\xi_0) = 0 \qquad Z_N'''\xi_0 = \frac{1}{4}\xi_0' \tag{4.90}$$

$$\tilde{a} = \xi_0' \tilde{t}^{1/4} \tag{4.91}$$

One can easy derive a rapidly convergent series for $Z_M(\xi)$:

$$Z_N(\xi) = c_2(\xi - \xi_0')^2 + \frac{\xi_0'}{24}(\xi - \xi_0')^3 + \frac{\xi_0'}{2\cdot 5!}c_2(\xi - \xi_0')^5 + \frac{1}{6!}\left(c_2 + \frac{1}{16}\xi_0'\right)^2(\xi - \xi_0')^6 + \ldots \tag{4.92}$$

The line separating the domains of the near-zone solution $Z_N(\xi)$ and the far-zone solution, $\tilde{t}^{1/2}Z(\xi)$, in the *rt* plane is determined by the continuity condition,

$$Z_N(\xi) = \tilde{t}^{1/2} Z(\xi) \tag{4.93}$$

The values of ξ_0 and ξ_0' can be found by equating the first and second derivatives of both solutions along the separation line.

From (4.91) it follows that the radius of the sintering neck grows as $t^{1/4}$ in the course of time if the mechanism of surface diffusion is dominant.

Vapor transport

The condensation/evaporation process has been also suspected to be the major mechanism in some cases[6]. Following Mullins[4] we derive the equation of vapor transport.

The Gibbs-Thompson formula

$$\ln \frac{p}{p_0} = K(s) \frac{\gamma \Omega}{kT} \tag{4.94}$$

provides the thermodynamic equilibrium condition of the vapor phase with pressure p and the solid phase surface of mean curvature K where p_0 corresponds to the flat surface when $K = 0$. When $\Delta p = p - p_0 \ll p_0$, Equation (4.94) becomes

$$\frac{\Delta p}{p_0} = K \frac{\gamma \Omega}{kT} \,. \tag{4.95}$$

The number of atoms θ emitted by a solid surface which is in equilibrium with its vapor, is given by the kinetic theory of gases as

$$\theta = \frac{p}{(2\pi MkT)^{1/2}} \tag{4.96}$$

per unit time per unit area, for total evaporation (M is the mass of a molecule).

From Eqs. (4.95) and (4.96) it follows that a segment of surface with curvature K emits a flux of atoms which exceeds the quantity θ_0 emitted by a plane by the amount

$$\theta - \theta_0 = \frac{\Delta p}{(2\pi MkT)^{1/2}} = K \frac{p_0 \gamma \Omega}{(2\pi M)^{1/2} (kT)^{3/2}} \tag{4.97}$$

For a plane being in equilibrium with its vapor, θ_0 atoms emanate from the plane and θ_0 atoms impinge on the plane. The latter flux remains the same for a curved surface, so that equation (4.97) gives the net loss of number of atoms per unit time per unit area from the curved surface. Hence, the surface element advances at the rate v_n along the normal to the surface at the point under consideration,

$$\mathrm{v}_n = -B_\mathrm{v} K, \tag{4.98}$$

where

$$B_\mathrm{v} = \frac{p_0 \gamma \Omega^2}{(2\pi M)^{1/2} (kT)^{3/2}} \ .$$

Because of (4.61) and (4.62) Equation (4.98) becomes

$$\frac{\partial z}{\partial t} = B_\mathrm{v} \frac{\partial^2 z}{\partial r^2} \left[1 + \left(\frac{\partial z}{\partial r} \right)^2 \right]^{-1} \tag{4.99}$$

or, in linear approximation,

$$\frac{\partial z}{\partial t} = B_\mathrm{v} \frac{\partial^2 z}{\partial r^2} \tag{4.100}$$

for the void profile $z = z(r,t)$ at $t > 0,\ r > a(t)$. In Equations (4.64) and (4.99) we neglected the curvature of the sintering neck $1/a$ compared to the curvature of sintering zone in a radial cross section.

According to Equation (4.98) the solid phase atoms evaporate from the convex part of a void surface (in the large far zone of small curvature) and condense on the concave part of the void surface (in the small near zone of large curvature). This is the vapor transport mechanism of sintering.

At the very beginning of sintering, when cohesion forces create the tensile zone and a parabolic profile in the near zone of the sintering front, the sintering rate is equal to

$$\frac{da}{dt} = -\mathrm{v}_n = \frac{2 B_\mathrm{v}}{l} = \frac{\pi E p_0 \Omega^2}{8 (2\pi M)^{1/2} (kT)^{3/2} (1 - \nu^2)} \tag{4.101}$$

based on Equations. (4.65) and (4.98). As a result of the short initial stage of sintering when the sintering rate is high (due to (4.101) or surface diffusion or grain boundary diffusion), the near zone is flattened, its size diminishes, and the sintering front neighbourhood degenerates to a specific point sink at $z = 0\ r = a(t)$. The number of atoms condensed in the sink per unit time per unit length equals $2 B_\mathrm{v} (\partial z / \partial r) \Omega^{-1}$ according to (4.98). All the atoms are deposited in the narrow layer of thickness δ along $z = 0\ r \le a(t)$, similar to a grain boundary layer, so that the mass conservation equation can be written as

$$\delta \frac{da}{dt} = 2 B_\mathrm{v} \frac{\partial z}{\partial r} \quad \text{at } z = 0\ r = a(t) \tag{4.102}$$

This is the equation controlling sintering by the vapor transport mechanism.

In equations (4.71) and (4.102) we neglected the mobility of atoms and the diffusion in the grain boundary layer.

Equations (4.99), (4.100) and (4.102) together with the evident initial condition, $z = r^2/(2R)$ at $t = 0$, compose the boundary value problem of the vapor transport sintering theory. Figure 4.5 shows the domain D of the problem with unknown boundary $r = a(t)$ in the r, t plane. In this theory, the dihedral angle at the sintering front profile is equal to $\delta(da/dt)/(2B_v)$ in view of (4.102). Besides, we need to satisfy the kinematic condition (4.73) at the sintering front.

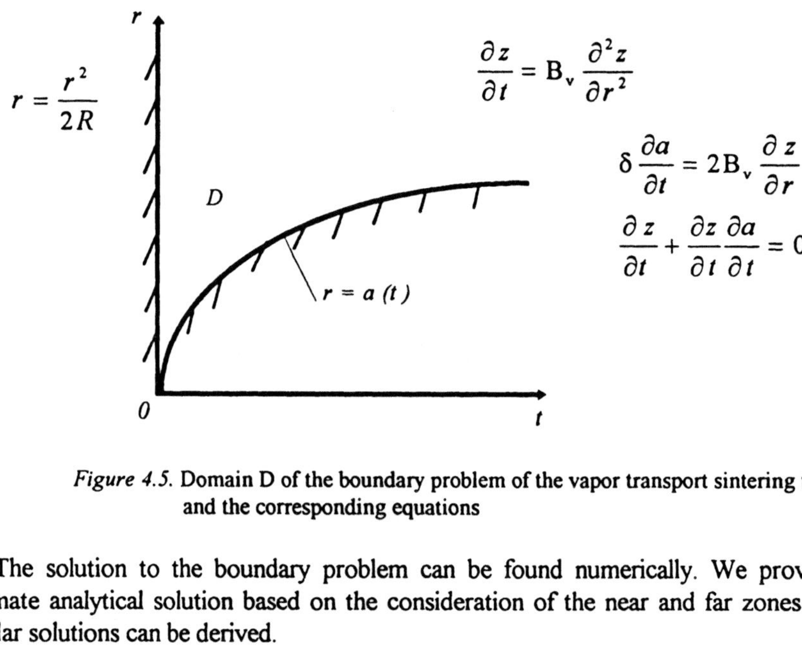

Figure 4.5. Domain D of the boundary problem of the vapor transport sintering theory and the corresponding equations

The solution to the boundary problem can be found numerically. We provide an approximate analytical solution based on the consideration of the near and far zones where self-similar solutions can be derived.

In the far zone the solution is self-similar and can be represented as

$$z(r,t) = \frac{B_v t}{R} Z(\xi) \quad \text{where} \quad \xi = \frac{r}{(B_v t)^{1/2}} \quad (\xi_0 < \xi < \infty) . \tag{4.103}$$

Here, $Z(\xi)$ should satisfy the following condition

$$Z''(\xi) + \frac{1}{2}\xi Z'(\xi) - Z(\xi) = 0 \qquad \text{for } \xi_0 < \xi < \infty; \tag{4.104}$$

$$Z(\xi_0) = 0; \qquad \text{for } \xi_0 \to \infty \quad Z = \frac{1}{2}\xi^2 \ ; \tag{4.105}$$

$$a = \xi_0 (B_v t)^{1/2} . \tag{4.106}$$

The solution to the boundary problem given by (4.104) and (4.105) can be written in the form of the convergent series,

$$Z(\xi) = \frac{1}{2}\xi^2 + 1 + F_0(\xi) - \frac{1}{2}F_1(\xi) + \frac{1}{4}F_2(\xi) - \frac{1}{8}F_3(\xi) + \ldots \qquad (4.107)$$

Here,

$$F_0(\xi) = -\left(\frac{1}{2}\xi_0^2 + 1\right)e^{-(\xi-\xi_0)} \quad \xi_0 < \xi < \infty \qquad (4.108)$$

$$F_n''(\xi) - F_n(\xi) = \xi E_{n-1}'(\xi) \quad \text{for} \quad n = 1,2,3,\ldots,$$

$$F_n(\xi_0) = 0; \quad F_n(\xi) = 0 \quad \text{for} \quad \xi \to \infty. \qquad (4.109)$$

The function F_n (x) is easily derived for any n from the recurrent equations (4.109).
In the near zone the solution can be represented in the following self-similar form:

$$z(r,t) = R\,Z_N(\xi) \quad \text{where} \quad \xi = \frac{r}{(B_v t)^{1/2}} \quad (\xi_0' < \xi < \infty) \qquad (4.110)$$

Here, Z_N (ξ) should satisfy the following equations:

$$2Z_N''(\xi) + \xi E_N'(\xi) = 0 \quad \text{for} \quad \xi_0' < \xi < \infty \; ; \qquad (4.111)$$

$$Z_N(\xi_0') = 0, \quad 4RZ_N'(\xi_0') = \delta\xi_0' \; ; \qquad (4.112)$$

$$a = \xi_0'(B_v t)^{1/2} \qquad (4.113)$$

The kinematic condition (4.73) at the sintering front is satisfied by both asymptotic solutions, (4.103) and (4.110). The solution to the boundary problem (4.111) and (4.112) is:

$$Z_N(\xi) = \frac{\delta}{4R}\xi_0' e^{\frac{(\xi_0')^2}{4}} \int_{\xi_0'}^{\xi} e^{\frac{-x^2}{4}} dx \qquad (4.114)$$

The latter integral is equal to $\sqrt{\pi}\left[\operatorname{erf}\left(\frac{\xi}{2}\right) \quad \operatorname{erf}\left(\frac{\xi_0'}{2}\right)\right]$ in terms of the tabulated function.

The line separating the near and far zones in the rt plane is given by the continuity equation,

$$B_v t Z(\xi) = R^2 Z_N(\xi) \; . \qquad (4.115)$$

In the zero approximation in the far zone, this equation has the form

$$2B_v t\left[\frac{r^2}{B_v t}+2-\left(\xi_0^2+2\right)exp\left(\xi_0-\frac{r}{\sqrt{B_v t}}\right)\right]= \\ =\delta R\xi_0' e^{\frac{(\xi_0')^2}{4}}\int_{\xi_0'}^{r(B_v t)^{-1/2}} exp\left(\frac{-x^2}{4}\right)dx \tag{4.116}$$

To find ξ_0 and ξ_0' it is necessary to equate the first and second time derivatives of both asymptotic solutions along the separation line. Analysis of (4.116) shows that $\xi_0' \geq \xi_0$ and the separation line approximately behaves as $r = \xi_0' \sqrt{t}(1-c_1 t)/(1-c_2 t)$ where c_1 and c_2 are some constants. This means that, for small times, one can use equation (4.114) as the solution of the sintering problem in the case under study.

According to the vapor transport theory, the sintering neck grows roughly as $t^{1/2}$ in the course of time. It is clear that the self-similar solutions constructed in this section lose their physical sense both for $t \to 0$ and $t \to \infty$.

4.5. Boundary Versus Lattice Diffusion

There exists a specific thin layer with a thickness of several interatomic spacings along the area of contact of two spheres. Diffusive conductivity in this layer can be six orders greater[5,18] than in bulk material (lattice). This layer is analogous to those on grain boundaries of polycrystalline materials. They create the main paths for material transport in solids. Similar thin layers of high diffusive conductivity exist on the solid surface. The material fluxes in such thin layers are proportional to concentration gradients and thus obey the usual Fick's law as in volume or lattice diffusion. However, the two-dimensional nature of this phenomenon called *boundary diffusion* creates both difficulties and opportunities for both theoreticians and users. One should clearly distinguish the boundary diffusion from the surface diffusion along curved solid surfaces and interfaces, which is driven by curvature gradients (see Section 4.4 for more detail). Volume diffusion proceeds also along some narrow channels which are probably dislocation lines always available in real lattices.

In large part, we follow Fisher's approach[5] to boundary diffusion, which is similar to Prandtl's approach to the flow of a viscous fluid in a thin boundary layer on a solid surface. Fisher studied a thin semi-infinite bundary layer between two quarter planes - the geometry of the sintering problem is different. Taking into account that the crack-like cavity between two sintering spheres is very narrow during almost all the time of the sintering process, we come at the problem of sintering two half-spaces along a disk-shaped area (as Hertz did). The latter area, along $z = 0$ $r < a(t)$ for the two identical spheres we study, is assumed to have a very small thickness d in the z direction and is called the boundary layer in what follows (Figure 4.6). This layer has much higher diffusive conductivity than the lattice because it contains a lot of free space for atom transport.

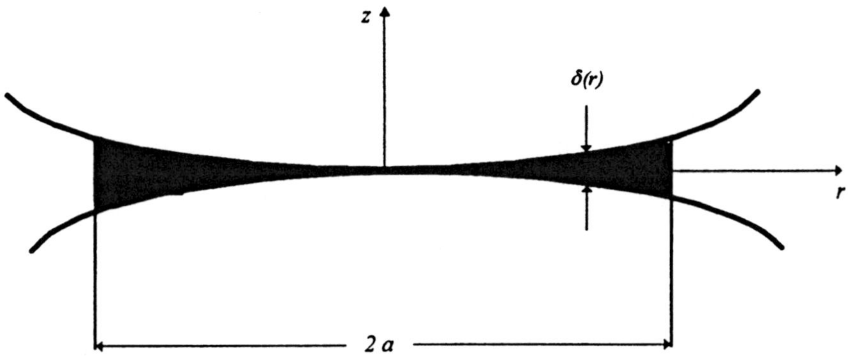

Figure 4.6. The boundary layer of sintering between two half-spaces

However, lattice diffusion influences the mass transport, too, because of much bigger volume of bulk material compared to that of boundary layer. Fisher suggested that δ is a constant equal to two interatomic spacings along a grain boundary. (See Fisher's problem in the axisymmetric formulation below in Section 4.8, Problems 4.6 and 4.7). In Fisher's problem, the grain boundary is fixed - in the sintering problem under study the boundary layer can propagate along the narrow cavity at $z = 0 \;\; r > a$ where $a = a(t)$. If one accepts that δ is constant, it can be proven that $a \sim t^{3/4}$ for large time (Problem 4.8). However, this result does not take into account a very important geometrical peculiarity of the sintering problem - that the cavity widens in the course of time. In the foregoing analysis we assume that

$$\delta = \eta \frac{r^2}{2R} \tag{4.117}$$

where η ia a fitting constant that should be chosen to comply with experimental data. In other words, we assume that δ widens in direct proportion to the distance between the oposite sides of the cavity to be sintered (Figure 4.6).

The equations of mass transport for the axisymmetric problem are:

$$\begin{aligned} \frac{\partial c}{\partial t} &= D_b\left(\frac{\partial^2 c}{\partial r^2} + \frac{1}{r}\frac{\partial c}{\partial r} + \frac{\partial^2 c}{\partial z^2}\right) \quad \text{for} \quad |z| < \frac{\delta}{2}, \quad r < a(t); \\ \frac{\partial c}{\partial t} &= D_L\left(\frac{\partial^2 c}{\partial r^2} + \frac{1}{r}\frac{\partial c}{\partial r} + \frac{\partial^2 c}{\partial z^2}\right) \quad \text{for} \quad |z| > \frac{\delta}{2}. \end{aligned} \tag{4.118}$$

Here, $c = c(r, z, t)$ is the concentration of mobile components (or, the material itself if all atoms may be mobile), D_b and D_L are the diffusive conductivity or diffusivity of the boundary layer and lattice respectively.

The concentration and normal flux of matter should be continuous along the surface

$z = \pm d/2 \;\; r < a(t)$ separating the bulk and boundary layer material. The sintering neck $a(t)$ should be found from the solution of the problem.

At the initial moment of time, the concentration was a given constant in the bulk material and zero in the boundary layer.

The sintering rate and normal flux of matter on the moving front are related by the mass conservation equation,

$$\frac{da}{dt} = -\Omega D_b \frac{\partial c}{\partial r} \qquad \text{for} \quad z = 0 \;\; r = a \tag{4.119}$$

where c is measured in number of atoms per unit volume (Ω, the atom volume). If $a = const$, the diffusion stops in a while after c is levelled everywhere.

There are two small parameters in this problem, δ/a and D_L/D_b, which are needed in order to apply the asymptotic methods of boundary layer theory.

Integrate the first equation (4.118) over z from $-\delta/2$ to $+\delta/2$ to obtain:

$$\frac{dc_*}{dt} = D_b\left(\frac{d^2 c_*}{d r^2} + \frac{1}{r}\frac{dc_*}{dr}\right) + q(r,t) \qquad \text{for} \quad r < a(t)\,. \tag{4.120}$$

Here,

$$c_* = \frac{1}{\delta}\int_{-\delta/2}^{\delta/2} c(r,z,t)dz\,, \qquad q = \frac{2}{\delta} D_L \left(\frac{\partial c}{\partial z}\right)\Bigg|_{z=\delta/2} \tag{4.121}$$

(c_*, the mean concentration in the boundary layer, and q, the input of matter from the bulk material per unit area per unit time).

Assume that :

(i) in the boundary layer

$$\frac{\partial c_*}{\partial t} << D_b \frac{\partial^2 c_*}{\partial r^2}\,, \qquad \text{and} \tag{4.122}$$

(ii) in the bulk material

$$\frac{1}{r}\frac{\partial c}{\partial r} << \frac{\partial^2 c}{\partial z^2}, \qquad \frac{\partial^2 c}{\partial r^2} << \frac{\partial^2 c}{\partial z^2}\,. \tag{4.123}$$

In other words, the field is assumed to be much steeper across the layer than along it.

These are the crucial assumptions based on our belief or guess that it should be so, because $r \sim a$, $t \sim t_s$ and $D_b >> a^2/t_s$ (in the boundary layer), so that

$$\frac{\partial c_*}{\partial t}\left(D_b \frac{\partial^2 c_*}{\partial r^2}\right)^{-1} \sim \frac{a^2}{t_s D_b} << 1$$

and because $r \sim a$, $z \sim$ d, so that (4.124)

$$\frac{\partial^2 c}{\partial r^2}\left(\frac{\partial^2 c}{\partial z^2}\right)^{-1} \sim \frac{\delta^2}{a^2} << 1$$

in the adjacent bulk material.

This kind of guess should be verified after the solution to the simplified problem is achieved. Fortunately, it usually appears to work remarkably well. This method may reduce the dimension of the problem or simplify the governing equations, allowing analytical solutions or numerical experiments on simpler problems. Practically, all the scientific literature considers only simplified problems. It is very seldom that an analytical solution is provided for all three spatial variables and time. Even if the full solution, numerical or analytical, would be found for a certain practical problem, it would be less useful for making a conclusion and decision because of the large amount of unnecessary information.

After the simplifications based on (4.120), (4.122) and (4.123), equations (4.118) become

$$\frac{\partial^2 c_*}{\partial r^2} + \frac{1}{r}\frac{\partial c_*}{\partial r} + \frac{q}{D_b} = 0 \quad \text{for } z = 0, \quad r < a(t); \tag{4.125}$$

Within the framework of this approximation, the value of d may be ignored in these equations, so that the boundary layer becomes a surface domain of zero thickness. Moreover, now we have $c_* = c_*(r)$ and $c = c(z,t)$, in terms of "steepest" variables.

Let us take the solution of equation (4.125) in the following form (see Section 4.4)

$$c = c_*\left(1 + \frac{c_0 - c_*}{c_*\sqrt{\pi}}\int_0^{\xi} e^{-\frac{x^2}{4}}dx\right), \quad \text{where} \quad \xi = \frac{z}{(D_L t)^{1/2}}. \tag{4.126}$$

According to (4.126) $c_* = c_*$ at $z = 0$ and $c = c_0$ at $z \to \infty$ (c_0 is the initial concentration in the unperturbed domain). Differentiating Equation (4.126) yields

$$\frac{dc}{dz} = \frac{c_0 - c_*}{(\pi D_L t)^{1/2}} \quad \text{at } z = 0. \tag{4.127}$$

Substitute Equation (4.127) into Eqs. (4.121) and (4.124) to obtain

$$\frac{d^2 c_\bullet}{d r^2} + \frac{1}{r}\frac{dc_\bullet}{dr} + \frac{\lambda^2}{r^2}(c_0 - c_\bullet) = 0 \tag{4.128}$$

where

$$\lambda = \left(\frac{4R D_L}{\eta D_b}\right)^{1/2} (\pi D_L t)^{-1/4} \tag{4.129}$$

The dimensionless number λ characterizes the relative role of lattice diffusion compared to boundary diffusion. For large λ, lattice diffusion is more important; for small λ, boundary diffusion dominates. As is seen, in the course of time the role of boundary diffusion increases.

The general solution to Equation (4.128) is given by

$$c_\bullet = c_0 + A_1 r^\lambda + A_2 r^{-\lambda} . \tag{4.130}$$

Here, A_1 and A_2 are some constants.

The quantity $c_\bullet$ should be bounded at $r \to 0$, hence $A_2 = 0$. The constant A_1 can be found from the condition that $c_\bullet = 0$ at $r = a$. (The latter condition is, probably, not satisfied at the end of sintering when the sintering rate is very small and the sintering front can be saturated by diffusive atoms). From here, it follows that

$$c_\bullet = c_0 \left(1 - \frac{r^\lambda}{a^\lambda}\right) . \tag{4.131}$$

According to this solution, $c_\bullet$ equals c_0 at $r = 0$. Equation (4.131) follows also from the assumption that the sintering is completed at $r = 0$ from the very beginning, so that $c_\bullet = c_0$ at $r = 0$ at any time.

Substitute Equation (4.131) into Equation (4.119) to derive

$$a\frac{da}{dt} = \lambda c_0 \Omega D_b . \tag{4.132}$$

Solve Equation (4.132) at the initial condition that $a = 0$ at $t = 0$ neglecting the small initial size of the neck. Take into account that λ depends on t due to Equation (4.129) to yield

$$a = 4(9\pi^{1/2}\eta)^{-1/4} (c_0 \Omega)^{1/2} (R D_b)^{1/4} D_L^{1/8}\, t^{3/8} \tag{4.133}$$

This is the main result of the present analysis.

4.6. Combined Diagram of Sintering

The analysis of Sections 4.2 - 4.5 shows that sintering is a complex process realized through many possible mechanisms and mass transport paths. The problem of identification of the major mechanism of sintering under prescribed conditions is very important for practitioners. Unfortunately, this problem stands now roughly as it was fifty years ago. At this point, theory has outstripped experiment and discovered many ways to explain the available test data. It is not clear whether the future theory can predict any test data and solve the identification problem. In any case, some additional refined experiments with radiotracers are necessary to distinguish the better models. However, it is possible that all the mechanisms contribute to the sintering process. Therefore, it is important to construct a logical chain or diagram of the many events or mechanisms leading to a new material.

First, one should distinguish whether the mechanisms operate parallel or in series. Two mechanisms are said to operate parallel if they work independently one from the other and, in principle, each can reach the final goal. Parallel mechanisms can either help or hinder one other, depending on specific conditions. Two mechanisms are said to operate in series if one of them can be realized only if the other works, so that the final goal can be reached only by the work of both.. Among several mechanisms in series, the major mechanism is that which provides the slowest transport rate - other mechanisms in series service the major. The rate of a chain of series mechanisms is equal to the rate of the major mechanism. Parallel mechanisms can be superposed and the major is that which provides the fastest transport rate, that is, the greatest contribution to the final goal (e.g. sintering rate or total time of sintering).

Table 4.3, which is based on the analysis of the sintering process presented in Sections 4.1 - 4.5, gives a block-diagram of the combined work of all the mechanisms involved. In Table 4.3, lattice diffusion was ignored as an independent parallel mechanism. However, low lattice diffusitivy can stop the boundary/lattice diffusion mechanism, so that the boundary layer can become a collector of atoms delivered by vapor transport and surface diffusion onto the sintering front. In this case, boundary diffusion can slow down sintering. Also, surface diffusion and vapor transport can interact and slow down each other and the boundary/lattice diffusion. Creep by pressure, probably, always helps sintering because it makes the cavity sides closer. Lattice diffusion is necessary to provide integrity to a solid material after the spheres or mixed powder is pressurized (creep/lattice diffusion mechanism). Needless to say, superposition does not mean a simple sum of sintering rates delivered by different parallel mechanisms working separately. One may not add da/dt values for different mechanisms (for example, those found in Sections 4.1 - 4.5) and consider the sum to be the total sintering rate - this is a mistake widely spread in the literature on sintering.

Table 4.3 allows one to construct the full system of partial differential equations governing the sintering process. It includes :

- creep equations (4.6), (4.7), and (4.10) or (4.22) - (4.30), in the volume domain D_v occupied by the bodies sintered, allowing one to determine the creep-induced component of velocity of a domain boundary point, $v_{CR} = v_{CR}(s,t)$ where s is the coordinate identifying the point (e.g. arc length);
- lattice diffusion equations of mobile components in D_v, Equations (4.4);

- boundary diffusion equations, Equations (4.120) or (4.124) in the sintering zone domain D_s interfacing the bodies sintered;
- the equation for the normal velocity $v_n = v_n(s,t)$ of the growth of the boundary surface,

$$v_n = v_{CR} + B_v K + B_s \frac{\partial^2 K}{\partial s^2} \tag{4.134}$$

synthesizing Eqs. (4.60) and (4.98) according to Table 4.3.

Table 4.3. Combined Diagram of Sintering

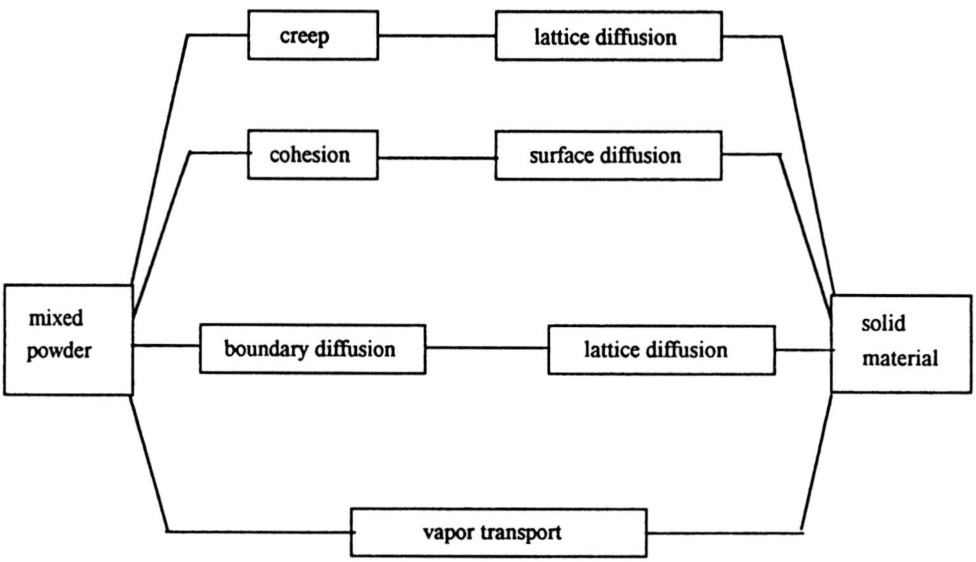

Temperature increase speeds up all the mechanisms.

Pressure increase enhances the creep mechanism.

Cohesion leads the process at the very beginning as the major mechanism, but stops working after a while, or remains as an optional servicing mechanism.

Cohesion or pressure is necessary to start surface diffusion of two spheres, this is the ignition mechanism.

Free surface tension is important only for vapor transport and surface diffusion.

The general equation for the sintering rate is:

$$\frac{da}{dt} = (\mathrm{v}_{CR})_0 + \frac{1}{\delta} B_v \theta_D + \frac{1}{\delta} B_s \left[\frac{\partial K}{\partial s} \right]_0 - \Omega D_b \left(\frac{\partial c_\bullet}{\partial r} \right)_0 \qquad (4.135)$$

Here, $(\mathrm{v}_{CR})_0$ is the creep component of material velocity at the sintering front, θ_D is the dihedral angle, $[\partial K/\partial s]_0$ is the discontinuity of $\partial K/\partial s$ on the sintering front (while coming across it), and $(\partial c_\bullet/\partial r)_0$ is the gradient of boundary diffusion across the sintering front.

Equation (4.135) is derived on the assumption that the sintering front is a singular geometrical line (of zero thickness) on the material surface moving in space and time, on which the state functions generally suffer a discontinuity. This is the line sink/source of atoms for the processes of condensation/vaporization (the second term), surface diffusion (the third term), and boundary diffusion (the fourth term). Stresses and strain rates at this line tend to

$$\frac{1}{E_{\bullet}}=\frac{1-\nu_1^2}{E_1}+\frac{1-\nu_2^2}{E_2}, \quad R_{\bullet}=\frac{R_1 R_2}{R_1+R_2} . \tag{4.136}$$

Here E_1, E_2, ν_1 and ν_2 are the Young's moduli and Poisson's ratios of the corresponding materials, and R_1 and R_2 are the radii of the corresponding spheres.

4.2. Consider a half-plane $y \geq 0$ of an elastic material with a certain surface tension, γ. According to Laplace's equation for surface tension in terms of curvature, a deformation of the boundary causes the normal load component which is equal to $\gamma\partial^2 u_y/\partial x^2$ where u_y is the normal displacement. Study the capillary waves propagating on the surface of the elastic body. Show that, if the propagation speed of the wave is considerably less than the speed of sound, the wave length is equal to $4\pi\gamma(1-\nu^2)E^{-1}$ (static wrinkling). This effect may be essential for materials like polymers and rubber, in which γ/E may be considerably greater than interatomic spacing.

4.3. Consider a traction-free opening mode crack along $y = 0$ $x > 0$ under plane strain. Assume that the elastic material possesses a surface tension γ which provides the normal load component $\gamma\partial^2 u_y/\partial x^2$ on the crack banks. All the stresses at infinity tend to zero. Prove that the solution to this problem can be reduced to the following first order differential equation for the Kolosov - Muskhelishvili potential $\Phi(z)$,

$$\frac{2\gamma(1-\nu^2)}{E} i\Phi'(z)+\Phi(z)=\frac{K_I}{2\sqrt{2\pi z}} , \tag{4.137}$$

valid in the half-plane $y > 0$, and $\Psi(z) = -z\Phi'(z)$. Derive that the crack tip has a cusp,

$$u_y=\frac{2K_I}{3\gamma\sqrt{2\pi}}x^{3/2} \quad \text{as} \quad y=0 \ \text{and} \ x\to 0+ \tag{4.138}$$

and the stress σ_y at the crack tip is equal to

$$\sigma_y=\frac{1}{8}\Gamma(1/2)K_I\left[\frac{E}{\pi\gamma(1-\nu^2)}\right]^{1/2} \quad \text{at } x=0 \ y=0 \tag{4.139}$$

in this model where $\Gamma(1/2)$ is the value of Gamma function $\Gamma(x)$ at $x = 1/2$.

4.4. Consider the harmonic function $\varphi(r,z)$ satisfying Laplace's equation for $z > 0$, whose normal derivative is equal to zero at $z = 0$ $r > a$ and

$$\frac{\partial\varphi}{\partial z}=2\pi\left(1-\frac{r^2}{a^2}\right)^{\lambda-(1/2)} \quad \text{at } z=0 \ r<a , \tag{4.140}$$

where λ is an integer or zero.

Prove that at $z = 0$ $r > a$

$$\varphi(r) = \pi a^2 \frac{\Gamma(\lambda + 1/2)}{\lambda \Gamma(\lambda)\Gamma(1/2)} \int_{r^2 - a^2}^{\infty} \left(1 - \frac{r^2}{a^2 + u}\right) \frac{du}{\sqrt{u}(a^2 + u)} \tag{4.141}$$

Use this result to derive Equation (4.16) and the shape of a deformed sphere everywhere outside the contact area in Hertz' problem of Section 4.3.

4.5. Consider the sintering problem of two identical spheres due to vapor transport (Figure 4.7). For small times, use the equation $z = Z_N(\xi)$ where $\xi = r(B_v t)^{-1/2}$ and Z_N is given by Equation (4.114). Find ξ_0' in the sintering rate equation, $a = \xi_0'(B_v t)^{1/2}$, using the mass conservation law, namely, that the volume of evaporated region E equals the volume of condensed region C in the equilibrium state of vapor.

Show that shrinkage and densification equal zero according to the vapor transport theory. Prove that the dihedral angle is equal to

$$\delta \xi_0' (4R)^{-1} (B_v t)^{-1/2} .$$

4.6. Consider an infinite cylinder, $r < r_0$ and $-\infty < z < +\infty$, made of two semiinfinite cylinders bonded along the ends at $z = 0$ by a thin bonding layer of thickness δ of high diffusive conductivity (this may be a grain boundary layer). Place the cylinder in an environment with concentration c_0 of a diffusive component. Find the concentration of the component in terms of time at the center, $z = 0$, $r = 0$.

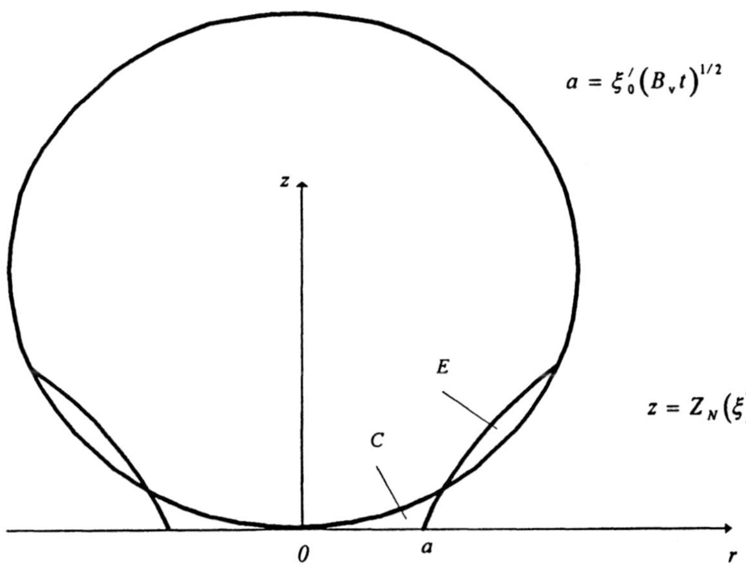

Figure 4.7. Sintering of two identical spheres by vapor transport for small times (E is the evaporated region , and C is the condensed region)

4.7. Consider an infinite plate, $|x| < x_0$ and $-\infty < z < +\infty$, made of two semiinfinite plates bonded along the ends at $z = 0$ by a thin bonding layer of thickness δ of high diffusive conductivity. Suppose one side of the plate is placed in contact with an environment containing a diffusive component of concentration c_0. Find the concentration of the component in terms of time on the other side of the plate. From here for $x_0 \to \infty$ derive Fisher's solution[5].

4.8. Consider the problem of Section 4.5 under the assumption that the thickness δ of the boundary layer is a constant. Prove that

$$c = c_*\left(1 + \frac{c - c_*}{c_*\sqrt{\pi}}\int_0^{\xi} e^{-\frac{x^2}{4}}dx\right) \quad \text{where} \quad \xi = \frac{z}{(D_L t)^{1/2}} \tag{4.142}$$

$$c_* = c_0\left[1 - \frac{I_0(\lambda r)}{I_0(\lambda a)}\right] \quad \text{where} \quad \lambda = \frac{\left(\frac{2D_L}{\delta D_b}\right)^{1/2}}{(\pi D_L t)^{1/4}} \tag{4.143}$$

$$\frac{da}{dt} = \lambda c_0 \Omega D_b \frac{I_1(\lambda a)}{I_0(\lambda a)} \tag{4.144}$$

in this case (I_0 and I_1 are modified Bessel's functions). Prove that $a = A_0 t^{3/4}$ for $t \to \infty$, where A_0 is a constant.

4.9. Consider the problem of Section 4.5 under the assumption that the thickness δ of the boundary layer is a very small constant (e.g equal to two interatomic spacings as in Fisher's analysis[5]). Assume in this case that the boundary layer grows along a fractal surface, so that the actual path of sintering is $a_{act} = A_1 a^{d_f}$ where A_1 is a constant, a is the radius of sintering neck in the scale of sphere radius R, and d_f is the fractal dimension of the sintering path (See Chapter 8 for more detail). Use the result of Problem 4.8 for this case and show that

$$a = A_3 t^{3/(4d_f)} \quad \text{for } t \to \infty \text{ where } A_3 \text{ is a constant.} \tag{4.145}$$

4.10. Consider[4] the small deformation of the free boundary of a crystal being in position $y = 0$ at time $t = 0$, due to surface diffusion and/or condensation / evaporation, if there is a grain boundary along $x = 0$ that maintains the dihedral angle of the constant slope, $dy/dx = \pm\gamma_b/(2\gamma_s)$, of the boundary $y = y(x,t)$ at $x = \pm 0$ (γ_b and γ_s is the surface tension of grain boundary and free surface respectively, and $\gamma_b \ll \gamma_s$). Show that this formulation of the problem makes sense if the specific time of observation is much greater than the relaxation time of the material creep.

4.11. Consider one-dimensional lattice interdiffusion of two different solid materials (denoted by subscripts 1 and 2) occupying the half-spaces $x < 0$ and $x > 0$ respectively, brought into close contact at $t = 0$. Denote the self-diffusion coefficients as D_{11} and D_{22} and the diffusion coefficient of atoms of the ith material into the kth material as D_{ik} where $i, k = 1,2$. Find the exact solution for the distribution of concentrations of both materials in terms of x and t, assuming that the interface is fixed at $x = 0$. Prove that at any time:

$$c_{10} = \frac{c_{1\infty}\sqrt{D_{11}}}{\sqrt{D_{11}} + \sqrt{D_{12}}}, \qquad c_{20} = \frac{c_{2\infty}\sqrt{D_{22}}}{\sqrt{D_{22}} + \sqrt{D_{21}}} \tag{4.146}$$

Here, c_{k0} is the concentration of the kth material on the interface, $x = 0$, and $c_{k\infty}$ is the original concentration of the kth material at infinity (at $x \to -\infty$ for $k = 1$ and $x \to +\infty$ for $k = 2$).

4.12. Consider Problem 4.11 of lattice interdiffusion in a more rigorous formulation taking into account that the interface is moving in the course of time. Prove that the solution to the problem is given by

$$c_{11} = A_1 \int_{-\infty}^{\xi} exp\left(-\frac{\xi^2}{4D_{11}}\right) d\xi + c_{10} \qquad \left(\xi = x/\sqrt{t} < \xi_0\right); \tag{4.147}$$

$$c_{12} = A_2 \int_{\xi}^{\infty} exp\left(-\frac{\xi^2}{4D_{12}}\right) d\xi \qquad \left(\xi = x/\sqrt{t} > \xi_0\right); \tag{4.148}$$

$$c_{22} = A_3 \int_{\xi}^{\infty} exp\left(-\frac{\xi^2}{4D_{22}}\right) d\xi + c_{20} \qquad \left(\xi = x/\sqrt{t} > \xi_0\right); \tag{4.149}$$

$$c_{21} = A_4 \int_{-\infty}^{\xi} exp\left(-\frac{\xi^2}{4D_{21}}\right) d\xi \qquad \left(\xi = x/\sqrt{t} < \xi_0\right); \tag{4.150}$$

Here, c_{ik} is the diffusion concentration of the ith material in the kth material ($i, k = 1, 2$), the interface position x_0 is determined by

$$x_0 = \xi_0\sqrt{t} \tag{4.151}$$

and the constants A_1, A_2, A_3, A_4, and ξ_0 are to be found from the following conditions at $x = x_0 t^{1/2}$:

$$c_{11} = c_{12}, \qquad c_{22} = c_{21} \tag{4.152}$$

$$D_{11}\frac{\partial c_{11}}{\partial x} = D_{12}\frac{\partial c_{12}}{\partial x}\,, \qquad D_{22}\frac{\partial c_{22}}{\partial x} = D_{21}\frac{\partial c_{20}}{\partial x} \tag{4.153}$$

$$D_{11}\frac{\partial c_{11}}{\partial x} = D_{22}\frac{\partial c_{22}}{\partial x} \tag{4.154}$$

Equations (4.152) physically express the condition of close contact of two material planes, equations (4.153) the mass conservation law for every material (which is equivalent to the assumption that there are no chemical reactions or transformations in the system), and equations (4.154) the physical nature of atom diffusion in solids through the diffusion of voids, namely, the vacant places left after the atoms of material 1 moved into material 2 should be replaced by the corresponding atoms of material 2. In other words, this is mass exchange without the creation of any new voids .

4.13. Show how the method of solution to Problems 4.11 and 4.12 can be employed for solution of the corresponding problems of non-linear diffusion when diffusivities D_{ik} are some given functions of concentrations.

4.14. Consider the strength and fracture toughness of an interface between two sintering spheres. Introduce the assumption that the interface strength and fracture toughness are determined by values of the concentrations of the mobile components on the interface. Based on the results of solution to Problems 4.11 and 4.12, show that the interface strength and fracture toughness do not depend on time of sintering.

4.15. Consider the sintering of two spheres of different radius, R_1 and R_2 , made of different materials. Assume that the sintering zone has the shape of a spherical segment with sphere radius R and radius a of the circular sintering front. Introduce spherical coordinate θ and study the problems of Sections 4.4 - 4.6 in terms of θ for different spheres of different materials. Determine R from the mass conservation integral.

References

1. G. C. Kuczynski (1949), *Transactions of the Metallurgical Society A. I. M. E.*, **185**, p. 169.

2. C. Herring (1951), Surface tension as a motivation for sintering, Chapter 8 in *The Physics of Powder Metallurgy*, T. E. Kingston (ed.), Mc Graw Hill, New York.

3. W. D. Kingery and M. Berg (1955), Study of the initial stage of sintering solids by viscous flow, evaporation-condensation, and self-diffusion, *J. Applied Physics*, **26**, p. 1205.

4. W. W. Mullins (1957), Theory of thermal grooving, *J. Applied Physics*, **28**(3), pp. 333-339.

5. J. C. Fisher (1950), Calculation of diffusion penetration curves for surface and grain boundary diffusion, *J. Applied Physics*, **22**(1), pp. 74-77.

6. F. B. Swinkels and M. F. Ashby (1981), A second report on sintering diagrams, *Acta Metallurgica*, **29**, pp. 259-281.

7. F. F. Lange and B. J. Kellett (1989), Thermodynamics of densification, *J. American Ceramic Society*, **72**(5), pp. 725-741.

8. J. C. Wang (1990), Analysis of early-stage sintering simultaneous surface and volume diffusion, *Metal Transactions*, **21A**, pp. 305-312.

9. J. Svoboda and H. Riedel (1995), *Acta Metallurgica*, **43**, pp. 499.

10. M. M. Mullins (1993) *Scripta Metallurgica Materialia*, **29**, p. 491.
11. H. E. Exner (1987), *Acta Metallurgica*, **35**, p. 587.
12. F. A. Nichols (1968), Theory of sintering of wires by surface diffusion, *Acta Metallurgica*, **16**, pp. 103-113.
13. H. Riedel (1990), *Ceramic Transactions*, **12**, p. 619.
14. D. L. Johnson and T. M. Clarke (1964), Grain boundary and volume diffusion in the sintering of silver, *Acta Metallurgica*, **12**, pp. 1173-1179.
15. K. L. Johnson, K. Kendall, and A. D. Roberts (1971), Surface energy and the contact of elastic solids, *Proceedings of the Royal Society* **A324**,p. 301
16. Alan Cottrell (1996), Continuum and atomistic theories in the mechanics of materials, in *Fracture: A Topical Encyclopedia of Current Knowledge*, G. Cherepanov (ed.), Krieger Publishing Co., Melburne, USA.
17. J. P. Stark (1980), *Solid State Diffusion*, John Wiley, New York.

CHAPTER 5

POINT DEFECTS IN SOLIDS

The properties of solid matter depend significantly on the distribution, mobility and mode of point defects like interstitial atoms, vacancies and holes. The methods of fracture mechanics enable us to derive the forces driving the point defects, study their motion, and thus predict the corresponding response of the solid matter. The technique of invariant integrals is extensively explored in this chapter for the derivation of the defect-driving forces, and the solution of some problems relating to the motion of the defects. The method of invariant integrals introduced earlier by the present author is treated here as an alternative approach to the basic classic theories of mathematical physics including gravitational theory, electromagnetic field theory, nonequilibrium thermodynamics, gas dynamics and elasticity theory. Arts of the formulation of well-posed governing equations of the physical problems and skills for their solution are also developed in this chapter.

5.1. Conservation Laws and Invariant Integrals

Energy, mass, charge, momentum, entropy, moment of momentum and other properties of a continuous medium may be subject to conservation laws. The law of conservation can be written in the form of the sum of the non-steady and convective components as

$$\int_V \frac{\partial F_v}{\partial t} dV + \int_\Sigma G_i n_i d\Sigma = 0 \qquad (i = 1,2,3). \tag{5.1}$$

Here, $F_v(x_1, x_2, x_3, t)$ and $G_i(x_1, x_2, x_3, t)$ are some functions depending on the field state variables specified in the reference frame $0(x_1x_2x_3)$ of Cartesian coordinates x_1, x_2, x_3 and time t, V is the volume domain in $0(x_1x_2x_3)$ bounded by the surface Σ, and n_1, n_2 and n_3 are the outer unit normal vector components on Σ. The coordinate frame $0(x_1x_2x_3)$ can be fixed or moving with a constant velocity with respect to a fixed coordinate frame. The repeated index denotes summation over the index (for example, $G_i n_i = G_1 n_1 + G_2 n_2 + G_3 n_3$). The quantities F and/or G_i can be some components of a vector and/or a tensor.

According to Gauss - Ostrogradsky's divergence theorem,

$$\int_\Sigma A n_i \, d\Sigma = \int_V A_{,i} \, dV \quad , \tag{5.2a}$$

$$\int_\Sigma A_i n_i \, d\Sigma = \int_V A_{i,i} \, dV \quad , \tag{5.2b}$$

$$\int_{\Sigma} A_{ik}\, n_i\, d\Sigma = \int_{V} A_{ik},_i\, dV \qquad (i = 1,2,3), \tag{5.2c}$$

and so on.

Here, A, A_i, and A_{ik} are the scalar, vector and second-rank tensor functions correspondingly. The comma with a succeeding index denotes the partial derivative with respect to the corresponding coordinate (for example, $A,_i = \partial A/\partial x_i$).

We apply the divergence theorem to the second term in Equation (5.1) and reduce this equation to

$$\int_{V} \left(\frac{\partial F_v}{\partial t} + G_{i,i} \right) dV = 0 \qquad (i = 1,2,3) \quad , \tag{5.3a}$$

and then derive the conservation law in the local form,

$$\frac{\partial F_v}{\partial t} + G_{i,i} = 0 \qquad (i = 1,2,3) \tag{5.3b}$$

because V is arbitrary. This is the common way governing equations are obtained.

This local formulation of the conservation law has some disadvantages, because the derivatives lose sense at the singular points, lines and surfaces of the physical field. Meanwhile, the study of the singularities is the principal interest for a theory and its applications. The singularities of the field and the corresponding singular solutions contain the basic information about the advantages and disadvantages of the theory, the limits of its applicability, and its opportunities in practice. Therefore, it is important to have other formulations of governing equations of mathematical physics which would hold at both regular and singular points of the field. Such a formulation can be provided by invariant integrals.

Introduce the following potential, $\Pi(x_1, x_2, x_3, t)$, by

$$\frac{\partial F_v}{\partial t} = \Pi_{i,i} \tag{5.4}$$

Substituting $\partial F_v/\partial t$ in Equation (5.1) by Equation (5.4) and applying the divergence theorem we find,

$$\int_{\Sigma} (G_i + \Pi_{,i})\, n_i\, d\Sigma = 0 \tag{5.5}$$

The integral in (5.5) is invariant with respect to Σ; therefore, it is called the invariant integral* .

Equation (5.5) is obviously equivalent to Equation (5.3b) at regular points, and to Equation (5.1) at any point. These are different mathematical formulations of one and the same law of conservation.

Consider a point singularity of the field at point 0 that can move in the reference frame $0(x_1x_2x_3)$ fixed at 0. Suppose that the singularity generates or absorbs the property which is conserved at the regular points according to the conservation law. Hence, the singular point can be considered as a source or sink of magnitude Γ,

$$\Gamma = \lim_{\Sigma\to O} \int_{\Sigma} (G_i + \Pi_i)\, n_i \, d\Sigma \quad , \tag{5.6}$$

according to Equation (5.5). In (5.6), closed surface Σ shrinks to 0.

Equations (5.5) and (5.6) are especially simple when the field is steady-state in the moving coordinate frame $0(x_1\, x_2\, x_3)$ or stationary in the fixed coordinate frame $0(x_1x_2x_3)$. In this case, $\Pi_i = 0$, and the G_i do not depend on t, so that Equations (5.5) and (5.6) are reduced to:

$$\int_{\Sigma} G_i n_i d\Sigma = 0 \quad \text{and} \quad \Gamma = \lim_{\Sigma\to 0} \int_{\Sigma} G_i n_i d\Sigma \tag{5.7}$$

Let us consider the basic fields of mathematical physics.

Gravitational field

Let the field of gravitation be defined by a potential $\varphi(x_1, x_2, x_3)$ satisfying the energy conservation law written in the form of the following invariant integrals

$$\frac{1}{4\pi f}\int_{\Sigma}\left(\frac{1}{2}\varphi,_i \varphi,_i n_k - \varphi,_i \varphi,_k n_i\right) d\Sigma = 0 \tag{5.8a}$$

$$\frac{1}{4\pi f}\lim_{\Sigma\to O}\int_{\Sigma}\left(\frac{1}{2}\varphi,_i \varphi,_i n_k - \varphi,_i \varphi,_k n_i\right) d\Sigma = \Gamma_k \ . \tag{5.8b}$$

where f is a constant, and Γ_k is the energy of the field spent to move the point singularity 0 unit length distance along the x_k axis, that is, Γ_k, is the kth component of the driving force.

* This name was introduced by the present author who suggested the general method. The path-independent integral is the name introduced by Eshelby for some integrals of this kind in elasticity theory.

We show that this formulation of the gravitation theory is equivalent to the Newtonian theory. By means of the divergence theorem, we transform Equation (5.8a) to

$$\int_{\Sigma} \left(\frac{1}{2} \varphi_{,i} \varphi_{,i} n_k - \varphi_{,i} \varphi_{,k} n_i \right) d\Sigma =$$

$$= \int_{V} \left[\frac{1}{2} (\varphi_{,i} \varphi_{,i})_{,k} - (\varphi_{,i} \varphi_{,k})_{,i} \right] dV = \tag{5.9}$$

$$\Gamma_k = M g_k \qquad (k = 1,2,3) \quad . \tag{5.13}$$

We apply Equation (5.13) to find the force of interaction between two point sources: M_1 at (0,0,0) and M_2 at (0,0,D). According to Equation (5.10), the field intensity created by M_1 at (0,0,D) is:

$$g_1 = \left(\frac{\partial \varphi}{\partial x_1}\right)_{x_1 = D} = \frac{f M_1}{D^2}, \quad g_2 = g_3 = 0 \tag{5.14}$$

Hence, based on Equations (5.13), the driving force from source M_1 applied to M_2 is:

$$F_1 = -\frac{f\, M_1 M_2}{D^2} \qquad (F_2 = F_3 = 0) \quad . \tag{5.15}$$

Let us choose M in Equations (5.10) - (5.15) to be equal to the inertial mass, M_{iN},

$$M = M_{iN} \quad . \tag{5.16}$$

The inertial mass is defined by Newton's law as

$$M_{iN} = \frac{F_1}{a} \quad , \tag{5.17}$$

where a is the acceleration produced by force F_1. Measuring a and F_1 (e.g., by elastic spring deformation) yields M_{iN}. From here, by the comparison of Equations (5.13) and (5.17), we deduce that g_k is the acceleration of the free motion of unit mass in the gravitation field, which can be measured. One measurement of the gravitational force between two point masses enables one to calculate f using Equation (5.15), which becomes the well-known Newton law of gravitation (f is the universal gravitational constant).

In the classic non-relativistic approach to gravitation, the identity of the inertial and gravitational masses is a separate postulate. In the present approach, this is the issue of our option of the gravitational constant f. From the viewpoint of energy conservation law, the term $\varphi_{,i}\varphi_{,i}/(4\pi f)$ represents the potential energy of the gravitational field per unit volume, and the term $\varphi_{,i}\varphi_{,k}n_i/(4\pi f)$ the work rate of gravitational stresses. From the viewpoint of the momentum conservation law, these terms can also be interpreted as the specific momentum and gravitation stress tensor, respectively.

An analogous approach is possible for the construction of the relativistic field theory of gravitation (Einstein's theory of gravitation is geometrical, therefore within its framework it is difficult to introduce invariant integrals).

Electromagnetic field

Consider the stationary electromagnetic field in a medium with zero conductivity, ignoring the medium deformation. The field equations can be specified in the form of the following invariant integrals expressing the energy and momentum conservation laws:

$$\int_{\Sigma}(Wn_k - D_i n_i E_k - H_k B_i n_i)d\Sigma = 0 \quad , \qquad (i,k = 1,2,3) \tag{5.18a}$$

$$\lim_{\Sigma\to 0}\int_{\Sigma}(Wn_k - D_i n_i E_k - H_k B_i n_i)d\Sigma = \Gamma_k \quad , \tag{5.18b}$$

and the constitutive equations,

$$H_i = \frac{\partial U}{\partial B_i}, \quad E_i = \frac{\partial U}{\partial D_i}, \quad U = U(D_i, B_i) \quad . \tag{5.18c}$$

Here **D, E, H** and **B** are the field vectors, U is the potential energy of the field per unit volume (only reversible thermodynamic processes are here considered), and W is the function defined by

$$W(E_i, H_i) = U(D_i, B_i) + E_i D_i + H_i B_i \quad , \tag{5.19}$$

so that

$$D_i = \frac{\partial W}{\partial E_i} \quad , \quad B_i = \frac{\partial W}{\partial H_i} \quad . \tag{5.20}$$

The quantity Γ_k in (5.18b) is the energy of the field spent to move the singularity 0 unit length along the x_k axis, that is, the kth component of the driving force. Let us show that Maxwell's equations follow from the invariant integrals (5.18a). To the end, convert (5.18a) using the divergence theorem:

$$\begin{aligned}
&\int_{\Sigma}(Wn_k - D_i n_i E_k - H_k B_i n_i)\,d\Sigma = \int_V \left[W_{,k} - (E_k D_i)_{,i} - (H_k B_i)_{,i}\right]dV = \\
&- \int_V \left[\frac{\partial W}{\partial E_i}E_{i,k} + \frac{\partial W}{\partial H_i}H_{i,k} - E_k D_{i,i} - E_{k,i} D_i - H_k B_{i,i} - H_{k,i}B_i\right]dV = \\
&= \int_V \left[D_i\left(E_{i,k} - E_{k,i}\right) + B_i\left(H_{i,k} - H_{k,i}\right) - E_k D_{i,i} - H_k B_{i,i}\right]dV = 0 \quad .
\end{aligned} \tag{5.21}$$

From here, because V, D_i, B_i, E_k and H_k are arbitrary, it follows that

$$E_{i,k} = E_{k,i}\ ; \quad H_{i,k} = H_{k,i}\ ; \quad D_{i,i} = 0\ ; \quad B_{i,i} = 0 \ ; \tag{5.22}$$

and thus

$$\text{curl}\ \mathbf{E} = 0,\ \text{curl}\ \mathbf{H} = 0,\ \text{div}\ \mathbf{D} = 0,\ \ \text{div}\ \mathbf{B} = 0\ . \tag{5.23}$$

These are Maxwell's equations in the stationary case.

For an electrostatic field in an isotropic linear dielectric, for which

$$\mathbf{H} = 0,\ \ \mathbf{D} = \varepsilon\mathbf{E} \tag{5.24}$$

where ε is the dielectric constant, Equations (5.18) reduce to

$$\varepsilon \int_{\Sigma} \left(\frac{1}{2} E_i E_i n_k - E_i n_i E_k \right) d\Sigma = 0 \quad , \tag{5.25a}$$

$$\varepsilon \lim_{\Sigma \to 0} \int_{S} \left(\frac{1}{2} E_i E_i n_k - E_i n_i E_k \right) d\Sigma = \Gamma_k \quad , \tag{5.25b}$$

where

$$E_i = -\varphi_{,i} \quad , \quad \varphi_{,ii} = 0 \quad , \qquad i,k = 1,2,3 \ . \tag{5.26}$$

(φ is the electrostatic potential).

Consider a singularity at 0. We write the singular solution of Laplace's equation in the form

$$\varphi = \frac{q}{4\pi\varepsilon r} + E_i x_i \quad , \tag{5.27a}$$

where q is a constant, and E_i is the intensity of the external electrostatic field at 0 based on (5.26). As with Equations (5.11) - (5.13), one can derive the force driving this singularity to be

$$\Gamma_k = q\, E_k \qquad (k = 1,2,3)\ . \tag{5.27b}$$

The quantity q (charge) can be measured using this equation. From here and Equation (5.27a), it follows that the interaction force between charge q_1 at (0,0,0) and q_2 at (L,0,0) is equal to $q_1\, q_2\, L^{-2}$, which is Coulomb's law.

Invariant integrals for a non-stationary electromagnetic field for irreversible processes in deformable media were given in references[1,2]. Invariant integrals for relativistic physical fields in the Minkovsky space-time were derived in reference[3] for ideal and viscous relativistic fluids and relativistic heat flow in fluids.

Nonequilibrium thermodynamics
Consider a nonequilibrium stationary thermodynamic field in which the entropy production rate, $\dot{S}$, can be written in the following form of a dot product of vectors

$$\dot{S} = \mathbf{X}^{\alpha}\,\dot{\mathbf{Q}}^{\alpha} \quad , \qquad (\alpha = 1,2,\ldots,N) \quad , \tag{5.28}$$

where the repeated superscript α denotes summation. Here $\mathbf{X}^{\alpha}$ and $\dot{\mathbf{Q}}^{\alpha}$ are the generalized thermodynamic force and flow, respectively; they are vectors in the physical space $0(x_1x_2x_3)$. The number N is equal to the number of different physical state functions describing the thermodynamical process under consideration. The typical state functions are temperature, concentrations of components taking part in diffusion and/or chemical reactions, pressure of the liquid or gaseous phase in a porous medium and so on. Equations (5.28) hold for the local equilibrium condition. Classical stationary fields of temperature or concentration in heat conductivity or diffusion can serve as examples where Equations (5.28) can be used.

In these fields, the entropy of any volume is stationary. This conservation law can be written in the form of the invariant integral

$$\int_{\Sigma}\left(Sn_k - X_i^{\alpha} n_i Q_k^{\alpha}\right) d\Sigma = 0 \quad , \qquad (i,k = 1,2,3) \tag{5.29a}$$

$$\lim_{\Sigma\to 0}\int_{\Sigma}\left(Sn_k - X_i^{\alpha} n_i Q_k^{\alpha}\right) d\Sigma = \Gamma_k \quad . \quad (\alpha = 1,2\ldots,N) \tag{5.29b}$$

Here Γ_k is the entropy flow from the external field spent (produced or absorbed) to move the field singularity 0 unit length along the x_k axis. The lower indices in X_i^{α} and Q_i^{α} denote the components of $\mathbf{X}^{\alpha}$ and $\mathbf{Q}^{\alpha}$ along the x_i axis.
Convert Equation (5.29a) using the divergence theorem:

$$\int_{\Sigma}\left(Sn_k - X_i^{\alpha} n_i Q_k^{\alpha}\right) d\Sigma = \int_{V}\left[S_{,k} - \left(X_i^{\alpha} Q_k^{\alpha}\right)_{,i}\right] dV = \tag{5.30}$$

$$= \int_{V}\left[\frac{\partial S}{\partial Q_i^{\alpha}} Q_{i,k}^{\alpha} - X_{i,i}^{\alpha}\, Q_k^{\alpha} - X_i^{\alpha}\, Q_{k,i}^{\alpha}\right] dV = \int_{V}\left[X_i^{\alpha}\left(Q_{i,k}^{\alpha} - Q_{k,i}^{\alpha}\right) - X_{i,i}^{\alpha}\, Q_k^{\alpha}\right] dV = 0$$

From here, because V, X_i^{α} and Q_k^{α} are arbitrary, it follows that

$$X_{i,i}^{\alpha} = 0 \quad , \quad Q_{i,k}^{\alpha} = Q_{k,i}^{\alpha} \quad , \qquad (i,k = 1,2,3) \tag{5.31}$$

and hence,

$$\text{curl } \mathbf{Q}^{\alpha} = 0, \; Q_i^{\alpha} = \text{grad } \varphi^{\alpha}, \qquad (\alpha = 1,2,\ldots,N) \tag{5.32}$$

where φ^{α} are some potentials.

For the linear law,

$$\mathbf{X}^{\alpha} = A_{\alpha\beta}\,\mathbf{Q}^{\beta} \qquad (\alpha,\beta = 1,2,\ldots,N) \tag{5.33}$$

the coefficients $A_{\alpha\beta}$ should satisfy Onsager's principle of symmetry

$$A_{\alpha\beta} = A_{\beta\alpha} \quad . \tag{5.34}$$

Expressing $\mathbf{X}^{\alpha}$ in terms of $\mathbf{Q}^{\beta}$ by using (5.33), and then (5.32), we may write (5.31) as

$$\left(A_{\alpha\beta}\,\varphi_{,i}^{\beta}\right)_{,i} = 0 \quad , \tag{5.35}$$

where $i = 1,2,3$ and α, $\beta = 1,2,\ldots,N$. This is the system of N partial differential equations governing the local-equilibrium thermodynamic process of the N coupled phenomena.

The motion of point sources or sinks of entropy is controlled by the driving force which can be derived from Equations (5.29b) and (5.35), just like the driving force on the gravitational mass or charge.

Gas dynamics

Consider irrotational stationary isoentropic flow of an inviscid compressible gas. In this case, the governing equations of gas dynamics can be written in the form of the following invariant integrals[4] :

$$\int_{\Sigma} \rho \mathrm{v}_i n_i \, d\Sigma = 0 \quad , \qquad (i,j,k = 1,2,3) \quad . \tag{5.36a}$$

$$\int_{\Sigma} \left(\rho \mathrm{v}_k \mathrm{v}_j n_j + p n_k\right) d\Sigma = 0 \quad , \tag{5.36b}$$

$$\lim_{\Sigma \to 0} \int_{\Sigma} \left(\rho \mathrm{v}_k \mathrm{v}_j n_j + p n_k\right) d\Sigma = \Gamma_k \quad , \tag{5.36c}$$

$$\frac{p}{p_{\infty}} = \left(\frac{\rho}{\rho_{\infty}}\right)^{\kappa} \quad . \tag{5.36d}$$

Here ρ, p and v_i are the density, pressure and velocity components of the gas, p_∞ and ρ_∞ are the pressure and density in the unperturbed flow, and $\kappa = c_p/c_v$ is the ratio of the heat capacities of the gas.

Equation (5.36a) expresses the mass conservation law, and Equations (5.36b) and (5.36c) express both momentum (per unit time) and energy (per unit path length) conservation laws. The quantity Γ_k represents the dissipation of energy of the external flow at 0 per unit path length of 0, that is, the force applied to 0 by the external flow. Equation (5.36d) expresses the process assumption that there is no entropy exchange

$$\mathrm{v}_i = \varphi,_i \; ; \qquad \varphi,_{ii} = 0 \; ; \qquad p = -\frac{1}{2}\rho \mathrm{v}_i \mathrm{v}_i + const. \quad , \tag{5.41}$$

where φ is the fluid potential.

Theory of elasticity

Consider the static process of small deformations of an elastic solid. The governing equations of the process (theory of elasticity) can be written in the form of the following invariant integrals:

$$\int_{\Sigma} \sigma_{ij} n_j \, d\Sigma = 0 \qquad (i, j, k = 1,2,3) \quad , \tag{5.42a}$$

$$\int_{\Sigma} \left(U n_k - \sigma_{ij} u_{i,k} n_j \right) d\Sigma = 0 \quad , \tag{5.42b}$$

$$\lim_{\Sigma \to O} \int_{\Sigma} \left(U n_k - \sigma_{ij} u_{i,k} n_j \right) d\Sigma = \Gamma_k \quad , \tag{5.42c}$$

$$\lim_{\Sigma \to O} \int_{\Sigma} \sigma_{ij} \, n_i \, d\Sigma = F_j \quad . \tag{5.42d}$$

Here u_i are the displacement vector components, σ_{ij} are the stress tensor components, U is the elastic potential of the solid per unit volume, F_i are the components of the concentrated force at 0, and Γ_i are the components of the energy flow rate from the external field into 0 (in other words, the change of the elastic energy of the system due to the motion of 0 along the x_i axis, respectively, per unit length). The quantities Γ_i are the components of the singularity-driving force of the point 0, if $F_1 = F_2 = F_3 = 0$.

Using the divergence theorem, one can derive from the invariant integrals (5.42) the partial differential equations of the theory of elasticity:

$$\sigma_{ij,j} = 0 \qquad (i, j = 1,2,3) \quad , \tag{5.43a}$$

$$\sigma_{ij} = \frac{\partial U}{\partial \varepsilon_{ij}} \quad , \tag{5.43b}$$

where $$\varepsilon_{ij} = \frac{1}{2}\left(u_{i,j} + u_{j,i} \right) \quad , \qquad U = U(\varepsilon_{ij}) \quad . \tag{5.43c}$$

When U is a quadratic function of strains ε_{ij}, Equations (5.43) represent the traditional linear theory of elasticity.

5.2. Point Inclusions

The interstitial atoms of carbon, nitrogen, nickel and other alloying elements left in the lattice of a parent metal after solidification of a liquid mixture, hydrogen atoms penetrating into the solid metal lattice, impurity atoms, etc. will be referred to as point inclusions, if the diameter of foreign atoms is greater than the interatomic spacing of the lattice. Point inclusions are able to move with respect to the lattice due to self-diffusion and external loading.

Point inclusions are modeled as centers of compression described by the following equations[5] :

$$u_R^S = \frac{1+\nu}{2E}\frac{qa^3}{R^2} \qquad (q>0), \tag{5.44a}$$

$$\sigma_R^S = -\frac{qa^3}{R^3}, \qquad \sigma_\psi^S = \sigma_\theta^S = \frac{qa^3}{2R^3}. \tag{5.44b}$$

Here R, ψ, and θ are spherical polar coordinates, u_R^S ,σ_R^S, σ_ψ^S and σ_θ^S are the displacement and stresses, respectively, a is the radius of the kernel of the compression center, q is the pressure of the kernel upon the elastic solid, and E and ν are Young's modulus and Poisson's ratio of the solid, respectively.

For $q < 0$ Equations (5.44) provide the singularity referred to as a "vacancy" in the physical literature. Both point inclusions and "vacancies" are traditionally called point defects. All the mathematical content of the present Section holds also for the "vacancies".

Physicists use the "vacancy" model for characterizing a small void in a lattice, for example, a vacant atomic volume not occupied by a parent atom. However, this model of small voids is physically incorrect because a void cannot create a tensile stress σ_R on the boundary (of the spherical void). The correct model of small voids referred to as point holes is given in Section 5.3 below. Probably, for $q < 0$ the "vacancy" model and the present theory can be applied to a foregn atom whose diameter is a little less than the void diameter so that the atom can create the cohesion force necessary to maintain the tensile stress σ_R on the void boundary.

It is easy to show, by means of Equations (5.42c) and (5.44), that the inclusion-driving force is zero for a uniform external stress field. Let us assume now that the unperturbed elastic field is non-uniform and is of the following form:

$$\begin{gathered} \sigma_{33}^0 = Ax_1, \quad u_1^0 = -\frac{\nu A}{2E}\left(x_1^2 - x_2^2 + \frac{1}{\nu}x_3^2\right), \\ u_2^0 = -\frac{\nu A}{E}x_1x_2, \quad u_3^0 = \frac{A}{E}x_1x_3 \end{gathered} \tag{5.45}$$

(A is a given constant).

Let us calculate the driving force which is, evidently, directed along the x_1 axis in this case. Using the invariance of the integral in Equation (5.42b), we take Σ as the parallelepiped formed by the faces: $x_3 = \pm\,\delta$, $x_1 = \pm\,L$, $x_2 = \pm\,L$ when $\delta/L \to 0$, $\delta \to \infty$, and $L \to \infty$. Using the rule of Γ-integration[6] yields

$$\Gamma_1 = -2\int_{-\infty}^{+\infty}\int_{-\infty}^{+\infty}\left(\sigma_{33}^{S}u_{3,1}^{0} + \sigma_{33}^{0}u_{3,1}^{S} + \sigma_{12}^{S}u_{2,1}^{0} + \sigma_{13}^{S}u_{3,1}^{0}\right)dx_1 dx_2 \quad (x_3 = \delta) \tag{5.46}$$

Here, we also used the symmetry with respect to the plane $x_3 = 0$ and the following relationships: $n_3 = 1$ for $x_3 = +\delta$, $n_3 = -1$ for $x_3 = -\delta$, $\sigma_{13}^{0} = \sigma_{23}^{0} = 0$ for $x_3 = \pm\delta$.
According to Equation (5.44), we have:

$$\sigma_{33}^{s} = \sigma_{z}^{s} = -\frac{qa^3}{4R^3}\left(1 + 3\cos 2\psi\right), \qquad \tau_{rz}^{s} = -\frac{3qa^3}{4R^3}\sin 2\psi \;,$$

$$\sigma_{13} = \frac{x_1}{r}\tau_{rz}, \qquad \sigma_{23} = \frac{x_2}{r}\tau_{rz}, \qquad u_{3,1}^{s} = -3\frac{1+\nu}{2E}qa^3\frac{x_1 z}{R^5} \;. \tag{5.47}$$

$$\left(z = x_3 \,,\, r^2 = x_1^2 + x_2^2 \,,\, R^2 = r^2 + z^2\right)$$

Here, r, θ and z are cylindrical coordinates (ψ is measured from the z axis).
Using the equations (for $x_3 = \delta$),

$$R^2 = r^2 + \delta^2, \quad \delta/R = \cos\psi, \quad r/R = \sin\psi, \quad x_1/r = \cos\theta,$$

$$r\,dr = R\,dR = R^2\tan\psi\,d\psi, \quad dx_1\,dx_2 = r\,dr\,d\theta,$$

calculate integral (5.46) by means of (5.45) and (5.47):

$$\Gamma_1 = -\frac{2A}{E}\int_0^{\infty}\int_0^{2\pi}\left(\delta\sigma_z^{s} - \nu r\tau_{rz}^{s} + Ex_1u_{3,1}^{s}\right)r\,dr\,d\theta =$$

$$= \pi\frac{A}{E}qa^3\int_0^{\pi/2}\left[\sin\psi\left(1 + 3\cos 2\psi\right) - 3\nu\sin\psi\;\tan\psi\;\sin 2\psi + \right. \tag{5.48}$$

$$\left. +3\left(1+\nu\right)\;\sin^3\psi\right]d\psi = 2\pi\frac{1-\nu}{E}Aqa^3$$

To verify and clarify this result, we performed the exact calculation of Γ_1, by means of (5.42), over the integration sphere Σ, $R = R_o$, in which the elastic field is the sum of the fields described by Equations (5.44) and (5.45). After some cumbersome calculations, we found the following equation:

$$\Gamma_1 = 2\pi \frac{1-\nu}{E} Aqa^3 \left\{ 1 + \frac{pR_o^3}{qa^3(1-\nu)} \left[1 - \frac{5}{8}\nu \left(1 + \frac{3\pi}{128} \right) \right] \right\} \tag{5.49}$$

As seen, if

Equation (5.55) may also be derived by the help of equations (5.5) and (8.9) in the work[9] by Eshelby. Thus, the inclusion-driving force is directly proportional to the gradient of the first invariant of the external stress tensor σ^0_{ij} (x_1, x_2, x_3). The elegant formula (5.55) is analogous to that of Peach-Koehler in the theory of dislocations and that of Irwin in the theory of cracks. Let us discuss some basic problems of point inclusions by means of Equation (5.55). In the studies of the trajectories of inclusions, we assume that there are no preferential canals for the inclusions in the solid.

Interaction between inclusions

Let one of the inclusions be at the origin of coordinates. According to Equation (5.44), we have $\sigma = 0$ in the entire space in this case. Therefore, it follows from Equation (5.55) that the inclusions do not interact in unbounded space.

Interaction of inclusions with a dislocation

Let the front of an edge dislocation coincide with the x_3 axis, the Burgers vector being directed along the x_1 axis. The σ for the dislocation field is equal to

$$\sigma = \frac{BEx_2}{2\pi(1-\nu)\left(x_1^2 + x_2^2\right)} \tag{5.56}$$

(B is the magnitude of the Burgers vector). Let us place an inclusion at point (x_1, x_2, 0) and calculate the driving force acting upon it by means of Equations (5.55) and (5.56):

$$\Gamma_1 = -\frac{2\Delta Bx_1x_2}{\left(x_1^2 + x_2^2\right)^2}, \qquad \Gamma_2 = \frac{\Delta B\left(x_1^2 - x_2^2\right)}{(1-\upsilon)\left(x_1^2 + x_2^2\right)^2} \quad . \tag{5.57}$$

The path of the moveable inclusion is a solution of the following equation:

$$\frac{dx_2}{dx_1} = \frac{\Gamma_2}{\Gamma_1} = \frac{x_2^2 - x_1^2}{2x_1\, x_2}. \tag{5.58}$$

The general solution of Equation (5.58) is :

$$x_1^2 + x_2^2 = Cx_1 \quad (C = const.) \quad . \tag{5.59}$$

The family of curves in Equation (5.59) is the set of circles centered on the x_1 axis and tangent to the x_2 axis at the origin. According to Equation (5.57), the inclusions move in a clockwise direction for $x_1 < 0$ and $B > 0$, and in a counterclockwise direction for $x_1 > 0$,

$B > 0$. Hence, the inclusions are attracted to the core of the edge dislocation in the stretched region.

If a continuous cloud of inclusions is distributed in the space with the same dislocation, then in accordance with Equation (5.57), the dislocation-driving force is

These are closed oval curves tangential to the x_1 axis at the origin and symmetrical with respect to the x_1 axis. The inclusions move along the ovals clockwise for $x_2 > 0$ and counterclockwise for $x_2 < 0$. Thus, the inclusions are attracted to the crack tip, namely to the stretched pre-fractured region of the crack's continuation. The inclusions near the crack front cause an increase of the crack-driving force, since

$$\Gamma_1 = \Gamma_r \cos\theta - \Gamma_\theta \sin\theta \ . \tag{5.65}$$

This means that inslusions embrittle the material. Futhermore, inclusions arrest dislocations near the crack tip which increases the embrittlement effect.

For a continuous cloud of inclusions in a material with the same crack, the crack-driving force is equal to

$$\Gamma_1 = \frac{1-\nu^2}{E} K_1^2 + \frac{\lambda\Delta(1+\nu) K_I}{E\sqrt{2\pi}} \iint \frac{\cos\frac{3\theta}{2} N(r,\theta)}{r^{1/2}} dr\, d\theta \tag{5.66}$$

Here, the first and second terms are, respectively, the Irwin force originated by external loads and the force induced by the cloud of inclusions.

Interaction of inclusions with a spherical cavity

Suppose an infinite elastic space with a spherical cavity of radius R_a is subjected to uniaxial extension by the stress $\sigma_z = p$. The center of the sphere is assumed to coincide with the origin of the cylindrical coordinate frame. The surface of the sphere is free of external loads. In this case, the first invariant of the stress tensor[5] is equal to

$$\sigma = p + \frac{5p(1+\nu)}{7-5\nu} R_a^3 \frac{\partial}{\partial z}\left(\frac{z}{R^3}\right) = p + 5p R_a^3 \frac{1+\nu}{7-5\nu} \frac{r^2-2z^2}{R^5} \tag{5.67}$$
$$(R^2 = r^2 + z^2).$$

Consider an inclusion at a point. According to Equations (5.55) and (5.67), the components of the inclusion-driving force are

$$\begin{aligned} \Gamma_r &= \frac{15\lambda p\Delta(1+\nu)}{E(7-5\nu)} \frac{rR_a^3(4z^2-r^2)}{R^7} , \\ \Gamma_z &= \frac{15\lambda p\Delta(1+\nu)}{E(7-5\nu)} \frac{zR_a^3(2z^2-3r^2)}{R^7} . \end{aligned} \tag{5.68}$$

(r and z are the coordinates of the inclusion).

The velocity of the moveable inclusion and the driving force have the same direction. Therefore, the path of the inclusion is an integral curve of the following equation:

$$\frac{dr}{dz} = \frac{r\left(4z^2 - r^2\right)}{z\left(2z^2 - 3r^2\right)} \quad . \tag{5.69}$$

The family of the integral curves of this equation is as follows:

$$\begin{aligned} &2CR^3 = \sin 2\psi \quad or \quad zr = CR^5 \\ &\left(R^2 = z^2 + r^2, \quad z/r = \cos\psi, \quad C = const\right) \quad . \end{aligned} \tag{5.70}$$

The trajectories of inclusions are qualitatively depicted in Figure 5.1. As seen, the inclusions travel to the zone of the highest tensile stresses near the cavity, which implies a hardening effect.

The continuum theory of inclusions

Consider a solid with a large number of point inclusions, that is, a cloud of inclusions. Within the framework of the asymptotic approach involved in Equation (5.55), the interaction of inclusions can be ignored. We study the self-diffusion and drift transport of the inclusions.

As generally accepted in the linear theory of diffusion[10], the drift rate of inclusions, $\mathbf{V}_d$, is directly proportional to the driving force, Γ,

$$\mathbf{V}_d = \eta\Gamma \quad , \tag{5.71}$$

where η is an empirical coefficient of mobility of inclusions.

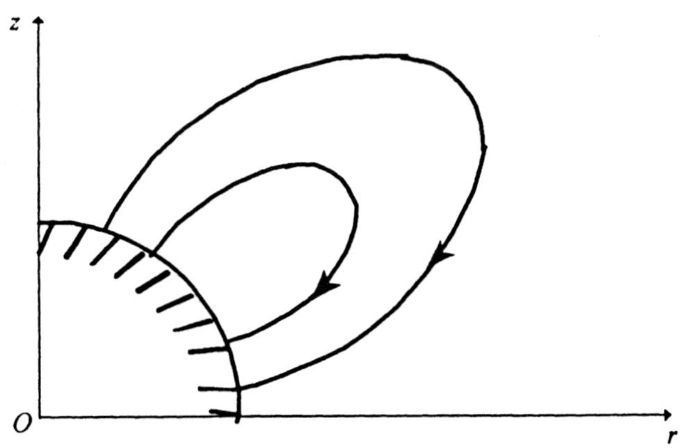

Figure 5.1. Schematic paths of moveable inclusions in the neighborhood of a fixed spherical hole

The inclusion transport equation will thus be

$$\frac{\partial \delta}{\partial t} = \left(D\delta_{,k}\right)_{,k} - \left(\eta \frac{\lambda \Delta}{E} \delta \sigma_{,k}\right)_{,k} \qquad (k = 1,2,3) \ . \qquad (5.72)$$

Here t is time, D is the self-diffusion coefficient, δ is the mass of inclusions per unit volume. The quantities η and D depend on temperature. For inclusions of the same type coefficients D, η and Δ are identical. Therefore, the total number of equations equals the number of different types of inclusions. The influence of other external fields (such as electric, heat and chemical ones) upon the transport of inclusions can be treated by methods using the relevant linear or non-linear governing equations of irreversible thermodynamics similar to Equation (5.71). The effect of new fields would result in additional terms in Equation (5.72).

As an illustration, we find the solution to Equation (5.72) for the strip $0 < x_1 < d$ subjected to pure bending by the moment M (per unit length). From the elastic solution, one obtains

$$\sigma_{,2} = \sigma_{,3} = 0, \qquad \sigma_{,1} = 12(1+\nu)Md^{-3}. \qquad (5.73a)$$

The stationary equilibrium concentration of inclusions in the strip under the action of bending is found from the solution of Equation (5.72) with $\partial\delta/\partial t = 0$,

$$\delta = C_o\left(e^{\mu d} - e^{\mu x_1}\right), \quad \text{where} \quad \mu = \frac{12\,(1+\nu)M\eta\lambda\Delta}{EDd^3} \ . \qquad (5.73b)$$

The constant C_0 is determined by the total mass m_0 of moveable inclusions:

$$C_0 = \frac{\mu m_0}{e^{\mu d} - 1}, \quad \text{where} \quad \int_0^d \delta \, dx_1 = m_0 \qquad (5.74)$$

This solution allows one to predict the distribution of material properties across the strip thickness, if it is known how the corresponding property depends on the concentration of inclusions.

According to the principle of superposition, the internal stresses induced by a continuous cloud of inclusions are as follows:

$$\sigma_{ik}\left(x_1, x_2, x_3\right) = \int \delta\left(x_1', x_2', x_3'\right)\sigma_{ik}'\left(x_1 - x_1',\ x_2 - x_2',\ x_3 - x_3'\right)dx_1'\,dx_2'\,dx_3' \ . \qquad (5.75)$$

Here, the stresses σ_{ik}' are determined by the stresses (5.44) in the corresponding coordinate system. The δ in this case is defined by (5.73) and (5.74).

5.3. Point Holes

Voids in a crystal lattice, small bubbles and cavities always present in materials are referred to as point holes. If they are sufficiently small or when there are some channels in the structure of the material, the holes can move relative to the lattice under the action of driving forces. Physically, the smallest point hole is a vacant atomic volume in a lattice not occupied by an atom.

Let us model a hole by a spherical cavity of radius r_0 with the surface of the cavity being free of external loads. The distance, R, between holes is assumed to be large compared to r_0. Actually, it is accurate to assume that $R > 4r_0$, that is, the porosity of the material involved should be less than 0.07. This assumption allows us to consider holes as point sources of asymptotically singular perturbation, or, in other words, as some quasi-particles[7,8]. Like any point source of perturbation, a hole is acted upon by a driving force, whose components are determined by Equation (5.42c).

The displacements u_r and u_z brought about by extension of an infinite elastic space with a spherical cavity of radius r_0 are[5] :

$$u_r = -\frac{\nu p}{E} r + \frac{p r r_0^3}{4GR^3}\left[-\frac{z^2}{R^2} - \frac{4-10\nu}{7-5\nu}\left(1-\frac{3z^2}{R^2}\right) + \right.$$
$$\left. + \frac{3r_0^2 - (2+5\nu)R^2}{(7-5\nu)R^2}\left(1-\frac{5z^2}{R^2}\right)\right] \qquad \left(R^2 = r^2 + z^2\right) , \tag{5.76a}$$

$$u_z = \frac{p}{E} z + \frac{p z r_0^3}{4GR^3}\left[\frac{2}{3} - \frac{z^2}{R^2} - \frac{4-10\nu}{7-5\nu}\left(1-\frac{3z^2}{R^2}\right) + \right.$$
$$\left. + \frac{(2+5\nu)R^2 - 3r_0^2}{(7-5\nu)R^2}\left(-3+\frac{5z^2}{R^2}\right)\right]. \qquad \left(G = \frac{E}{2(1+\nu)}\right) \tag{5.76b}$$

Here, r and z are cylindrical coordinates whose origin is at the center of the sphere and, p is the tensile stress σ_z at infinity. The stresses are determined from (5.76) by Hooke's law.

Using the invariance of Γ_k with respect to Σ, we contract Σ onto the surface of the spherical cavity. Making use of Equations (5.42) and (5.76), we can prove that $\Gamma_k = 0$.

We now assume that the unperturbed external field has a non-uniform component $\sigma^0_{33} = Ax_1$, and that $Ar_0 << p$. The following displacements correspond to this component:

$$u_1^0 = -\frac{\nu A}{2E}\left(x_1^2 - x_2^2 + \frac{1}{\nu}x_3^2\right), \quad u_2^0 = -\frac{\nu A}{E}x_1 x_2,$$
$$u_3^0 = \frac{A}{E}x_1 x_3 \qquad \left(x_3 = z, \quad x_1^2 + x_2^2 = r^2\right). \tag{5.77}$$

Using the invariance of Γ integrals in Equation (5.42), we take integration surface Σ, to be the parallelepiped $x_1 = \pm L$, $x_2 = \pm L$, $x_3 = \pm \delta$, with $\delta/L \to 0$, $\delta \to \infty$, $L \to \infty$. From this, taking account of the rule of Γ integration[6], we have [1-3]

$$\Gamma_1 = -2\int_{-\infty}^{+\infty}\int_{-\infty}^{+\infty}\left(\sigma_{33}^{0} \; u_{3,1}^{s} + \sigma_{33}^{s} \; u_{3,1}^{0} + \sigma_{23}^{s} \, u_{2,1}^{0} + \sigma_{13}^{s} \, u_{1,1}^{0}\right)dx_1 \, dx_2 \,, \tag{5.78}$$

along $x_3 = \delta \to \infty$.

The singular field of perturbation (with the superscript s) is determined with the help of Equation (5.76). The calculation yields the following value for the hole-driving force:

$$\Gamma_1 = \frac{\lambda_2 pA}{E} r_0^{\,3} \quad (\Gamma_2 = \Gamma_3 = 0), \tag{5.79a}$$

where

$$\lambda_2 = \frac{2\pi}{3} \; \frac{59 - 86\nu - 25\nu^2}{7 - 5\nu} \tag{5.79b}$$

In the general case of an arbitrary non-uniform field of external stresses σ^0_{ij} (x_1, x_2, x_3) satisfying the condition,

$$\left|\sigma_{ij}^{0}\right| >> r_0 \left|\sigma_{ij,k}^{0}\right| \,, \tag{5.80}$$

the stresses in the singular solution of the hole-type singularity are directly proportional to the stresses σ^0_{ij} of the hole-unperturbed field. According to the Γ integration rule[6], the hole-driving force is directly proportional to the stresses σ^0_{ij} and their gradients, $\sigma^0_{ij,k}$. Therefore, the interaction energy, U, of the hole with the external stress field is a positive quadratic function of stresses, so that, in the general case of anisotropic solids and holes of arbitrary shape, U and Γ_k have the form:

$$U = C_{ijmn}\sigma_{ij}^{0}\sigma_{mn}^{0} \,, \tag{5.81a}$$

$$\Gamma_k = -\frac{\partial U}{\partial x_k}, \tag{5.81b}$$

where the constants C_{ijmn} satisfy

$$C_{ijmn} = C_{jimn} = C_{ijnm} = C_{mnij} \qquad (i, j, k, m, n = 1,2,3). \tag{5.81c}$$

For a spherical hole and isotropic elastic solid, U depends only on the first and second invariants, σ and I, of the external stress tensor:

$$U = r_0^3 E^{-1}\left(\alpha\sigma^2 + \beta I\right), \qquad \Gamma_k = -\frac{\partial U}{\partial x_k}, \tag{5.82a}$$

where

$$\begin{aligned}\sigma &= \sigma_{11}^0 + \sigma_{22}^0 + \sigma_{33}^0, \\ I &= \sigma_{11}^0\sigma_{22}^0 + \sigma_{22}^0\sigma_{33}^0 + \sigma_{11}^0\sigma_{33}^0 - \left(\sigma_{12}^0\right)^2 - \left(\sigma_{13}^0\right)^2 - \left(\sigma_{23}^0\right)^2.\end{aligned} \tag{5.82b}$$

Here, α and β are some material constants to be defined.

For omni-directional uniform extension,

$$\sigma_{ij} = \frac{1}{3}\sigma\delta_{ij} \quad \text{where } \sigma > 0, \tag{5.83}$$

the hole is the compression center defined by Equation (5.44), whose interaction energy, according to Equation (5.55), is equal to

$$-\lambda\frac{\sigma^2 r_0^3}{6E}. \tag{5.84}$$

From Equations (5.82) and (5.84), it follows that

$$3\alpha + \beta = -\pi(1-\nu). \tag{5.85}$$

For another example, take $\sigma_{33}^0 = p$ and $\sigma_{33,1}^0 = A$, and the remaining stresses equal zero. The interaction energy given by Equation (5.79) is $-\lambda_2\, x_1 pAr_0^3 E^{-1}$ and $-2\alpha pA\, x_1\, r_0^3\, E^{-1}$ given by Equation (5.82). Comparison of both expressions yields $2\alpha = -\lambda_2$, and matching this result with (5.85), one can find α and β:

$$\alpha = -\frac{1}{2}\lambda_2, \quad \beta = \frac{3}{2}\lambda_2 - \pi(1-\nu), \qquad (\alpha < 0, \beta > 0), \tag{5.86}$$

where λ_2 is given by (5.79b) in terms of ν.

According to the basic equation (5.82a), the hole-driving force is directed to move the hole to a more stressed zone, independent of the sign of the stresses. Therefore, the hole behaves absolutely different, even qualitatively, from the classical vacancy described by Equations (5.44) and (5.55) for q <0. For example, when the external field is close to the omni-directional extension, the hole behaves rather like an inclusion.

The continuum theory of holes

Suppose an elastic medium contains a very large number of small holes which may be described by a continuous distribution of a cloud of holes. The movement of holes is assumed to be occurring under the action of external loads and heat fluctuations

The flux of holes is obviously equivalent to the flux of matter in the opposite direction. Therefore, the hole mass transport equation is:

$$\frac{\partial \varepsilon}{\partial t} = \left(D\varepsilon_{,i}\right)_{,i} + \left(\eta \varepsilon U_{,i}\right)_{,i} \quad , \tag{5.87a}$$

where

$$\Gamma_i = -U_{,i} \; , \quad \varepsilon = 1 - \frac{\rho}{\rho_o} \; , \qquad \mathbf{V}_d = \eta \Gamma \; . \tag{5.87b}$$

Here ε is the porosity, $\rho\,(x_1, x_2, x_3, t)$ is the sought-for density of the material, D is the self-diffusion coefficient, η is the mobility coefficient of holes, and U is defined by Equation (5.82a). The second term in Equation (5.87a) is written using the assumption that the drift rate of holes is directly proportional to the hole-driving force[7,8].

Equation (5.87) should be supplemented with the equations of the theory of elasticity

$$\sigma^o_{ij,j} = 0, \quad \frac{1}{2} E\left(u_{i,j} + u_{j,i}\right) = (1+\nu)\sigma^o_{ij} - \nu\sigma^o_{kk}\delta_{ij} \; . \tag{5.88}$$

For porous materials, it is natural to assume the following rule of mixture:

$$E = E_{max} \frac{\rho}{\rho_{max}} \tag{5.89}$$

(E_{max} and ρ_{max} correspond to zero porosity).

The closed system of Equations (5.87) - (5.89) allows one to study evolution of the cloud of holes and the progressive formation of weakened crack-like zones of high porosity. It may be important for studying creep and fracture of a material under high temperature.

Interaction between two holes

Consider a fixed hole of radius r_o with its center at the coordinate origin in an infinite space uniformly stretched along the z axis by stress $\sigma_z = p$. Let another hole of radius r_1 be placed in the field of singular perturbation of the first hole at a point with spherical coordinates R, ψ and θ ($r_1 << R$, ψ is measured from the z axis). Because of the non-uniformity of the total field, the latter hole is acted upon by the driving force with components Γ_R and Γ_ψ ($\Gamma_\theta = 0$).

After some calculations, Equations (5.76) and (5.82a) lead to the following simple equations

$$\Gamma_R = \frac{45\lambda p^2 r_o^3 r_1^3}{2(5-7\nu)ER^4}\left(-1+2\nu-6\nu\cos^2\psi+5\cos^4\psi\right) ,$$
$$\Gamma_\psi = \frac{15\lambda p^2 r_o^3 r_1^3}{(7-5\nu)ER^4}\sin 2\psi\left(-1-\nu+5\cos^2\psi\right) . \tag{5.90}$$

Let a cloud of small moveable holes be in the field of a large fixed hole at the coordinate origin, the interaction of small holes being ignored. According to Equation (5.90), the moveable holes are attracted, under the action of uniaxial extension-compression, to the cone $\psi = \psi_\bullet$ where

$$\psi_\bullet = arc\cos\sqrt{(1+\nu)/5}, \quad \text{so that} \quad \psi_\bullet = 60 \pm 4^o \ . \tag{5.91}$$

and then travel along this cone into the fixed hole (Figure 5.2). The exact paths of holes are determined if one integrates the following equation:

$$\frac{V_R}{V_\psi} = \frac{d\ln R}{d\psi} = \frac{\Gamma_R}{\Gamma_\psi} , \tag{5.92}$$

where V_R and V_ψ are components of the velocity of holes.

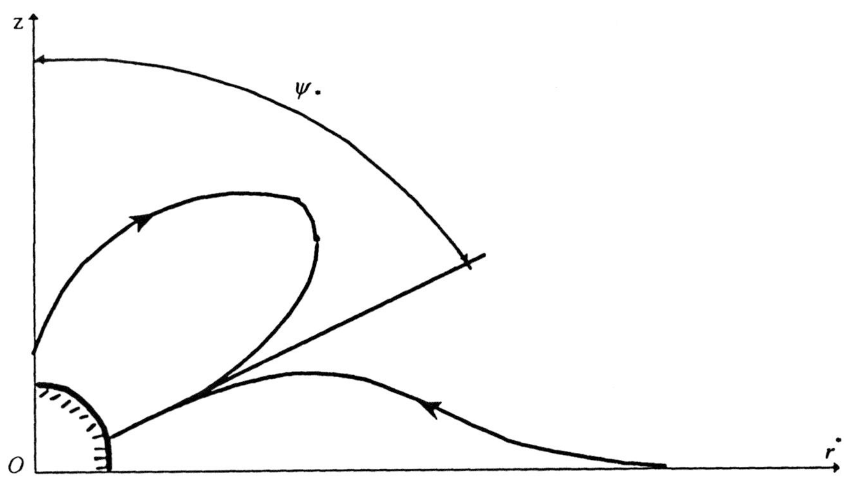

Figure 5.2. A schematic path of a moveable hole in the vicinity of a fixed spherical cavity

Because of the attraction of movable holes to the cone $\psi = \psi_*$, the planes $\psi = \psi_*$ far from the big hole become the place where the movable holes accumulate and form a loosen zone which develops to a shear fracture (under uniaxial load $\sigma_z = -\sigma$). The angle $\psi = \psi_*$ thus can be interpreted as the angle of internal friction of the material characterized by the friction coefficient

$$f = \frac{\tau_n}{\sigma_n} = cotan\,\psi_*$$

where $\tau_n = \sigma \sin\psi_* \cos\psi_*$, $\sigma_n = \sigma \sin^2\psi_*$.

So, we have predicted the friction coefficient to be $f = 0.6 \pm 0.1$ which is the value of f which is most frequently observed in compression tests of various brittle rocks.

Interaction of holes with a crack front

Let a hole of radius r_o be in the vicinity of an opening mode crack along $\theta = \pi$, $0 < r < \infty$, $-\infty < z < \infty$ where r, θ and z are cylindrical coordinates. The external stress field is as follows:

$$\sigma_r^0 = \frac{K_I}{\sqrt{2\pi r}}\left(\cos\frac{\theta}{2} + \frac{1}{2}\sin\theta\,\sin\frac{\theta}{2}\right),$$
$$\sigma_\theta^0 = \frac{K_I}{\sqrt{2\pi r}}\left(\cos\frac{\theta}{2} - \frac{1}{2}\sin\theta\,\sin\frac{\theta}{2}\right), \qquad (5.93)$$
$$\tau_{r\theta}^0 = \frac{K_I}{\sqrt{2\pi r}}\sin\theta\,\cos\frac{\theta}{2}, \quad \sigma_z = \frac{2\nu K_I}{\sqrt{2\pi r}}\cos\frac{\theta}{2}.$$

(K_I is the stress intensity factor).

According to Equation (5.82a), the components of the driving force are:

$$\Gamma_r = -\frac{\partial U}{\partial r}, \quad \Gamma_\theta = -\frac{1}{r}\frac{\partial U}{\partial \theta},$$
$$U = \frac{r_0^3 K_I^2}{2\pi E r}\cos^2\frac{\theta}{2}\left[4\alpha\,(1+\nu)^2 + \beta\left(4\nu + \cos^2\frac{\theta}{2}\right)\right],$$
$$\sigma = \frac{2\,(1+\nu)}{\sqrt{2\pi r}}\;K_I\cos\frac{\theta}{2}, \qquad (5.94)$$
$$I = \frac{K_I^2}{2\pi r}\cos^2\frac{\theta}{2}\;\left(4\nu + \cos^2\frac{\theta}{2}\right).$$

Taking Equations (5.93) and (5.94) into account, one can find

$$\Gamma_r = \frac{r_0^3\, K_I^2}{2\pi E r^2} cos^2 \frac{\theta}{2}\left[4\alpha\,(1+\nu)^2 + \beta\left(4\nu + cos^2\frac{\theta}{2}\right)\right] ,$$
$$\Gamma_\theta = \frac{r_0^3\, K_I^2}{2\pi E r^2} sin\,\theta\left[2\alpha\,(1+\nu)^2 + \beta\left(2\nu + cos^2\frac{\theta}{2}\right)\right] . \qquad (5.95)$$

Here r and θ are the hole coordinates.

The values of Γ_r and Γ_θ are negative for all θ in the range $0 < \theta < \pi$, that is, the hole is always attracted to the crack tip. The path of moveable holes is depicted qualitatively in Figure 5.3. The family of the paths is governed by the following equation:

$$\frac{V_r}{V_\theta} = \frac{dlnr}{d\theta} = \frac{\Gamma_r}{\Gamma_\theta} . \qquad (5.96)$$

The travel of holes into the crack tip leads to a reduction of fracture toughness and a subcritical creep-like growth of the crack. The holes concentrate on the continuation of the crack; their condensation at the crack tip allows one to predict the kinetics of slow crack growth due to the mechanism of migration/diffusion of holes in the material.

In addition, the holes change the crack-driving force. Let a cloud of holes be distributed in a material, so that there are N holes in unit volume, with porosity equal to $\varepsilon = 4\pi N r_0^3 / 3$. Adding up forces induced by external loads and cloud of holes, we get

$$\Gamma = \frac{1-\nu^2}{E} K_I^2 - \int_{\delta_0}^{L} \int_{-\pi}^{\pi} N\left(\Gamma_r\, cos\,\theta - \Gamma_\theta\, sin\,\theta\right) r\, dr\, d\theta . \qquad (5.97)$$

Here, L is a characteristic length of a crack (or a body). The first term in Equation (5.97) represents the Irwin force originated by external loads, the second being the force induced by holes. For the case of uniform distribution of holes when N and ε are independent of r and θ, the second term in Equation (5.97) vanishes because of (5.95).

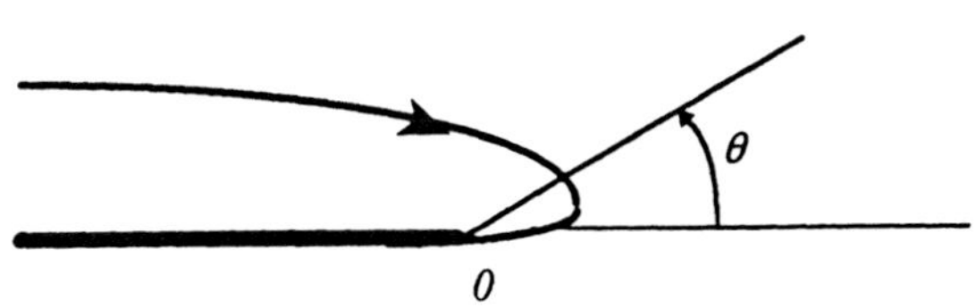

Figure 5.3. A schematic path of a moveable hole in the neighborhood of a fixed opening mode crack

Interaction of holes with a dislocation

Let a hole of radius r_o be near the front of an edge dislocation, so that

$$\sigma_r^o = \sigma_\theta^o = \frac{BE\,sin\theta}{4\pi\,(1-v^2)r}, \quad \tau_{r\theta}^o = -\frac{BE\,cos\theta}{4\pi\,(1-v^2)\,r}, \quad \sigma_z^o = 2v\sigma_r^o \quad , (5.98)$$

where B is the value of the Burgers' vector. By means of Equations (5.94) and (5.98), we calculate the components of the hole-driving force

$$\Gamma_r = -\frac{r_o^3\,B^2E}{8\pi^2\,(1-v^2)^2r^3}\left[2(1+2v)\,(2\alpha+\beta)\,sin^2\theta-\beta\right] \ ,$$

$$\Gamma_\theta = -\frac{r_o^3\,B^2E}{8\pi^2\,(1-v^2)^2r^3}\,sin2\theta\left[2\alpha\,(1+v)^2+\beta\,(1+2v)\right] \ . \qquad (5.99)$$

(r and θ are coordinates of the hole). From (5.99), it follows that the hole is always attracted to the core of the dislocation. The path of a moveable hole is qualitatively depicted in Figure 5.4.

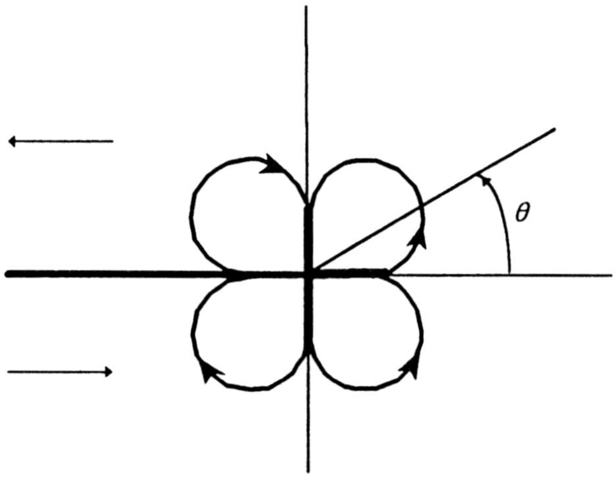

Figure 5.4. A schematic path of a moveable hole in the neighborhood of a fixed edge dislocation

If holes are continuously distributed in a material, the force Γ driving an edge dislocation is equal to

$$\Gamma = B\tau_\infty - \int_{r_d}^{L}\int_{-\pi}^{\pi} rN\left(\Gamma_r\,cos\theta - \Gamma_\theta\,sin\theta\right)drd\theta \quad . \qquad (5.100)$$

Here r_d is the radius of the dislocation core, L is a characteristic length of the body, and τ_∞ is an external shear stress on the plane of dislocation. The terms in Equation (5.100), represent the Peach-Koehler force and the force induced by a cloud of holes, respectively. For a uniform distribution of holes, the second term vanishes.

5.4. Conclusion

The invariant integral appears to be a very efficient tool for the study of driving forces and motion of miscellaneous point defects in solids.

The inclusion-type defects are important for the understanding of the hardening effect of a basic metal by alloying elements and their influence on fracture toughness. The inclusions were proved to move to stretched zones.

The hole-type defects are necessary for the understanding of creep in metals. The following features are proved to be characteristic of holes:

(*i*). Attraction and absorption of small holes by bigger ones;

(*ii*). Tendency to get collected along some planes that are inclined at an angle to the direction of uniaxial extension/compression;

(*iii*). Indifference with respect to the sign of external stress (extension or compression).

Driving forces constitute one side of the problem. The other side is the measurement of drag experienced by a point defect moving in a lattice. Knowledge of the threshold value of the driving force that is necessary for a point defect to overcome the resistance and move in the lattice is very important for the prediction of the response of a solid material to loading and temperature variations.

Some conceptual points of view treated in the present chapter are near to those developed in the recent books by Maugin[11-13].

The following problems constitute an essential part of the chapter and are designed for those who want to make creative use of the methods treated.

5.5. Problems

5.1. Find the invariant integrals for the stationary flow of viscous incompressible fluid. Derive the Navier-Stokes equations using the invariant integrals.

5.2. Find the drag of a slender semi-infinite penetrator in an elastic medium moving along its axis with zero friction. Consider both cracking and no cracking conditions.

5.3. Using the invariant integral, prove the following law of action and counteraction: if a field singularity, A, induces the force Γ acting upon another field singularity, B, then a force of the same magnitude, but opposite direction, $-\Gamma$, acts from B upon A.

5.4. Prove that the cavity-driving force in an elastic medium is equal to zero for a cavity of arbitrary shape, if the stress field is uniform, that is, does not vary in space. This is an analogue of Euler's paradox in hydrodynamics.

5.5. Prove that the resultant force (drag) applied to a solid body moving steadily in an ideal compressible gas is equal to zero if there are no vortices, mass sinks and sources, cavities, and shock waves (Euler's paradox).

5.6. Consider the plane flow of an ideal compressible gas of density ρ_∞ over a profile moving at constant speed V_∞. Prove that the propelling force,

$$\Gamma_1 = 2\pi\rho_\infty V_\infty^2 q, \tag{5.101}$$

appears when the profile surface absorbs the gas of mass q per unit length.

Prove that the lift force of a plane profile is equal to

$$\Gamma_2 = 2\pi\rho_\infty V_\infty^2 I_v \quad , \tag{5.102}$$

where I_v is the circulation (vortex intensity) at infinity produced by the profile (the Joukovsky-Chaplygin theorem).

5.7. Solve Problems 5.5 and 5.6 for an incompressible fluid.

5.8. Prove the transformation,

$$U_{,k}=(\sigma_{ij}\, u_{i,k})_{,j} \;=\; \sigma_{ij,j}\, u_{i,k} + \frac{1}{2}\sigma_{ij}\,(u_{i,jk} + u_{j,ik}) \;, \tag{5.103}$$

and derive Equation (5.43) from the invariant integrals (5.42).

5.9. Consider a thin flattened rigid penetrator moving in an elastic medium at a constant speed. Prove that the drag is equal to zero, if: (*i*) there is no friction on the body surface; and (*ii*) there is no cracking or fracturing in the elastic medium.

5.10. As distinct from the conditions of Problem 5. 9, consider a crack formed by the flattened penetrator, but closed at a finite distance behind the penetrator. Prove that the drag is also equal to zero in this case.

5.11. Consider the steady-state motion of a semi-infinite cylindrical rigid body along its axis in gas or fluid. Show that Euler's paradox is not valid in this case and the drag is equal to

$$\Gamma_1 = \frac{1}{2}\rho V^2 \Sigma_c, \tag{5.104}$$

where V is the velocity, and Σ_c is the cross-section area of the cylinder.

5.12. Study interaction between a point hole and a point inclusion.

References

1. G.P. Cherepanov (1977), Invariant Γ-integrals and some of their applications in mechanics. *Applied Mathematics and Mechanics (PMM)*, **43**(3), pp. 399-412.

2. G.P. Cherepanov (1981), Invariant Γ-integrals, *Engineering Fracture Mechanics*, **14**(1), pp. 39-58.

3. G.P. Cherepanov (1990), On invariant integrals in continuum mechanics, *Prikladnaia Mekhanika* (Soviet Applied Mechanics), **26**(7), pp. 3-16.

4. G.P. Cherepanov (1993), Introduction to singular integral equations in gas dynamics, in *Method of Discrete Vortices* by S.M. Belotserkovsky and I.M. Lifanov, CRC Press, Boca Raton, Florida.

5. A.E.H. Love (1927), *A Treatise on the Mathematical Theory of Elasticity*, 4th Ed, Cambridge University Press, Cambridge.

6. G.P. Cherepanov (1974), *Mekhanika Khrupkogo Razrushenia*, Nauka, Moscow; English edition, *Mechanics of Brittle Fracture*, R. de Wit and W. C. Cooley (eds.), McGraw-Hill, New York, 1979 .

7. G.P. Cherepanov (1985), Point defects in solids, pp. 605-623 in *Fundamentals of Deformation and Fracture*, B. Bilby, K. Miller and J. Willis (eds.), Cambridge University Press, (Eshelby Memorial Volume).

8. G.P. Cherepanov (1986), Motion of point defects in elastic solids, *Applied Mathematics and Mechanics (PMM)*, **50**(3), pp. 407-426.

9. J.D. Eshelby (1956), The continuum theory of defects, *Solid State Physics*, **3**, pp. 79-124.

10. J.P. Stark (1976), *Solid State Diffusion*, John Wiley, New York.

11. G.A. Maugin (1993), *Material Inhomogeneities in Elasticity*, Chapman and Hall, London.

12. G.A. Maugin (1992), *The Thermomechanics of Plasticity and Fracture*, Cambridge University Press, Cambridge.

13. G.A. Maugin (1988), *Continuum Mechanics of Electromagnetic Solids*, North-Holland, Amsterdam.

CHAPTER 6

DISLOCATION EMISSION

Dislocation emission induced by a crack tip is probably the most important unsolved physical problem of the theory of dislocations. This problem is addressed in the Chapter using the approach named by John Gilman the nanofracture mechanics. To simplify mathematics and focus on the physical aspects of the problem, we study: (*i*) small deformations of a single cubic crystal (or nanoscale strains of a grain in a polycrystalline material), and (*ii*) small scale yielding caused by screw or edge dislocations which are collinear to the rectilinear crack front (plane model, that is, plane strain or plane stress or plane shear).

The methods involved in the Chapter are:

- mathematical modeling of dislocation generation vs. subcritical crack growth based on the principle of adequacy and Mach's principle of "economy of thought";
- methods of the thermofluctuation theory of fracture for analysis of dislocation nucleation;
- analytical methods of the theory of functions of a complex variable (Carleman's method of solving Riemann's problem, multi-valued functions, and using general representation equations);
- method of singularity driving forces;
- analysis of solid matter states (including superplastic and amorphous states) from the viewpoint of nanofracture mechanics;
- numerical methods including the iterative and minimization methods applied to strongly non-linear systems with many unknowns governing dislocation emission and crack growth.

Using nanofracture mechanics, A. Richter and V. Sutyrin, independently conducted numerical experiments using different methods while studying the same problem of emission of many dislocations by a crack tip. Some results of these experiments are treated in Sections 6.3, 6.11 and 6.12.

6.1. Introduction

Brittle fracturing and plastic yielding are known as the main competing phenomena in polycrystalline materials that are subject to high loading. The phenomenological half-empirical theories relating to these phenomena are known as fracture mechanics and the theory of plasticity; they are based on experiments with macrospecimens, cracked or yielded, and have been investigated extensively. The nanoscale of both phenomena embracing distances from nanometer to several micrometers (grain size) has had limited study. The nanoscale physical theory of materials would predict the brittle or ductile behavior of a crystal, the superplasticity of a specimen, or the amorphous state of polycrystalline materials.

Within the framework of the contemporary theory of dislocations, there has been a reasonable understanding of the mechanics of single dislocation and very simple dislocation configurations; however, there has been no formal bridge between the dislocation theory[1-5] and fracture mechanics[6] or the theory of plasticity[7-8] (for more details, see Bullough and Matthews[9]). The difficulties in constructing the general theory are rooted, mainly, in two challenging problems, namely those of the mechanism of dislocation nucleation, and the interaction of many free and bound dislocations. Both of the problems are closely connected and form the combined problem of dislocation emission. It is easy to find stresses and strains if the number and positions of dislocations are given. However, it is not easy to determine the number and positions of dislocations emanated from a crack-tip, in terms of loading, and that is why this problem has not been solved until recently. Dislocation generation theory should provide an operable algorithm for any number of generated dislocations; the theory must secure, at least, the practical existence and uniqueness of the solution. It is not easy because a system of even small number of dislocations has many mathematical solutions (equilibrium positions), and the number of solutions grows very fast as new dislocations are born. For a large number of generated dislocations, the theory should agree with the generally accepted solutions of continuum theory of plasticity and non-linear fracture mechanics.

It is a matter of fact that if all the preexisting dislocations were simply driven out of a crystal by an applied load, the resultant plastic strain would be quite small, so that some mechanisms for dislocation generation that provide the plastic flow of crystalline materials must exist. The well-known Frank-Read source is not the most prevalent source of dislocations[9]. The most important sources of dislocation generation are various internal stress concentrations, such as surface imperfections and grain boundary corners[9], which can be modeled as microcracks[6]. The precise mechanisms involved in dislocation emission by the latter sources have not, as yet, been identified[9]. During the past 20 years, great efforts of many researchers, including Argon, Armstrong, Cherepanov, Clarke, Cottrell, Kobayashi, Li, Lin, Ohr, Rice, Thomson, Weertman, and others were made to solve the problem of dislocation emission from a crack-tip.

There are basically two approaches in studies of dislocation emission. The first one suggested by Rice and Thomson[10] was developed in[11-21] (in[14], about 25 papers of this direction were cited). The other approach put forward by Cherepanov[22,23] was developed in[24-29]. The latter has been called nanofracture mechanics because it has simultaneously addressed the problem of subcritical nanocrack growth due to dislocation emission. A comparison between these approaches to dislocation emission is given in the diagram.

Comparative diagram of two approaches to dislocation emission

	Rice/Thompson/Anderson/Beltz/ Wang/Sun/Truskinovsky	Cherepanov
1. What temperature is necessary for the approach to be valid?	Zero absolute temperature $T = 0$	Any temperature below the melting point
2. Where is a nucleated dislocation located?	At infinity	At a certain distance from the crack-tip
3. Can the approach describe the yielding of a lattice?	No	Yes
4. Can the approach describe the sub-critical crack growth?	No	Yes
5. What information is necessary to apply the approach?	Some additional constants describing dislocation core cutoff are necessary, or the Peierls approximation of interatomic forces	Only fundamental classical constants of a crystal lattice are used
6. Can the approach describe an accumulation of dislocations near a crack-tip?	No	Yes
7. What material model is used in the approach?	Cracked elastic continuum with special modeling of a crack-trip cohesion zone	Cracked elastic continuum with elementary actual and virtual dislocations while the cohesion zone is zero
8. Can the approach describe the fracture toughness?	No	Yes
9. Does the approach study the nucleation process within some interatomic spacings from a crack-tip?	Yes, only the nucleation process in considered	No, only the motion and stability of nucleated dislocations are considered
10. What is the value of the stress intensity factor at which the first stable dislocation is born?	100 $KPa\ m^{1/2}$ for the Al crystal ($T = 0°K$)	3.27 $KPa\ m^{1/2}$ for the Al crystal at $T = 0°K$
11. What is the value of the stress intensity factor at which the first elementary dislocation is nucleated ?	100 $KPa\ m^{1/2}$ (for the Al, at $T = 0$)	100 $Pa\ m^{1/2}$ at $T = 4°K$ (for Al) due to the mechanism of thermal fluctuations

The goal of the present Chapter is to investigate the combined process of dislocation emission and crack growth in the plane model which has served as the generally recognized mathematical laboratory for testing dislocation theories. In Chapter 2, we gave an estimate of the critical stress intensity factor at which a dislocation can be nucleated by the mechanism of a thermal fluctuation. Section 6.2 briefly describes the nanofracture mechanics approach to dislocation emission used in this Chapter, as well as the main modeling assumptions. Emission of screw dislocations along the crack continuation is studied in Section 6.3 for any number of dislocations. The analytical method for edge dislocations is given in Section 6.4. Generation of the first pair of edge dislocations (Section 6.5), the second pair of edge dislocations (Section 6.9) and any number of edge dislocations (Section 6.10) is analytically investigated. For many edge dislocations, the different numerical methods of iteration and optimization are used, and some results of calculations are compared and provided in graphs and Tables (Sections 6.11 and 6.12). The solutions are utilized for explanation of brittle vs. ductile behavior of crystals (Section 6.5), superplastic and amorphous state of solid matter (Sections 6.6 and 6.7). Some conclusions and problems are summarized in Sections 6.13 and 6.14.

It is noteworthy that Armstrong[30] and Kelly, Tyson and Cottrell[31] were the first to advance the viewpoint of brittle versus ductile response as the competition between Griffith cleavage and plastic shear at a crack-tip. Later, it was suggested by Cherepanov[6] that a dislocation emanates from a crack tip at a certain critical stress intensity factor k_{Ic} in super-fine scale structure (free dislocation zone). Recently, Rice[14] introduced the notion of stacking energy γ_{stack} for the corresponding released energy rate simply expressed in terms of k_{Ic} by the formula, $\gamma_{stack} = (1-\nu^2)\, k_{Ic}^2/(2E)$ where E and ν are Young's modulus and Poisson's ratio. Nanofracture mechanics [22-29] allows one to calculate k_{Ic} and γ_{stack} in terms of fundamental constants. The papers[32-35] on the mathematics of the dislocation-crack interaction are also related to the topic of the present Chapter.

Dislocation emission from a crack-tip has been directly observed by the use of several experimental techniques. Burns[36] employed etch pit techniques to observe edge dislocations on slip planes that emanated from a crack that had been cut parallel to the {110} planes in lithium fluoride. Michot and George[37], and Chiao and Clarke[38] used X-ray topography to carry out similar observations in silicon. Ohr[39] observed dislocation emission *in situ* using the electron microscope technique. Thin films of several materials, including f.c.c. and b.c.c. metals, were investigated with a high resolution, typically on the order of a few angstroms. The conditions of the latter experiments are probably close to those specified in the screw plane model of the present Chapter. Indeed, a through crack-tip in a thin film emits screw dislocations along the direction of the crack (with a slip line inclined to the film plane under 45 degrees). The direct experimental observation of a dislocation-free zone by Ohr[39], and Kobayashi and Ohr[40], also provides important evidence for understanding the present approach.

Dislocation emission at a crack-tip and cleavage decohesion along an interface as two competing processes were studied by Wang and Anderson[17], in their work on the symmetric tilt bicrystal of copper. In this work, a directional effect on the toughness of the grain boundary in a $\Sigma 9$ [110] $(2\bar{2}1)$ bicrystal was observed, in which two specimens were cut and notched along the boundary such that a crack would run in the opposite directions $[\bar{1}14]$ and $[1\bar{1}\bar{4}]$, respectively. The specimens were fatigued under a cyclic mode I loading of increasing amplitude. The specimen with the $[1\bar{1}\bar{4}]$ cracking direction broke along the interface when the maximum normal stress reached 28.1 *MPa* corresponding to the effective surface energy, 28 J/m^2. An intergranular fracture surface with cleavage "tongues" was observed. The other specimen, with a cracking direction of

[$\bar{1}$14], was loaded under identical conditions and eventually fractured at a normal stress of 76.7 *MPa*. The fracture surface contained large regions of ductile transgranular fracture and plastic tearing, so that the value of effective surface energy, more that 210J/m^2, was beyond the reliably measurable range of fracture mechanics. It was concluded that the cause of this behavior was the difference in ease with which dislocations could be emitted at each crack-tip, because the only difference between the two mentioned specimens was the cracking direction. Beltz and Wang[18] have performed experiments on copper crystals bonded on the same {221} face to sapphire, to form a layered beam subjected to four-point bending. Again, a directional dependence of toughness was observed. In their experiment, the ductile direction was observed to be [1$\bar{1}\bar{4}$], the opposite of the ductile direction with the Wang and Anderson[17] bicrystal specimen. The discussion of the above experiments within the framework of the Thomson - Rice model of dislocation nucleation is given in[19-20]. Also, it is necessary to mention the first paper on nanofracture by Gilman and Tong[41].

A dislocation can nucleate at an atomistically sharp crack-tip only after a critical stress intensity factor, k_{Ic}, is achieved[6,14]. The mechanism of thermal fluctuations[42] provides for a very low threshold of dislocation nucleation for ductile crystals, see Chapter 2 for detail. For example, for aluminum, we have: according to Honeycomb[4] τ_0=0.75 *MPa*, and a = 0.285nm, so that from Equation (2.26) it follows that k_{Ic} = 100.4 $Pa\ m^{1/2}$. The energy release rate Γ at the crack-tip corresponding to k_{Ic}, is equal to Γ_{min} = 0.2 $\mu J/m^2$ because for aluminum, ν = 0.33 and E = 70 *GPa*. This means that when K>k_{Ic}, (or $\Gamma > \Gamma_{min}$), dislocations are nucleated at a crack-tip by thermal fluctuations. The frequency of dislocation nucleation grows very fast with the increase of temperature and the stress intensity factor. The question is whether a newborn dislocation can find an equilibrium position or not.

(The Rice-Thomson model provides quite different predictions for the dislocation nucleation threshold. According to the calculations based on this model, which are summarized on page 266 in Rice[14], the value of Γ_{min} for pure aluminum is equal to 2.4J/m^2, that is, on seven orders greater than that following from the theory of thermal fluctuations. The Rice-Thomson model relates to theoretical strength which has never been realized in practice.)

6.2. An Outline of Nanofracture Mechanics

Nanofracture mechanics[22-29] considers any solid material as an ideally elastic medium with dislocations, cracks, and point defects. The nucleation, motion, and interaction of these singularities of an elastic field creates what we observe as inelastic deformation and fracturing. Let us confine ourselves by the dislocation emission at a crack-tip and the subcritical crack growth during loading. In this case, the method of nanofracture mechanics is comprised of two fundamental steps:

1. An elementary virtual dislocation is introduced to move along a physically determined slip plane subject to the Peach-Koehler driving force of an elastic field and the friction, Schmid, force. We study equilibrium motion of the virtual dislocation and

assume that **the actual elementary dislocation is born at the stable position at the lowest load, if there is a stable equilibrium position for the virtual dislocation.** If there are several positions of stable equilibrium, it is assumed that **the actual elementary dislocation is born at the stable position that is nearest to the crack-tip.**

2. Just after an actual dislocation is born, **the crack-tip advances by a certain amount, which is specific for the given crystal lattice (roughly equal to the interatomic spacing).**

These postulates launch the self-consistent, self-organized mechanism of the combined process of dislocation generation and crack growth. The method is reduced to a study of a system of non-linear equation system, see Sections 6.4, 6.5, 6.8, and 6.9.

Physically, a virtual dislocation can be understood to be a real dislocation that emanates from a crack-tip because of a thermal fluctuation. If a stable equilibrium position exists, the dislocation settles at the position; if not, the dislocation returns to the crack-tip. The second postulate describes the physical crack growth as a result of a blunting deformation of the crack-tip after a dislocation has been emitted.

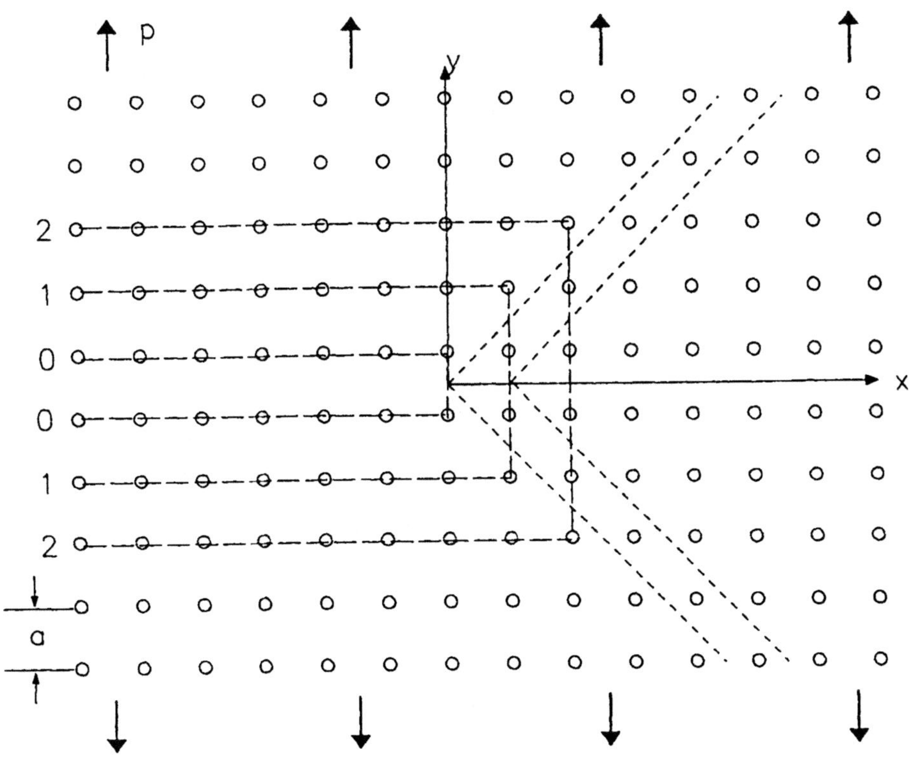

Figure 6.1. A scheme of dislocation emission/crack growth in the cubic lattice of atoms

We examine this approach as applied to a cubic lattice of atoms under plane strain conditions as shown in Figure 6.1. The symmetry planes of the crystal are assumed to be parallel to the x and y axes of the Cartesian coordinate system 0xy with the origin at the initial crack tip. The slip planes in the crystal are inclined at $\pm$ 45° to the x axis. The characteristic constants of the crystal lattice are interatomic spacing a, and Schmid's constant of friction along a slip plane, τ_0.

We investigate the neighborhood of the opening mode crack tip under small scale yielding conditions, when a crack can be considered as a rectilinear semi-infinite cut along the line $y = 0$, $x < 0$, and the loads are sufficiently small or the crack length is sufficiently great. The faces of the crack are free of traction, and form a semi-infinite rectangle $y = \pm a/2$, $x \leq 0$ as shown in Figure 6.1.

The only loading parameter in this problem is the stress intensity factor K_I. The crystal is assumed to be isotropic and homogeneous to simplify further calculations.

An increase in the external load K_I is accompanied by an increase in the stress concentration of interatomic bonds near the crack tip. After the threshold of dislocation nucleation $K = k_{Ic}$ is achieved, the regime of dynamic dislocation motion becomes possible until there is no stable position for the emitted dislocations[22]. During a thermal fluctuation the crack tip emits a dislocation pair at a near-sonic speed, the dislocations reach the limiting position under the influence of the external stress field, and then return to the crack tip intercepted by the attraction force of interaction[22]. With a further increase in external load, the stable position of the first pair of dislocations becomes possible at $K_I = K_I^{(1)}$, and eventually two elementary edge dislocations nucleated by a thermal fluctuation at the crack tip move along their slip planes, achieve the stable position and stay there. So, the first dislocation pair is generated. Simultaneously one atomic bond is ruptured ahead of the crack tip, and the crack advances one interatomic spacing a from state 0 with $l_C = 0$ to state 1 with $l_C = a$ and the crack faces along $y = \pm 3a/2$, $x \leq a$ where l_C is the crack growth length (Figure 6.1). The value of the critical stress intensity factor $K_I^{(1)}$ and the coordinates of the stable position of dislocations must be determined.

Thus for nanoscale cracks, the initiation of plastic deformation or dislocation emission occurs simultaneously with a crack advancement. The newborn dislocations cause the stress relaxation in the vicinity of the crack tip which is analogous to plastic deformation in macroscopic volumes.

With a further increase in K_I, the dislocations slowly move away from the crack tip along their slip planes, and the local stress concentration increases until a stable position of the second pair of dislocations becomes possible. This critical state is characterized by the value $K_I = K_I^{(2)}$. At this state, the next atomic bond ahead of the crack tip is ruptured, the second pair of dislocations is emanate, and the crack advances the distance a again. Simultaneously, the system transits from the state 1 to the state 2 with $l_C = 2a$ and the crack faces along $y = \pm 5a/2$, $x \leq a$ (Figure 6.1).

The second pair of dislocations moves along the slip planes $y = a \pm x$, while the first pair moves along the planes $y = \pm x$. The value of $K_I = K_I^{(2)}$ corresponding to the emission of the second pair of dislocations and the stable position of the two pairs of dislocations at this moment are to be determined. The quantitative analysis of state 2 is more complicated than of state 1 because of the interaction of four dislocations with the crack and between themselves.

As a result of the new stage of stress relaxation near the crack tip a further increase in the external stress field is necessary before the next critical state, $K_I = K_I^{(3)}$, is reached. The emission of the third pair of dislocations, the movement of the first and second pairs of dislocations, and the advancement of the crack tip are qualitatively similar to those at the previous stage. Similarly, we can examine the generation of new dislocations and the transition to states 4, 5, 6, ..., N during monotonic loading.

At a certain stage the initial pairs of dislocations get into the less loaded region behind the crack-tip and become "frozen". If we decrease K_I, the interaction of the dislocations with the crack may lead to the generation of reverse dislocations located closer to the crack tip.[25]

The state of the elastic system under study develops continuously with an increase of load (stress intensity factor) between the instants of the birth of two neighboring dislocations; however, at the instant of the birth of a new dislocation, the elastic system moves discontinuously to a new equilibrium state. The discontinuous development through a succession of equilibrium states will be rigorously analyzed below in the plane model. The frequent fluctuation process of the nucleation of dislocations at the crack-tip (most of which dynamically move away from the crack-tip and then return to the crack-tip because of the absence of a proper stable equilibrium position) creates the background of this discontinuous development.

A critical level at which a dislocation is emitted is analogous to a critical level at which the quantum mechanical system emits a photon or electron. The system under consideration develops by quanta through a series of successive critical ("quantum") levels like a quantum mechanical system.

It is necessary to emphasize the following points:

1. **The crack is** assumed to be a mathematical slit **of zero thickness**. This means that the distance of the dislocations from the crack-tip is large compared to the opening of the tip, and that all strains are small. For simplicity in the calculations, we assume small scale yielding when the initial crack may be considered semi-infinite. The role of the initial crack can be played not only by the inevitable defects and sharp grain boundaries, but also by depressions and cavities in the natural surface roughness, having the order of 10^{-1}-$10^{2}\mu m$. The role of the initial crack can also be played by interatomic layers consisting of weak, low-melting impurities at the grain boundaries[22]. Such impurities of low-melting metals like lead, antimony, tin, and bismuth, as well as sulfur and nickel sulfide, can form a low-melting eutectic with parent metal.

2. The forces controlling the motion of a dislocation near a crack front are forces of interaction between the dislocations and the crack, **friction**, and the forces of interaction among the dislocations. The fundamental role of the friction must be emphasized. This force always acts in a direction opposite to the direction of dislocation movement and is equal to $\tau_0 b$, where b is the Burgers vector of the screw or edge dislocation. The quantity τ_0 is the critical shear stress for the given slip plane in the crystal. According to Schmid's empirical law[1], τ_0 is a physical constant of the given crystal and it depends on the slip

plane, temperature, and the purity of the crystal. The last dependence reflects the effect of point defects such as holes, inclusions, interstitial atoms, vacancies, etc. on dislocation movement. The attraction of points defects toward the core of a dislocation and the formation of an "atmosphere" of such defects near the front of the dislocation is an important factor in slowing dislocations. The quantity τ_0 has a value of the order of 1-100*MPa*. Friction imparts an irreversible character to the process of dislocation movement; it is only with allowance for frictional forces that actual plastic deformations in crystalline materials can be described by the dislocation theory.

3. In annealed metals, there are usually about 10^5 - 10^8 initial, **preexisting dislocations** per cm^2 of a metallographic section. There are up to 10^{12} dislocations per cm^2 in work-hardened metals. From here, it follows that the mean distance between the nearest dislocations is 10^{-2} - 10^2 μm. Meanwhile, according to the succeeding Sections 6.9 - 6.11, when a crack advances over this distances, its tip generates 10^3 - 10^6 dislocations, that is, the density of generated dislocations 10^3 to 10^6 times that of preexisting dislocations. Thus, the effect of initial, preexisting dislocations on the crack growth is negligible. In the range of 0°*K* to 150°*K*, the value of τ_0 decreases (roughly as a linear function) with an increase in temperature. For example[4], τ_0 for aluminum at 100° *K* is nearly 20 percent lower than at 0°*K*.

4. Allowing for elastic strains does little to alter the pattern near the crack-tip depicted in Figure 6.1, since these strains are small and the plastic-strain maximum at the tip reaches values of the order of 100%. At the same time, these strains are relatively small, so that we can ignore the **local strain-hardening** of the crystal because the plastic strains can reach 1000% in good crystals, and strain-hardening during the second stage becomes appreciable at strains of several hundred percent. This observation is valid only for crystals that are sufficiently pure. The presence of alloying elements significantly lowers ductility and accelerates strain-hardening. In super-plastic alloys, conversely, impurity atoms can facilitate plastic deformation. Such impurities can be considered anti-alloying elements.

5. Some authors consider **superplasticity** as a consequence of intensive void formation. The present approach proves that, on the contrary, the formation and growth of ductile voids is a consequence of the ductility properties of the given crystalline lattice[22-29]. As is known, there are also crystals that are quite incapable of any plastic deformation. Examples are diamond and mica. The characterization of the relative capability of a crystalline lattice to undergo plastic growth of cavities or of its susceptibility to brittle fracture is one of the important problems that can be solved by means of nanofracture mechanics [22-29].

6.3. Emission of Screw Dislocations

A screw dislocation is a simpler mathematical object than an edge one, and we start just with this case.

A dislocation near a crack tip
Let an elastic plane subject to screw-shear-strain conditions contain a semi-infinite rectilinear slit along $y = 0, x < 0$. The slit is free of tractions. The plane also contains an elementary screw dislocation with Burgers vector b_s at point $x = x_1$ on the continuation of the crack. We use the Cartesian coordinate frame with the center at the crack-tip position at the start of loading.

The screw-shear elastic field can be represented as follows[6]

$$u_z = Re\, f(z), \quad \tau_{xz} + i\tau_{yz} = \mu \overline{f'(z)} \quad (z = x + iy) \tag{6.1}$$

Here, u_z is the only nonzero component of the displacement vector, τ_{xz} and τ_{yz} are the shear components of the stress tensor (remaining components equal zero), $f(z)$ is an analytic function of a complex variable z, and μ is the shear modulus.

The elastic field of a single screw dislocation is described by

$$f(z) = \frac{b_s}{2\pi i} log(z - z_0), \quad \tau_{xz} - i\tau_{yz} = -\frac{\mu b_s}{2\pi i}\frac{1}{z - z_0}, \tag{6.2}$$

where b_s is the absolute value of the Burgers vector and z_0 is the complex position vector of a dislocation in the z plane. For an elementary dislocation, $b_s = a$ for a cubic lattice.

The $f(z)$ has to satisfy the following conditions:

At the crack, $y = \pm 0$, $x < 0$, $$\tau_{yz} = -Im\, f'(z) = 0 \tag{6.3}$$

At the dislocation, $z \to x_1$, $$f(z) = \frac{b_s}{2\pi i} log(z - x_1) \tag{6.4}$$

At infinity, $z \to \infty$, $$f'(z) = -\frac{iK_{III}}{\mu\sqrt{2\pi z}}. \tag{6.5}$$

Here, K_{III} is the screw-crack stress intensity factor; and the square root, $\sqrt{z}$, is assumed to be positive at $Im\, z = 0$, $Re\, z > 0$. The K_{III} is the only loading parameter for the small scale yielding considered. This formulation describes the neighborhood of the front of any screw-shear-crack in a body of arbitrary shape with other cracks, voids, or inclusions.

The solution of the problem stated by Equations (6.1) - (6.5), is given by the following equation:

$$f'(z) = -\frac{iK_{III}}{\mu\sqrt{2\pi z}} - \frac{b_s \sqrt{x_1}}{2\pi i\sqrt{z}(x_1 - z)}. \tag{6.6}$$

The shear stress τ_{yz} on the slip plane of the dislocation at $z \to x_1$ has the form,

$$0\left(|z-x_1|^{-1}\right)+\tau^0_{yz}+0\left(|z-x_1|\right), \tag{6.7}$$

where τ^0_{yz} is the regular component of τ_{yz} defining the driving force of the dislocation in accordance with the Peach-Koehler equation (Chapter 5).

Calculation of the driving force, F, provides:

$$F(x_1)=\lim_{z\to x_1} Im\left\{\mu\overline{f'(z)}-\frac{b_s\ \mu}{2\pi i(x_1-\bar{z})}\right\}=$$
$$=\frac{K_{III}}{\sqrt{2\pi x_1}}-\frac{\mu b_s}{2\pi}\lim_{z\to x_1} Re\left\{\frac{1}{\sqrt{\bar{z}}}\frac{\sqrt{x_1}-\sqrt{\bar{z}}}{x_1-\bar{z}}\right\}=\frac{K_{III}}{\sqrt{2\pi x_1}}-\frac{\mu b_s}{4\pi x_1}. \tag{6.8}$$

This expression consists of two terms. The first term is due to the interaction between a dislocation and the background elastic field; this force propels the dislocation from the crack. The second term is the self-induction force due to the interaction between the dislocation and crack, which, conversely, drives the dislocation back to the crack.

We study the equilibrium positions of a virtual dislocation in the neighborhood of a crack-tip, after the dislocation is nucleated at the crack-tip in accordance with any mechanism. Taking into account the friction force resisting the removal of the dislocation from the crack-tip, and using Equation (6.8) we find:

$$\frac{K_{III}}{\sqrt{2\pi x_1}}-\frac{\mu b_s}{4\pi x_1}=\tau_0, \tag{6.9}$$

or

$$K_{III}=\tau_0\sqrt{2\pi x_1}+\frac{\mu b_s}{2\sqrt{2\pi x_1}}. \tag{6.10}$$

This equation defines the possible equilibrium positions of a virtual dislocation in terms of the stress intensity factor. In dimensionless variables,

$$K=\frac{K_{III}}{\sqrt{2\mu b\tau_0}} \quad \text{and} \quad \bar{x}_1=\frac{4\pi\tau_0}{\mu b}x_1, \tag{6.11}$$

Equation (6.10) thus becomes:

$$K=\frac{1}{2}(\bar{x}_1^{1/2}+\bar{x}_1^{-1/2}). \tag{6.12}$$

The graph of this function is drawn in Figure 6.2. It has two branches: the left one, for $\bar{x}_1<1$, corresponds to the unstable equilibrium states of a dislocation, and the right one, for $\bar{x}_1 > 1$, corresponds to the stable equilibrium states of the dislocation. The minimum point ($K = 1$, $\bar{x}_1 = 1$) at which

$$K_{III} = K_{III}^{(1)} = \sqrt{2\mu b \tau_0} \quad \text{and} \quad x_1 = x_1^{(1)} = \frac{\mu b}{4\pi\tau_0}, \tag{6.13}$$

determine the minimum values of the stress intensity factor and the dislocation coordinate, at which a stable state is possible. From the diagram of Figure 6.2, it follows that there are no stable equilibrium positions of a dislocation while the K_{III} grows from zero to $K_{III}^{(1)}$. This means that any dislocation nucleated at a crack-tip in this range, returns to the crack-tip. The minimum stable position may be realized, only when K_{III} achieves the value of $K_{III}^{(1)}$. *The basic principle of nanofracture mechanics is that an actual dislocation is always born at this state* due to any mechanism of nucleation (e.g., the thermal fluctuations). A dynamic dislocation nucleated at a crack-tip propagates with the speed of sound so that the self - induction force lags and, at first, only the instantaneous forces of friction and the background field of K_{III} act upon the dislocation.

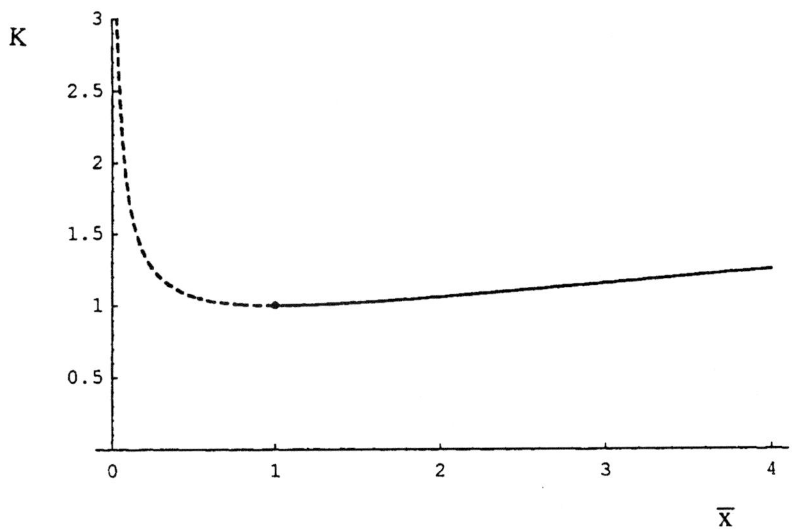

Figure 6.2. The first dislocation birth diagram, where $\bar{x} = x_1 / x_1'$ and $K = K_{III} / K_{III}^{(1)}$. The values of $x_1' = \mu b / (4\pi\tau_0)$ and $K_{III}^{(1)} = \sqrt{2\mu b\tau_0}$ determine the minimum quantum level at which the stable equilibrium dislocation can exist. (K_{III} is the stress intensity factor, and x_1 is the equilibrium position of screw shear dislocation.)

The birth of an actual dislocation at the minimum point becomes less probable for a very high loading rate, when the frequency of the thermally fluctuated dislocations is not sufficient to supply this rate. At very high loading rates, $\dot{K} > Kt_*^{-1}$, where t_* is the mean time between thermal emission of two successive dislocations, a dislocation is born at an metastable state on the stable branch of diagram of Figure 6.2.

For example, we apply Equation (6.13) to aluminum. According to reference[4] $\mu = 25GPa$, $\tau_0 = 0.75\ MPa$, $b_s = 0.285\ nm$ in this case, and from (6.13), it follows: $K_{III}^{(1)} = 3.27KPa\ m^{1/2}$ and $x_1^{(1)} = 756\ nm$. In the same aluminum, dislocations emanate due to thermal fluctuations at $K > K_E = 100.4\ Pa\ m^{1/2}$ (See the end of Section 2.4).

As seen, the threshold of the birth of the first dislocation in aluminum is 33 times greater than the K_E corresponding to the threshold of dislocation nucleation at the crack-tip due to thermal fluctuations. (However, the value of $K_{III}^{(1)}$ is still a hundred times less than that provided by the Rice-Thomson model: or, in energy release rate, 10^4 times less compared to the Rice-Thomson model). Generally, we have:

$$\frac{K_{III}^{(1)}}{K_E} = \sqrt{\frac{\mu}{10\pi\tau_0}}, \tag{6.14}$$

according to Equations (2.26) and (6.13).

Any number of dislocations near a crack-tip

Consider the state of the elastic system after N dislocations have been emitted by the crack tip one by one. The crack increment equals $l_C = Na$, at this stage.

From Equation (6.6), it follows that the elastic field for any number of dislocations is:

$$f'(z) = -\frac{iK_{III}}{\mu\sqrt{2\pi(z-l_C)}} - \frac{b_s}{2\pi i\sqrt{z-l_C}}\sum_{n=1}^{N}\frac{\sqrt{x_n - l_C}}{x_n - z}, \tag{6.15}$$

where N is the total number of elementary screw dislocations, and x_n is the position of the nth dislocation.

The force driving the mth dislocation is given by the following equation:

$$F(x_m) = \frac{K_{III}}{\sqrt{2\pi(x_m - l_C)}} - \frac{\mu b_s}{4\pi(x_m - l_C)} - \frac{\mu b_s}{2\pi\sqrt{x_m - l_C}}\sum_{\substack{n=1\\ n\neq m}}^{N}\frac{\sqrt{x_n - l_C}}{x_n - x_m} \tag{6.16}$$

$$(m = 1,2,\ldots N)$$

Assuming that in the loading process all dislocations withdraw from a crack-tip to their equilibrium positions, and taking into account the friction force, we find:

$$F(x_m) = \tau_0 \qquad (m = 1,2,\ldots N), \tag{6.17}$$

where $F(x_m)$ is given by Equation (6.16)

The nonlinear equation system (6.17) serves to determine the stable equilibrium positions, $x_1, x_2, \ldots x_N$, of N dislocations. This system can be written in the following dimensionless form:

$$\frac{2K}{\sqrt{\bar{x}_m}} - \frac{1}{\bar{x}_m} - \frac{2}{\sqrt{m}} \sum_{\substack{n=1 \\ n \neq m}}^{N} \frac{\sqrt{\bar{x}_n}}{\bar{x}_n - \bar{x}_m} = 1 \qquad (m = 1,2,..N) \tag{6.18}$$

Here,

$$\bar{x}_m = \frac{4\pi\tau_0}{\mu b_s}(x_m - l_C); \quad K = \frac{K_{III}}{\sqrt{2\mu b_s \tau_0}}. \tag{6.19}$$

Equation (6.18) defines the equilibrium position of (N-1) actual dislocations and the Nth virtual dislocation. The number N plays the role of a specific time.

It is noteworthy that, for $N > 1$, the solution to Equations (6.18) is non-unique, and the number of solutions rapidly grows as N increases. To choose the physically correct solution and achieve the practical convergence, a numerical method must comply with the successive dislocation generation process, in nanofracture mechanics.

The following iteration procedure corresponds to the physical process described in Section 6.2 above. At first, let us number the coordinates of successive dislocations as follows:

$$x_N < x_{N-1} < \ldots < x_2 < x_1 \tag{6.20}$$

Let us examine the force driving the first, nth, and Nth dislocations, on the basis of Equations (6.18). The qualitative plot of the force driving the nth dislocation versus its position while the positions of the remaining dislocations are fixed, is shown in Figure 6.3 ($a = x_{n-1}$ and $b = x_{n+1}$ in this figure). As seen from the figure, there is always a unique equilibrium position for the nth dislocation, and it can be easily found using the Newton - Raphson method. The unique stable equilibrium position of the first dislocation always lies in the semi-infinite line $x \geq a = x_2$ (see Figure 6.4 where $a = x_2$). However, the virtual, Nth, dislocation may or may not have an equilibrium position, depending on the value of K_{III} and the positions of other dislocations (see Figure 6.5, where $b = x_{N-1}$ and 0 corresponds to the crack-tip).

The iteration procedure is to calculate the equilibrium position of each dislocation, one by one, until all the equations are satisfied with the prescribed accuracy. The main cycle of iteration is as follows:

Suppose the solution for a certain N corresponding to a K_{III} is known, so that there are N actual dislocations. We increase the K_{III} by a small value and try to find the solution with (N + 1) dislocations, including the virtual, (N + 1)th one. If the solution does not exist, we correct the position of N dislocations to the new, greater load, K_{III}, by running

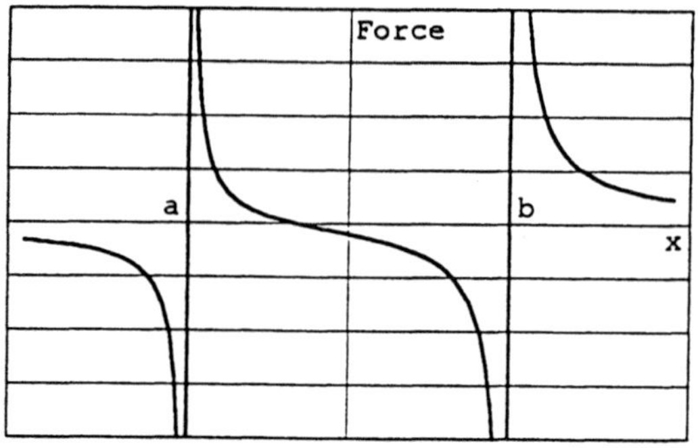

Figure 6.3. Qualitative plot of the force driving the *n*th dislocation (*a* and *b* are the positions of neighboring dislocations.)

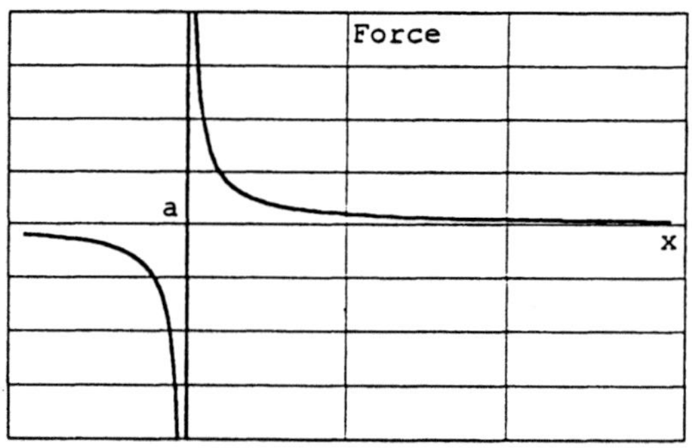

Figure 6.4. Qualitative plot of the force driving the first dislocation (*a* is position of the second dislocation.)

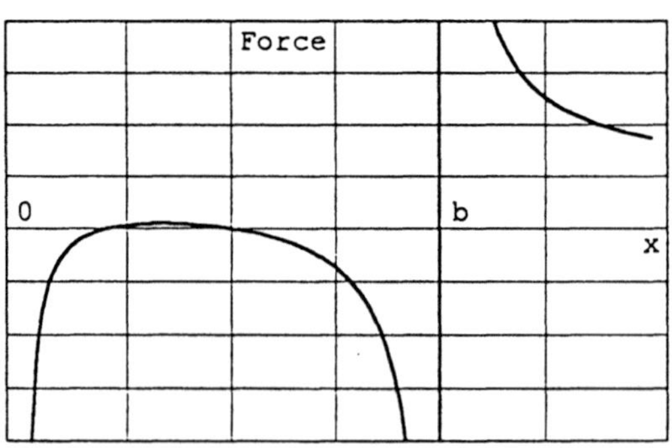

Figure 6.5. Qualitative plot of the force driving the virtual dislocation (*b* is the position of the *(N-1)*th dislocation , and 0 is the position of the crack-tip.)

load, K_{III}, again and try to find the equilibrium position of the (N+1)th dislocation. We repeat the process until there is an equilibrium position for the (N+1)th dislocation; then, the main iteration procedure with the system of (N+1) dislocations is run in order to find a new equilibrium position for all (N+1) dislocations. The cycle is finished at this point.

This procedure provides excellent convergence. The equation system (6.18) has been solved for up to N=150 using PC 486. Some results are given in Table 6.1 in the following designation: N is the number of generated dislocations, $\bar{x}_N = 4\pi\tau_0 x/(\mu b_s)$ is the dimensionless coordinate of the Nth newborn dislocation, $\bar{k}_N = k_{III}(2\mu b_s\tau_0)^{-1/2}$ is the dimensionless superfine scale stress intensity factor at the moment of the birth of the Nth dislocation, $\bar{K}_N = K_{III}(2\mu b_s\tau_0)^{-1/2}$ is the dimensionless stress intensity factor. The error in $\bar{x}_N$ and $\bar{k}_N$ is less than four percent. The error in $\bar{K}_N$ is much less than one percent.

Table 6.1. Quantum levels of dislocation births up to 12 dislocations

N	1	2	3	4	5	6
$\bar{x}_N$	1	0.819477	0.745282	0.701085	0.670582	0.647703
$\bar{k}_N$	1	0.99995	0.999785	0.99861	0.998775	0.99818
$\bar{K}_N$	1	1.41421	1.73205	2	2.23607	2.44949
N	7	8	9	10	11	12
$\bar{x}_N$	0.630082	0.615193	0.603058	0.592298	0.583294	0.575432
$\bar{k}_N$	0.99778	0.99716	0.996465	0.996285	0.995665	0.99502
$\bar{K}_N$	2.64575	2.82843	3	3.16228	3.31663	3.4641

The analytical theory for many dislocations

Consider a large number of dislocations as a one-dimensional continuum dislocation distribution along $l_C < x < l_D$ on the crack continuation. Let ξ_n be the Lagrange coordinate that characterizes the dislocations as follows:

$$\xi_n = 1 - \frac{n-1}{N} \qquad (0 \le \xi_n \le 1). \tag{6.21}$$

Equations (6.17) can be rewritten using $x(\xi)$ as:

$$\begin{aligned}\tau_0 = &\frac{K_{III}}{\sqrt{2\pi x(\xi_m) - l_C}} - \frac{\mu b_s}{4\pi x(\xi_m) - l_C} - \\ &-\frac{\mu b_s}{2\pi\sqrt{x(\xi_m) - l_C}} \sum_{\substack{n=1\\ n\ne m}}^{N} \frac{\sqrt{x(\xi_n) - l_C}}{x(\xi_n) - x(\xi_m)}\end{aligned}, \tag{6.22}$$

where $x_n = x(\xi_n)$.

By passing to the limit, $N \to \infty$, in Equations (6.22), we obtain:

$$\tau_0 = \frac{K_{III}}{\sqrt{2\pi(x(\xi)-l_C)}} - \frac{\mu b_s N}{2\pi\sqrt{x(\xi)-l_C}}\int_0^1 \frac{\sqrt{x(\xi')-l_C}}{x(\xi'_n)-x(\xi)}\,d\xi' \quad . \tag{6.23}$$

Introducing the density of the Burgers vector,

$$B(\xi) = Nb_s \left|\frac{dx}{d\xi}\right|^{-1}, \tag{6.24}$$

and considering $B(\xi)$ as a function of x, one can get:

$$\tau_0 = \frac{K_{III}}{\sqrt{2\pi(x-l_C)}} - \frac{\mu}{2\pi\sqrt{x-l_C}}\int_{l_C}^{l_D} \frac{B(x')\sqrt{x'-l_C}}{x'-x}\,dx' \tag{6.25}$$

Here, $l_D = x_1$ is the coordinate of the first dislocation.
Notice the evident condition:

$$\int_{l_C}^{l_D} B(x)dx = Nb \quad . \tag{6.26}$$

The solution of the integral equation (6.25) should satisfy Equation (6.26). The function $f(z)$can be expressed through B(x) by the following equation[6,43,44]:

$$f'(z) = \frac{K_{III}}{\sqrt{2\pi(z-l_C)}} - \frac{\mu}{2\pi\sqrt{z-l_C}}\int_{l_C}^{l_D} \frac{B(x')\sqrt{x'-l_C}}{x'-z}\,dx' \tag{6.27}$$

$$B(x) = \frac{d[u_z]}{dx} = Re\left[f'(z)\right]_-^+, \tag{6.28}$$

where $[u_z]$ means the discontinuity as follows:

$$[u_z] = \lim_{z\to x+i0} u_z - \lim_{z\to x-i0} u_z \tag{6.29}$$

It can be shown[6,43,44] that the integral equation (6.25), with the condition (6.26), can be reduced to the following boundary value problem of the theory of functions of a complex variable:

At the crack, $y = 0$, $x < l_C$: $Im\, f'(z) = 0$, (6.30)

At the dislocation array, $y = 0$, $l_C < x < l_D$: $Im\, f'(z) = -\frac{\tau_0}{\mu}$, (6.31)

At the infinity, $z \to \infty$: $f'(z) = -\frac{iK_{III}}{\mu\sqrt{2\pi z}}$ (6.32)

We can write the solution to this problem in the following form[44]:

$$f'(z) = -i\frac{\tau_0}{\mu}\left[1 - \frac{1}{\pi i}\log\frac{i\sqrt{l_D - l_C} - \sqrt{z - l_D}}{i\sqrt{l_D - l_C} + \sqrt{z - l_D}}\right] \tag{6.33}$$

Using Equations (6.24) and (6.28), one can obtain[44]:

$$B(x) = -\frac{2\tau_0}{\pi\mu}\log\frac{i\sqrt{l_D - l_C} - \sqrt{x - l_D}}{i\sqrt{l_D - l_C} + \sqrt{x - l_D}} , \tag{6.34}$$

$$\frac{d\xi}{dx} = -\frac{2\tau_0}{\pi\mu\, bN}\log\frac{i\sqrt{l_D - l_C} - \sqrt{x - l_D}}{i\sqrt{l_D - l_C} + \sqrt{x - l_D}} , \tag{6.35}$$

$$\xi = \left(1 - \sqrt{1 - \frac{x - l_C}{l_D - l_C}}\right) - \frac{x - l_C}{2(l_D - l_C)}\log\frac{i\sqrt{l_D - l_C} - \sqrt{x - l_D}}{i\sqrt{l_D - l_C} + \sqrt{x - l_D}} , \tag{6.36}$$

where

$$l_C = Na \tag{6.37}$$

$$l_D = l_C + \frac{\pi K_{III}^2}{8\tau_0^2} = l_C + \frac{\pi\, b\mu}{4\tau_0}N \tag{6.38}$$

$$N = \frac{K_{III}^2}{2b\mu\tau_0} . \tag{6.39}$$

From Equations (6.37) - (6.39), it follows that:

$$K_{III} = \sqrt{\frac{2b\mu\tau_0}{a}}\sqrt{l_C} \tag{6.40}$$

$$K_N = \sqrt{2b\mu\tau_0}\sqrt{N} \tag{6.41}$$

$$l_D - l_C = \frac{\pi K_{III}^2}{8\tau_0^2}. \qquad (6.42)$$

Comparison of the analytical solution with the exact numerical one (see Table 6.1) shows that the N-K_N diagrams coincide for any N = 1,2,3,... . Equation (6.40) provides the subcritical crack growth diagram or R-curve. Equation (6.42) coincides with that derived from the Dugdale solution[45] for the corresponding elastic-plastic problem. Equation (6.42) derived here from nanofracture mechanics has been thoroughly verified in numerous experiments[45,6], and nanofracture mechanics is thus circumstantially confirmed by the test evidence. Equations (6.33), (6.34) and (6.42) coincide with the exact solution of the corresponding elastic-plastic problem[6,44,45].

Continuum theory cannot describe the near neighborhood of the crack-tip, particularly the interval, $l_C < x < x_N$, between the crack-tip, $x = l_C$, and the last dislocation, $x = x_N$ (the free dislocation zone).

Superfine-scale stress intensity factor in the dislocation-free zone.

In the small neighborhood of a crack-tip $|z - l_C| = \varepsilon << |x_n - l_C|$, inside the dislocation-free zone, the elastic field has the following form[6]:

$$f'(z) = -\frac{ik_{III}}{\mu\sqrt{2\pi(z - l_C)}}. \qquad (6.43)$$

Here, k_{III} is the superfine-scale stress intensity factor[6]. From Equations (6.15) and (6.43), it follows that

$$k_{III} = K_{III} - \frac{b\mu}{\sqrt{2\pi}}\sum_{n=1}^{N}\frac{1}{\sqrt{x_n - l_C}}, \qquad (6.44)$$

or

$$k = K - \sum_{n=1}^{N}\bar{x}_n^{-1/2}, \qquad (6.45)$$

where

$$k = \frac{k_{III}}{\sqrt{2\mu b\tau_0}}, \quad K = \frac{K_{III}}{\sqrt{2\mu b\tau_0}}, \quad \bar{x}_n = \frac{4\pi\tau_0}{\mu b}(x_n - l_C). \qquad (6.46)$$

In particular, $k_{III} = K_{III}$ when N=0.

Figures 6.6 and 6.7 show the numerical results based on Equations (6.18) and (6.44), namely the birth and motion of dislocations, and the superfine-scale stress intensity factor, in terms of K. The figures can serve as some illustrations of the nanofracture mechanics approach. In the figures, the characteristic length and stress intensity factor units are $x_1^{(1)}$ and $K_{III}^{(1)}$ defined by Equation (6.13):

$$\bar{x} = \frac{x}{x_1^{(1)}} \quad , \quad K = \frac{K_{III}}{K_{III}^{(1)}} \quad . \tag{6.47}$$

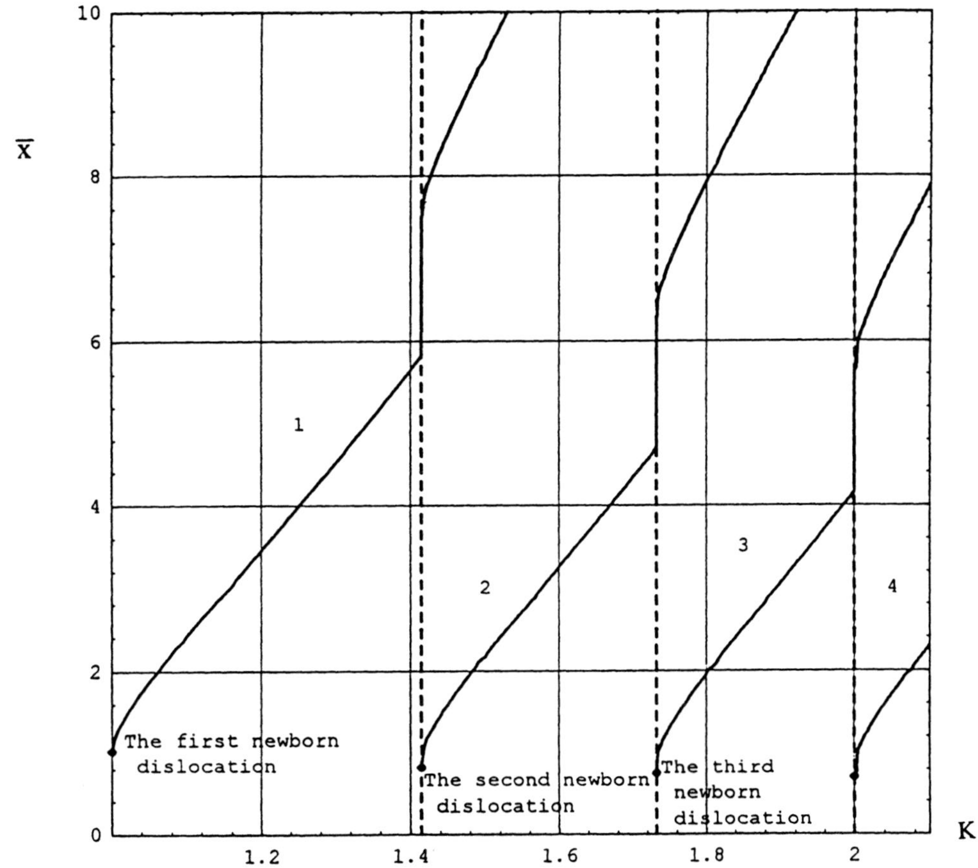

Figure 6.6. Discontinuous process of dislocation birth and motion: $\bar{x}$ is the dimensionless coordinate, and K is the dimensionless load, namely $\bar{x} = 4\pi\tau_0 x/(\mu b)$ and $K = K_{III}/\sqrt{2\mu b \tau_0}$. Numbers 1,2,3 and 4 mark the trajectories of corresponding dislocations.

The quantities $x_1^{(1)}$ and $K_{III}^{(1)}$ identify the first quantum level at which the first dislocation is born.

One can observe that, by increasing K, the system under study runs through an infinite series of critical states or quantum levels, where the system behavior is discontinuous. Between any two adjacent quantum levels, the system changes its state continuously. At a critical state, a new dislocation is born in the dislocation-free zone at a certain distance from a crack-tip, and the crack-tip advances by the quantum, a. Any other parameter of the system suffers a discontinuity at that moment, too. The K can be considered as a specific time, because we may increase K directly proportional to the real time. The system is similar to a stationary quantum mechanical system subject to a continuous input of an external energy.

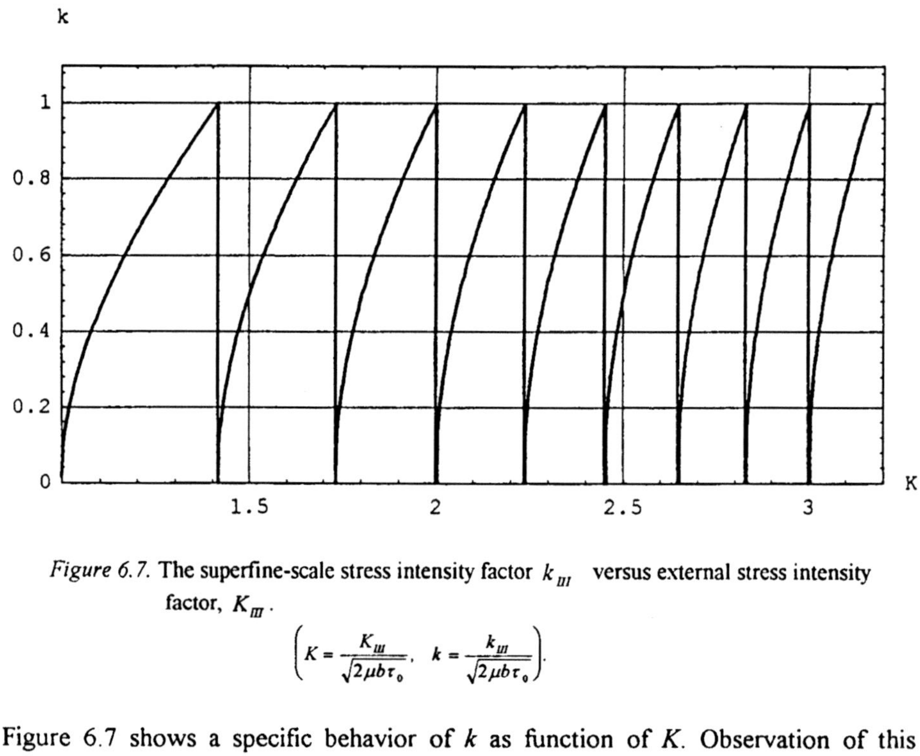

Figure 6.7. The superfine-scale stress intensity factor k_{III} versus external stress intensity factor, K_{III}.

$$\left(K = \frac{K_{III}}{\sqrt{2\mu b \tau_0}}, \quad k = \frac{k_{III}}{\sqrt{2\mu b \tau_0}}\right).$$

Figure 6.7 shows a specific behavior of k as function of K. Observation of this behavior is important because the k_{III} plays the role of a watchdog or judge controlling the competition of dislocation emission or yielding versus a brittle cleavage advance of a screw crack. Thus a screw crack starts and grows without emitting a dislocation, after k_{III} achieves the value of $2\sqrt{\mu\gamma}$ specifying the brittle cleavage propagation of the screw-crack along its line. If k_{III} is less than $2\sqrt{\mu\gamma}$, the system demonstrates only a tough behavior of a crack emitting dislocations. According to Figure 6.7, the k increases from zero to one while K is growing, inside an interval of continuous development between two adjacent quantum levels, which means that the screw-shear rectilinear crack behavior is either ideally tough (without any cleavage stage), if $K_{III}^{(1)} < 2\sqrt{\mu\gamma}$, or ideally brittle (without any dislocation emission), if $K_{III}^{(1)} > 2\sqrt{\mu\gamma}$. Figures 6.8 and 6.9 showing the comparison between the analytical and numerical solutions are self-explanatory.

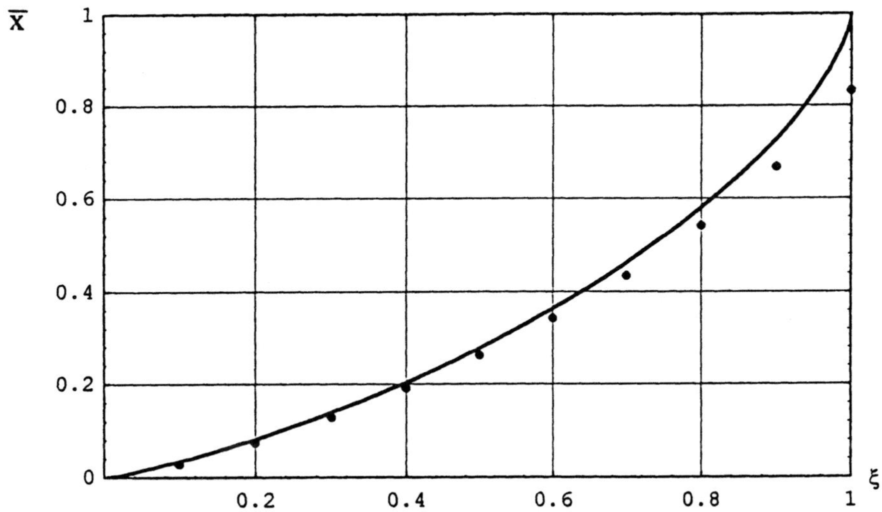

a). N = 10

Figure 6.8. The positions of dislocations vs their Lagrange coordinate: the solid line corresponds to the analytical theory, and the dots correspond to computerized results (both are based on nanofracture mechanics). Here, $\bar{x} = 4\tau_0 x / (b\mu\pi N)$ and $\xi = \xi_n = 1-(n-1)/N$ for the nth dislocation, where N is the number of dislocations emitted .

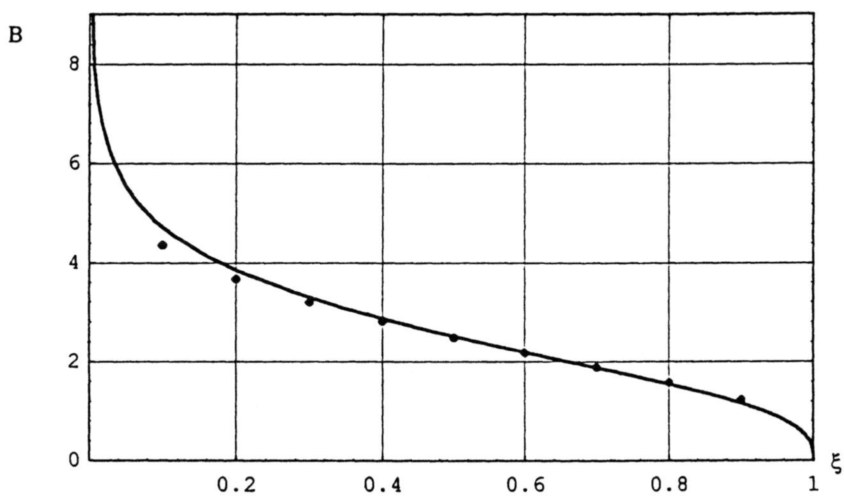

a). N = 10

Figure 6.9. Density of the Burgers vector vs the Lagrange coordinate: the solid line corresponds to the analytical theory, and the dots to computerized results (both are based on nanofracture mechanics). Here, $\bar{B}(\xi) = \pi\mu(2\tau_0)^{-1} Nb|dx/d\xi|^{-1}$ is the Burgers vector density, ($d\xi/dx$ is defined by Equation (6.35)) and $\xi = \xi_n = 1-(n-1)/N$ for the nth dislocation, where N is the number of dislocations emitted.

6.4. Edge Dislocations Near a Crack Tip

An efficient mathematical method is necessary for calculating the crack stress and displacement field with any number of dislocations in a concise closed form. The Muskhelishvili solution cannot be used because it holds only for single-valued displacements. The method of reference[10] is applicable only for a single dislocation, while the general approach[32] is too cumbersome.

Consider the elastic plane with a semi-infinite crack along $y=0$, $x < 0$ and N pairs of symmetrical edge dislocations at points z_k and $\bar{z}_k$, k = 1,..., N (Figure 6.10). The boundary conditions are:

$$\sigma_y = \tau_{xy} = 0 \quad \text{when} \quad y = 0, x < 0 \ . \tag{6.48}$$

Two polar coordinate systems, $r\ \theta$ and $\rho\ \varphi$, are used, the former associated with the crack tip, and the latter with the slip plane (Figure 6.10),

$$z_k = x_k + iy_k = a_k + \rho_k e^{i\varphi_k} = r_k e^{i\theta_k} \ . \tag{6.49}$$

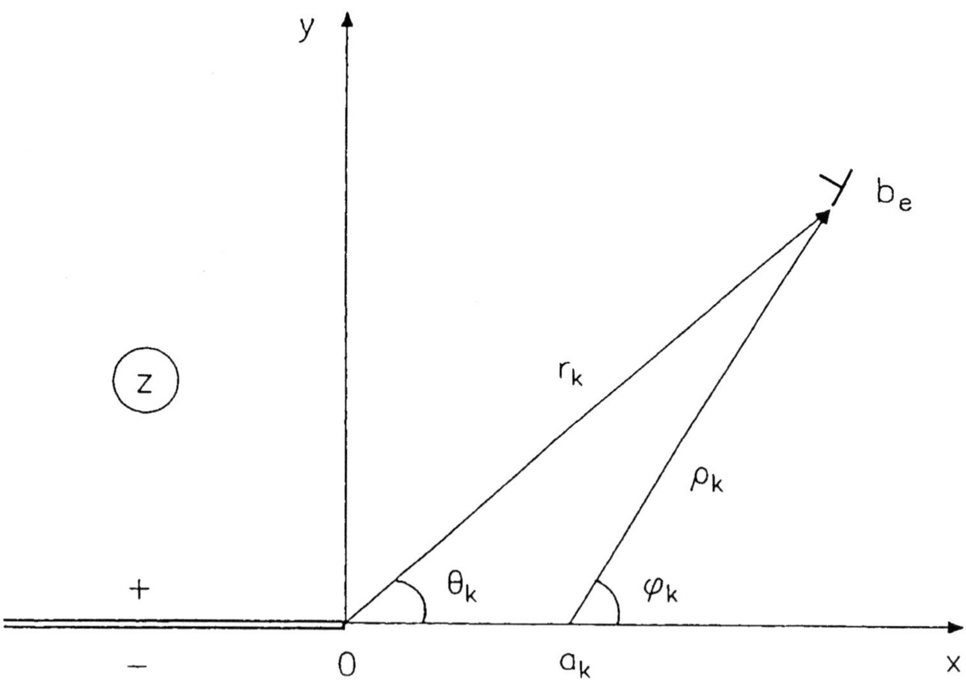

Figure 6.10. An edge dislocation near a crack tip

The complex Burgers vectors of the kth pair of dislocations are given by

$$B_{e+} = b_e^{(k)} e^{i\varphi_k}, \quad B_{e\_} = -b_e^{(k)} e^{-i\varphi_k}, \tag{6.50}$$

where + and - correspond to the upper and lower half-planes. The elastic field can be represented by the Kolosov-Muskhelishvili potentials $\Phi(z)$ and $\Psi(z)$, which are analytic functions of the complex variable $z = x + iy$ in the domain occupied by the body:

$$\begin{aligned}
&\sigma_x + \sigma_y = 2\left[\Phi(z) + \overline{\Phi(z)}\right] \qquad (z = x + iy) \\
&\sigma_y - \sigma_x + 2i\tau_{xy} = 2\left[\bar{z}\Phi'(z) + \overline{\Psi(z)}\right] \\
&2\mu(u_x + iu_y) = \kappa\phi(z) - z\overline{\phi'(z)} - \overline{\psi(z)} \\
&\Phi(z) = \phi'(z), \quad \Psi(z) = \psi'(z), \quad \kappa = 3 - 4\nu.
\end{aligned} \tag{6.51}$$

where u_x and u_y are the components of the displacement vector along the x and y axes respectively, σ_x, σ_y and τ_{xy} are the components of the stress vector, and μ and ν are the shear modulus and Poisson's ratio, respectively.

The subscripts ρ, φ and r, θ are used for the corresponding components of the displacement vector and stress tensor in the polar coordinates. The transformation formulas are given by

$$\begin{aligned}
&u_x + iu_y = (u_\rho + iu_\varphi)e^{i\varphi} = (u_r + iu_\theta)e^{i\theta} \\
&\sigma_y + \sigma_y = \sigma_\rho + \sigma_\varphi = \sigma_r + \sigma_\theta \\
&\sigma_y - \sigma_x + 2i\tau_{xy} = e^{-2i\varphi}(\sigma_\varphi - \sigma_\rho + 2i\tau_{\rho\varphi}) = e^{-2i\theta}(\sigma_\theta - \sigma_r + 2i\tau_{r\theta})
\end{aligned} \tag{6.52}$$

At first, study the multivalued potentials:

$$\begin{aligned}
&\phi(z) = C_1 \ln z, \quad \psi(z) = C_2 \ln z \\
&\chi(z) = C_3 \ln z, \quad \psi(z) = \chi'(z),
\end{aligned} \tag{6.53}$$

where C_1, C_2 and C_3 are certain complex constants, the physical meaning of which will be clarified by studying the singular point $z = 0$.

The principal vector of forces (X, Y) applied to the arc AB passed from A to B and the moment M of these forces relative to the coordinate origin can be expressed in terms of the analytic functions as follows,

$$X + iY = -i\left[\phi(z) + z\overline{\phi'(z)} + \overline{\psi(z)}\right]_A^B$$

$$2\mu[u_x + iu_y]_A^B = \left[\kappa\phi(z) - z\,\overline{\phi'(z)} - \overline{\psi(z)}\right]_A^B \quad ,$$
$$M = Re[\chi(z) - z\psi(z) - z\bar{z}\phi'(z)]_A^B \quad , \tag{6.54}$$

where the brackets $[D]_A^B$ denote the jump $D(B)$ - $D(A)$ in any quality D enclosed in the brackets. Consider an arbitrary closed arc AB encircling the coordinate origin $z = 0$ and traveled in the counterclockwise direction. Substitution (6.53) into (6.54) gives

$$C_1 = \frac{2\mu B_e + Y - iX}{2\pi i(1+\kappa)}, \; C_2 = \frac{-2\mu\bar{B}_e + \kappa(Y + iX)}{2\pi i(1+\kappa)}, \; C_3 = -\frac{M + iC_0}{2\pi i} \quad , \tag{6.55}$$

where B_e is the discontinuity of the complex displacement vector on the arc AB (the Burgers vector of the dislocation), and C_0 is a real constant which describes a point source of compression (if $C_0 < 0$) or extension (if $C_0 > 0$). Thus, potentials (6.53) describe a combined point source including concentrated force, concentrated moment, dislocation and compression center. The problem under consideration involves only dislocation, so that $X = Y = M = C_0 = 0$, and the potentials of a dislocation source are:

$$\Phi_0(z) = \frac{\mu_0 b_e}{4\pi i}\frac{e^{i\varphi}}{z}, \quad \Psi_0(z) = -\frac{\mu_0 b_e}{4\pi i}\frac{e^{-i\varphi}}{z} \quad , \tag{6.56}$$

where $\mu_0 = \mu/(1-\nu)$.

The stress field at $z = 0$ can be translated to $z = z_1$ using the equations following from (6.51):

$$\Phi(z, z_1) = \Phi_0(z - z_1)$$
$$\Psi(z, z_1) = \Psi_0(z - z_1) - \bar{z}_1\Phi_0'(z - z_1) \tag{6.57}$$

From (6.56) and (6.57) it follows that as $z \to z_k$ $\left(B_e = b_e^{(k)} e^{i\varphi_k}\right)$,

$$\Phi(z, z_k) = \frac{\mu_0 b_e^{(k)} e^{i\varphi_k}}{4\pi i(z - z_k)} \quad ,$$
$$\Psi(z, z_k) = -\frac{\mu_0 b_e^{(k)} e^{-i\varphi_k}}{4\pi i(z - z_k)} + \bar{z}_k \frac{\mu_0 b_e^{(k)} e^{i\varphi_k}}{4\pi i(z - z_k)^2} \quad , \tag{6.58}$$

and as $z \to \bar{z}_k$ $\left(B_e = -b_e^{(k)} e^{-i\varphi_k}\right)$

$$\Phi(z,z_k) = -\frac{\mu_0 b_e^{(k)} e^{-i\varphi_k}}{4\pi i(z-\bar{z}_k)} ,$$
$$\Psi(z,z_k) = \frac{\mu_0 b_e^{(k)} e^{i\varphi_k}}{4\pi i(z-\bar{z}_k)} - z_k \frac{\mu_0 b_e^{(k)} e^{-i\varphi_k}}{4\pi i(z-\bar{z}_k)^2} . \tag{6.59}$$

Introduce the potential,

$$\Omega(z) = \overline{\Phi}(z) + z\overline{\Phi}'(z) + \overline{\Psi}(z) . \tag{6.60}$$

The limits of $\Omega(z)$ as $z \to z_k$ and $z \to \bar{z}_k$ can be obtained using (6.58) - (6.60)

$$\Omega(z,z_k) = -\frac{\mu_0 b_e^{(k)}}{4\pi i}\left[\frac{e^{-i\varphi_k}}{z-z_k} + \frac{z_k - \bar{z}_k}{(z-z_k)^2} e^{i\varphi_k}\right], \quad as\ z \to z_k \tag{6.61}$$

$$\Omega(z,z_k) = \frac{\mu_0 b_e^{(k)}}{4\pi i}\left[\frac{e^{i\varphi_k}}{z-\bar{z}_k} + \frac{\bar{z}_k - z_k}{(z-\bar{z}_k)^2} e^{-i\varphi_k}\right], \quad as\ z \to \bar{z}_k \tag{6.62}$$

The boundary conditions (6.48) can be rewritten as

$$F_1^+ = -F_1^-, \qquad F_2^+ = F_2^-, \qquad \text{when} \quad y = \pm 0,\, x \le 0 , \tag{6.63}$$

in terms of new analytic functions,

$$F_1(z,z_k) = \Phi(z) + \Omega(z), \qquad F_2(z,z_k) = \Phi(z) - \Omega(z) \tag{6.64}$$

In accordance with Equations (6.58), (6.59), (6.61) and (6.62), the analytic functions F_1 (z,z_k) and F_2 (z,z_k) have prescribed second-order poles at points z_k and $\bar{z}_k$. The functions F_1 and F_2 behave as $O(z^{-1/2})$ for $z \to 0$, and $O(z^{-3/2})$ for $z \to \infty$. The corresponding potentials of the external loads have the order $O(z^{-1/2})$ at $z \to \infty$. The potentials corresponding to the external loads should be considered given since the problem under study belongs to class N of the problems of the theory of elasticity, which do not obey St. Venant's principle (See[6], for more details).

The general solution of the Riemann boundary value problem (6.63) in the given class of functions has the form (Problem 6.16) ,

$$F_1(z,z_k)=\frac{2}{\sqrt{z}}\left[\frac{A_3}{z-z_k}+\frac{A_6}{(z-z_k)^2}+\frac{A_7}{z-\bar{z}_k}+\frac{A_8}{(z-\bar{z}_k)^2}\right]$$
$$F_2(z,z_k)=\frac{2A_1}{z-z_k}+\frac{2A_2}{(z-z_k)^2}+\frac{2A_4}{z-\bar{z}_k}+\frac{2A_5}{(z-\bar{z}_k)^2} \tag{6.65}$$

where $A_1,\ldots, A_8$ are some constants.

The potentials $\Phi(z,z_k)$ and $\Omega(z,z_k)$ following from (6.64) and (6.65) are:

$$\Phi(z,z_k)=\frac{A_1}{z-z_k}+\frac{A_2}{(z-z_k)^2}+\frac{\bar{A}_1}{z-\bar{z}_k}+\frac{\bar{A}_2}{(z-\bar{z}_k)^2}$$
$$+\frac{1}{\sqrt{z}}\left[\frac{A_3}{z-z_k}-\frac{A_2\sqrt{z_k}}{(z-z_k)^2}+\frac{\bar{A}_3}{z-\bar{z}_k}-\frac{\bar{A}_2\sqrt{\bar{z}_k}}{(z-\bar{z}_k)^2}\right], \tag{6.66}$$

$$\Omega(z,z_k)=-\frac{A_1}{z-z_k}-\frac{A_2}{(z-z_k)^2}-\frac{\bar{A}_1}{z-\bar{z}_k}-\frac{\bar{A}_2}{(z-\bar{z}_k)^2}$$
$$+\frac{1}{\sqrt{z}}\left[\frac{A_3}{z-z_k}-\frac{A_2\sqrt{z_k}}{(z-z_k)^2}+\frac{\bar{A}_3}{z-\bar{z}_k}-\frac{\bar{A}_2\sqrt{\bar{z}_k}}{(z-\bar{z}_k)^2}\right], \tag{6.67}$$

The potential $\Psi(z,z_k)$ can be found from Equations (6.60), (6.66), and (6.67) as

$$\Psi(z,z_k)=-\frac{2A_1}{z-z_k}-\frac{2A_2}{(z-z_k)^2}-\frac{2\bar{A}_1}{z-\bar{z}_k}-\frac{2\bar{A}_2}{(z-\bar{z}_k)^2}-\bar{z}\Phi'(z) \tag{6.68}$$

We calculate the constants $A_1,\ldots,$ and A_8 using Equations (6.58), (6.59), (6.61), (6.62), (6.66), and (6.67). They are given by

$$A_1=\frac{\mu_0 b_e^{(k)}\cos\varphi_k}{4\pi i},\quad A_2=\frac{\mu_0 b_e^{(k)}e^{i\varphi_k}}{8\pi i}\left(z_k-\bar{z}_k\right),$$
$$A_3=\frac{\mu_0 b_e^{(k)}}{4\pi i}\sqrt{z_k}\left(i\sin\varphi_k-\frac{z_k-\bar{z}_k}{4z_k}e^{i\varphi_k}\right), \tag{6.69}$$

$$A_4=\bar{A}_1,\quad A_5=\bar{A}_2,\quad A_6=-A_2\sqrt{z_k},\quad A_7=\bar{A}_3,\quad A_8=\bar{A}_6 \tag{6.70}$$

Equations (6.66) and (6.68) provide the potentials of a pair of dislocations at $z=z_k$ and $z=\bar{z}_k$ near the crack tip. For N dislocations the potentials take the form:

$$\Phi(z)=\sum_{k=1}^{N}\Phi(z,z_k),\quad \Psi(z)=\sum_{k=1}^{N}\Psi(z,z_k) \tag{6.71}$$

The solution of the problem can also be generalized for an arbitrary number of finite and infinite cracks along the real axis[23].

According to the Peach-Koehler equation, the force driving the kth dislocation is equal to $F = b_e^{(k)}\tau_{\rho\varphi}^0(z_k)$, where $\tau_{\rho\varphi}^0(z_k)$ is the stress $\tau_{\rho\varphi}(z_k)$ minus the singular component of this stress at $z = z_k$ (the latter describes the self field of the dislocation).

Applying this definition we can calculate $\tau_{\rho\varphi}^0$ from (6.51), (6.52), (6.66) and (6.68) as

$$\tau_{\rho\varphi}^0(z_k) = Im\left(P.V.\lim_{z\to z_k}\left[\frac{1}{2}e^{+2i\varphi_k}(\sigma_y - \sigma_x + 2i\tau_{xy})\right]\right) \tag{6.72}$$

$$= Re\left[\frac{-\bar{A}_1}{z_k - \bar{z}_k} - \frac{5A_2(z_k - \bar{z}_k)}{16z_k^3} - \frac{3A_3(z_k - \bar{z}_k)}{8z_k^{5/2}} - \frac{\bar{A}_2\bar{z}_k^{1/2}(5z_k - \bar{z}_k)}{2z_k^{3/2}(z_k - \bar{z}_k)^2} + \frac{\bar{A}_3(3z_k - \bar{z}_k)}{2z_k^{3/2}(z_k - \bar{z}_k)}\right]$$

The stress intensity factor induced by the elastic field of dislocations is denoted by k_{Id}. Since K_I is equal to lim $\left[2\sqrt{2\pi z}\Phi(z)\right]$ as $z \to o$, we can use (6.66) to find

$$k_{Id} = -2\sqrt{2\pi}\sum_{k=1}^{N}\left[\frac{A_3}{z_k} + \frac{A_2\sqrt{z_k}}{z_k^2} + \frac{\bar{A}_3}{\bar{z}_k} + \frac{\bar{A}_2\sqrt{\bar{z}_k}}{\bar{z}_k^2}\right], \tag{6.73}$$

where the constants are defined by Equations (6.69).

6.5. Generation of the First Edge Dislocation

Consider a crack in a cubic lattice of atoms under plain strain conditions as shown in Figures 6.1 and 6.10. Initially, the crack is situated along $y = 0$, $x < 0$. The first pair of dislocations moves along slip planes $y = \pm x$, so that

$$z_1 = \rho e^{i\varphi_1} = re^{i\theta_1}, \quad \theta_1 = \varphi_1 = \pi/4, \quad B_e = b_e e^{i\pi/4}, \quad b_e = a\sqrt{2} \tag{6.74}$$

Due to the symmetry of the problem only the upper half-plane will be considered.

Let us introduce a virtual dislocation on slip plane $y = x$ following the algorithm described in Section 6.2. The forces upon the virtual dislocation are the forces produced by external load, self-induction and friction.

The force from the external field is equal to $b_e\ \tau_{\rho\varphi}^e$, where $\tau_{\rho\varphi}^e$ is the shear stress on the slip plane[6],

$$\tau_{\rho\varphi}^e = \frac{K_I}{2\sqrt{2\pi r}}\sin\theta_1\cos\frac{\theta_1}{2} + \frac{1}{2}\left(\sigma_y^0 - \sigma_x^0\right) + O\left(\sqrt{r}\right) \tag{6.75}$$

and K_I is the stress intensity factor, σ_x^0 and σ_y^0 are the regular components of σ_x and σ_y at the crack tip.

For example, for an isolated crack of finite length with the traction-free sides and the stresses $\sigma_x^\infty, \sigma_y^\infty, \tau_{xy}^\infty = 0$ at infinity, $\sigma_x^0 = \sigma_x^\infty - \sigma_y^\infty$ and $\sigma_y^0 = 0$. Unless otherwise noted, it is assumed that $K_I >> \sqrt{2\pi x}\left|\sigma_y^0 - \sigma_x^0\right|$. Under this condition the shear stress on the slip plane due to the external load can be written as

$$\tau_{\rho\varphi}^e = \frac{K_I}{\sqrt{2\pi r}} \frac{\sin\theta_1 \cos(\theta_1/2)}{2} \tag{6.76}$$

From (6.74) and (6.76), it follows that

$$\tau_{\rho\varphi}^e \approx \frac{0.327 K_I}{\sqrt{2\pi r}} \tag{6.77}$$

We see that the force of the external field repulses the dislocation from the crack tip.

The value of $\tau_{\rho\varphi}^0$ induced by the dislocation pair can be calculated from (6.72) using (6.69) and (6.74) as

$$\tau_{\rho\varphi}^0 = -\frac{\mu_0 b_e}{\sqrt{4\pi r}} \tag{6.78}$$

This is the self-induction force which attracts the dislocation to the crack tip.

The friction force on the dislocation is equal to $b_e\tau_0$ where τ_0 is the Schmid constant of the lattice. This force acts in the direction opposite to the dislocation movement. Since the emission of a dislocation from the crack tip is considered, the friction is the force attracting the dislocation to the crack tip.

The balance of forces on the virtual dislocation thus has the form

$$\frac{0.327}{\sqrt{2\pi r}} K_I = \tau_0 + \frac{\mu_0 b_e}{4\pi r} \tag{6.79}$$

In the dimensionless quantities,

$$\bar{r}_1 = \frac{4\pi\tau_0}{\mu_0 b_e} r, \qquad \bar{K}_I = \frac{K_I}{\sqrt{2a\mu_0\tau_0}} \quad , \tag{6.80}$$

Equation (6.79) is written as

$$\bar{K}_I = 1.82\left(\sqrt{\bar{r}_1} + \frac{1}{\sqrt{\bar{r}_1}}\right) \tag{6.81}$$

The curve of $\overline{K}_I$ vs $\bar{r}$ is shown in Figure 6.11. The minimum of $\overline{K}_I$ $(\bar{r})$ is achieved at

$$\bar{r} = 1, \quad \overline{K}_I = \overline{K}_I^{(1)} = 3.641 \tag{6.82}$$

No equilibrium dislocation position exists for $\overline{K}_I < 1$.

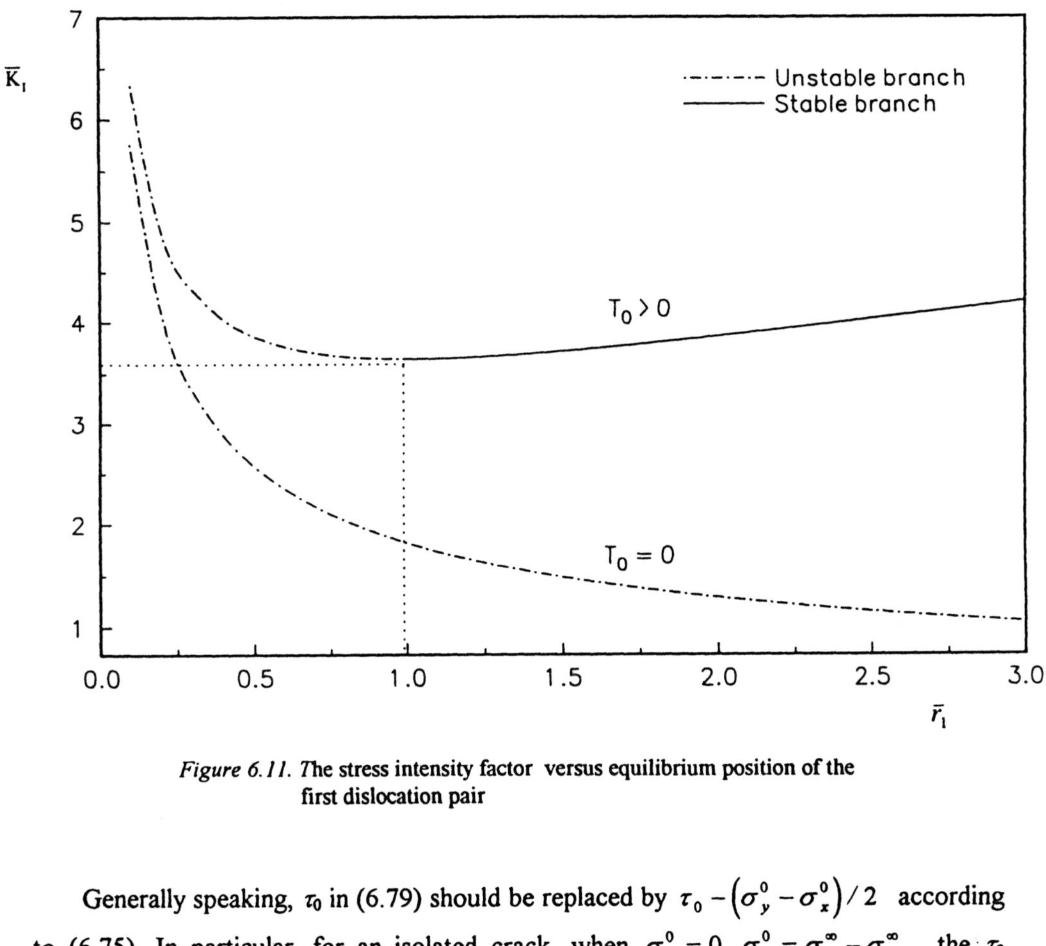

Figure 6.11. **The stress intensity factor versus equilibrium position of the first dislocation pair**

Generally speaking, τ_0 in (6.79) should be replaced by $\tau_0 - \left(\sigma_y^0 - \sigma_x^0\right)/2$ according to (6.75). In particular, for an isolated crack, when $\sigma_y^0 = 0$, $\sigma_x^0 = \sigma_x^\infty - \sigma_y^\infty$, the τ_0 should be replaced by $T_0 = \tau_0 - \left(\sigma_y^0 - \sigma_x^0\right)/2$.

The case $T_0 < 0$ is impossible since the condition of elastoplastic equilibrium at infinity, $\tau_{max} = \left|\sigma_y^\infty - \sigma_x^\infty\right|/2 \le \tau_0$, is not satisfied. The extreme case is $T_0 = 0$ which corresponds to yielding at infinity. In the latter case there is no stable position for a dislocation, so that a dislocation pair emitted from the crack tip moves away until it encounters a boundary or an obstacle.

We analyze the equilibrium equation (6.79) for τ_0. Note that the rate of propagation of the dislocation interaction is finite and is equal to the speed of sound (the speed of longitudinal waves for edge dislocations, and transverse waves for screw dislocations), so that dislocation interaction or self-induction acts with a certain delay. The rate of propagation of dislocations is limited to the speed of transverse waves (rather than to the speed of Rayleigh waves, as is the case for cracks). Meanwhile, the force of interaction of the dislocation with the external field and the friction force are instantaneous. This delay in self-induction force makes possible the regime of dynamic dislocation propagation at the values of K_I less than $K_I^{(1)}$.

Since dislocations do not have mass, the equilibrium equation for the regime of dynamic propagation is given by

$$0.327\frac{K_I}{\sqrt{2\pi r}} = \tau_0, \quad \text{or,} \quad \overline{K}_I = 1.82\sqrt{\overline{r}} \quad . \tag{6.83}$$

The self-induction force is added to this equation only after the perturbation from the crack tip overtakes the moving dislocation.

Thus, the process of slow monotonic loading (increase in K_I) is accompanied by the dynamic process of dislocation emission due to thermal fluctuations. During one fluctuation a dislocation pair propelled by the external field moves away from the crack tip at near-sonic speed, until it achieves the limiting position,

$$r_1 = \frac{(0.327)^2}{2\pi}\frac{K_I^2}{\tau_0^2}, \tag{6.84}$$

where the dynamic dislocation pair stops intercepted by the self-induction force. This force drives the dislocation pair to the crack tip, if $K_I < K_I^{(1)}$, where $K_I^{(1)}$ is given by

$$K_I^{(1)} = 3.641\sqrt{2a\mu_0\tau_0}\,. \tag{6.85}$$

If $K_I = K_I^{(1)}$ the self-induction force propels the dislocation pair to the stable equilibrium position at $r = r_1^{(1)}$, which is given by

$$r_1^{(1)} = \frac{\mu_0 b_e}{4\pi\tau_0}\quad . \tag{6.86}$$

Simultaneously, the atomic bond ahead of the crack tip is ruptured, and the crack advances one interatomic spacing a. With a further increase in K_I, the dislocation pair moves along the stable branch of the curve $\overline{K}_I(\overline{r}_1)$ shown in Figure 6.11. Equations (6.85) and (6.86) determine the moment of the crack growth initiation and the generation of the first pair of elementary edge dislocations as well as the stable positions of the dislocations. The delay of the self-induction force also provides some opportunities for metastable positions of the dislocation pair at $K_I > K_I^{(1)}$, that is, on the right branch of the curve of Figure 6.11.

6.6. Brittle vs Ductile Behavior of Crystals

There are two possible mechanisms of crack growth in a crystal, with and without dislocation generation. Ductile crack growth is accompanied by the emission of dislocations at the crack tip. Brittle crack growth occurs in a crystal if the crack advances without emission of dislocations, i.e. the atomic bond ahead of the crack tip is ruptured with no dislocations generated. The brittle fracture is characterized by the critical value of the stress intensity factor K_{Ic}^0 which is given by

$$K_{Ic}^0 = 2\sqrt{\mu_0 \gamma} \ , \tag{6.87}$$

where γ is the surface energy of the crystal along the crack plane.

Introduce the dimensionless quantity η,

$$\eta = \frac{K_I^{(1)}}{K_{Ic}^0} \ , \tag{6.88}$$

called the brittleness number of the crystal[22].

If $K_I^{(1)} < K_{Ic}^0$, that is $\eta < 1$, dislocation emission starts before the brittle crack propagation becomes possible. On the other hand, if $K_I^{(1)} > K_{Ic}^0$, that is, $\eta > 1$, the crack grows without dislocation generation. In other words, if $\eta < 1$, the crystal is ductile, and if $\eta > 1$, the crystal is brittle. Using (6.85) the brittleness number can be written as

$$\eta = 2.165\sqrt{\frac{\tau_0 b_e}{\gamma}} \tag{6.89}$$

The smaller η, the more ductile is the crystal.

A crystal is absolutely or ideally brittle, if $\eta > 1$, that is,

$$b_e \, \tau_0 > 0.213 \, \gamma \tag{6.90}$$

Table 6.2 shows the results of the calculations based on Equations (6.85), (6.86) and (6.89) of the present approach. It contains the values of $K_I^{(1)}, K_I^0, r_1^{(1)}$, and the brittleness number for several common crystals. The values of a, μ and ν are taken from the work of Rice and Thomson[10], while the values of τ_0 for pure crystals are taken from the book by Honeycomb[4].

It is difficult to determine experimentally the value of τ_0 in brittle crystals. In this case, it may be necessary to make use of a theoretical estimate. The theoretical estimate based on the Peierls-Nabarro formula was used to calculate τ_0 for diamond and salt in Table 6.2.

Table 6.2. Ductile versus brittle behavior of crystals

Crystal	μ $10^4 \times MPa$	ν	a $10^{-10} \times m$	γ $Pa \times cm$	τ_0 MPa	$K_{Ii}^{(1)}$ $MPa \times m^{1/2}$	$r_1^{(1)}/b_e$	K_{Ic}^{0} $MPa \times m^{1/2}$	η
Aluminum	2.5	0.35	2.85	84/60	0.55-1	$1.3 - 1.7 \cdot 10^{-2}$	$5.6 - 3.1 \cdot 10^3$	0.36/0.30	0.02-0.04
Copper	4	0.32	2.55	170/98	0.9-1	$1.9 - 2.0 \cdot 10^{-2}$	$5.2 - 4.7 \cdot 10^3$	0.63/0.48	0.02-0.03
Gold	2.4	0.41	2.87	142/140	0.5	$1.2 \cdot 10^{-2}$	$6.47 \cdot 10^3$	0.48	0.02
Nickel	7.5	0.28	2.49	172/101	3.3-7.5	$4.8 - 7.2 \cdot 10^{-2}$	$2.5 - 1.1 \cdot 10^3$	0.85/0.65	0.04-0.06
Silver	2.6	0.35	2.88	114/62	0.4-0.7	$1.1 - 1.4 \cdot 10^{-2}$	$7.9 - 4.5 \cdot 10^3$	0.43/0.32	0.01-0.02
Zinc	3.8	0.33	2.74	58/69	0.3	$1.1 \cdot 10^{-2}$	$15.0 \cdot 10^3$	0.36/0.39	0.03
α-Iron	6.9	0.29	2.49	198/120	15	$9.8 \cdot 10^{-2}$	$0.52 \cdot 10^3$	0.82/0.64	0.08-0.1
Diamond	50.9	0.07	2.52	540/550	$\sim 10^4$	~ 5		3.47	~ 1.5
Salt	1.6	0.25	3.99	25/135	$\sim 10^3$	~ 0.5		0.15/0.34	~ 1.0

It is also difficult to measure the reversible surface energy of ductile crystals in experiments. The empirical value of γ is always higher than the actual value due to the emission of dislocations. More accurate data for surface energy of a crystal can be found by extrapolating on temperature the empirical data for the corresponding liquid (see Chapter 1). Table 6.2 uses two values of γ - the higher, empirical value was taken from the work by Rice and Thomson[10], and the lower, theoretical estimate was obtained by Cherepanov[6].

It is noted that the theoretical value of γ of $NaCl$ is several times higher than the empirical value. The reason of this discrepancy is explained by a humid environment effect[6]. The value of $K_I^{(1)}$ appears to be in the range 0.01 - $0.1 MPa\sqrt{M}$ for metals, which is a hundred of times less than the value of $K_{I\Gamma}$ that corresponds to the threshold of macroscopic fatigue crack growth under cyclic load[6].

The present approach shows that there is practically no latent period in the growth of a nanoscale fatigue crack (short crack). The crack grows almost from the moment of load application even if the loading level is extremely low. The fundamental difference between the long and short fatigue cracks is that the growth rate of the short microcrack is two orders greater than the growth rate of the long macrocrack for the same value of the stress intensity factor $K_I < K_{I\Gamma}$. This difference is due to the high resistance to the growth of the long crack because of grain boundaries saturated with alloying elements.

6.7. Superplastic State of Crystals

Some authors consider superplasticity as a consequence of intensive void formation. The present approach proves that on the contrary, the formation and growth of ductile voids is the consequence of ductile properties of a given crystalline lattice. The growth of ductile voids in a crystalline lattice occurs by the same slip mechanism as the subcritical crack growth. We define *superplasticity as a capability of a crystal for unlimited strain.*

The superplastic state of a crystal can be explained in terms of nanofracture mechanics by two factors: (*i*) a low friction τ_o on slip planes, and (*ii*) a small size of a specimen. In Section 6.5, we calculated the first stable position of edge dislocations in the cubic lattice at $r_1^{(1)} = \mu_0 b_e / (4\pi\tau_0)$. From here, it follows that the lower the friction τ_0 on a slip plane, the larger is the value of $r_1^{(1)}$. This means that the first pair of dislocations settles farther from the crack tip in the crystals with lower friction, and is driven out if the specimen is sufficiently small.

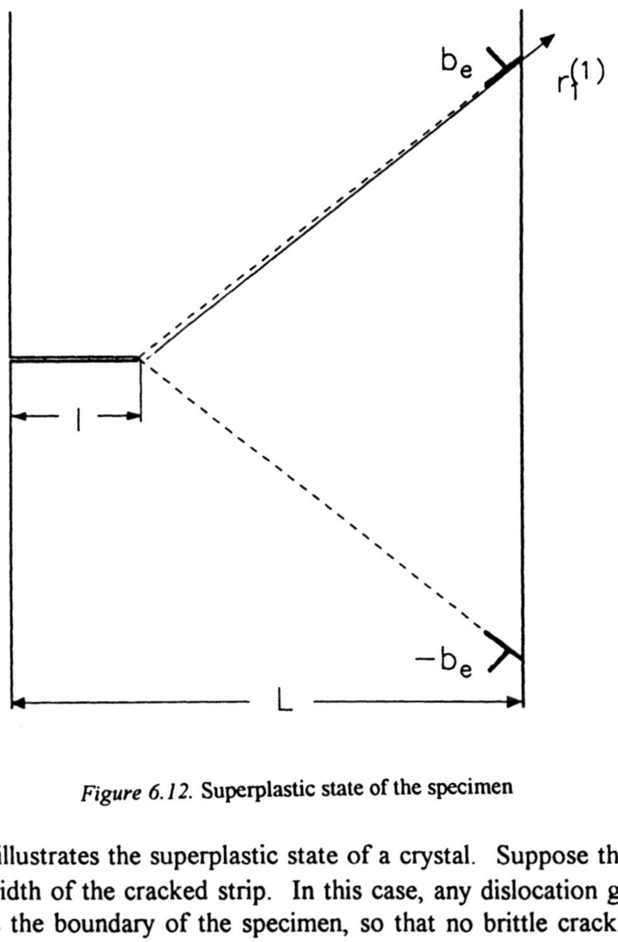

Figure 6.12. Superplastic state of the specimen

Figure 6.12 illustrates the superplastic state of a crystal. Suppose that $L < r_1^{(1)} / \sqrt{2}$ where L is the width of the cracked strip. In this case, any dislocation generated at the crack tip reaches the boundary of the specimen, so that no brittle crack propagation is possible. This corresponds to the superplastic state of the cracked crystal. The condition of the superplastic state of a specimen with the characteristic dimension L is given by

$$\frac{L}{a} \leq \frac{\mu_0}{4\pi\tau_0} \tag{6.91}$$

So, there can be a superplastic state in any crystal, if the size of the crystal is sufficiently small.

Figure 6.13 demonstrates superplastic behavior of a strip-shaped specimen with an initial atomically sharp crack subjected to extension (the numerical experiment). The width of the specimen is taken equal to $L = 10^6\, a$ while the length of the initial crack to $l = 10^5\, a$, where a is the interatomic spacing of the lattice. At a certain value of the loading parameter $K_I = K_I^{(1)}$ characteristic of the given crystal, the crack tip emits the first pair of dislocations. The dislocations reach the boundary of the specimen, and the crack advances one interatomic spacing a. The crystal domains above and below the slip planes move as a whole along these planes. Then the next pair of dislocations is generated at the crack tip, and so on. Figure 6.13 shows the calculated shape of the specimen after 10^5 and $2{\cdot}10^5$ loading steps. The initial crack opens and the upper and lower parts of the specimen are shifted along the slip planes when the dislocations are emitted. Deformation of the superplastic specimen in the process zone may be very great (in this theoretical model, the superplastic deformation is unlimited).

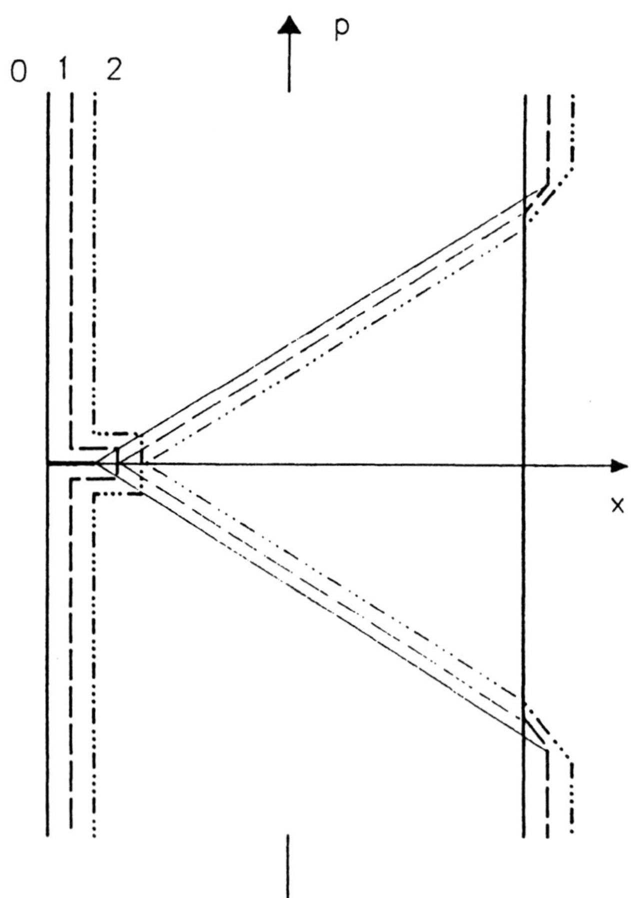

Figure 6.13. Deformation of a superplastic specimen under extension

6.8 Amorphous State of Polycrystalline Materials

The amorphous state is another extreme state of a solid matter apart from the superplastic state described above. *No free dislocations are possible in the solid in the amorphous state.* We consider this property as a definition of the amorphous state. The structure of metals in an amorphous state is built of many small clusters or crystalline grains of ordered atoms, each cluster containing some hundreds of atoms.

According to the nanofracture mechanics, this state corresponds to the situation where the clusters are too small for a dislocation to reach a stable position. Any pair of dislocations emitted from the atomically sharp crack tip inside a cluster returns to the crack tip if the material satisfies the condition

$$d < r_1^{(1)} \quad or \quad \frac{d}{b_e} < \frac{\mu_0}{4\pi\tau_0}, \tag{6.92}$$

where d is the characteristic dimension of the cluster as shown in Figure 6.14.

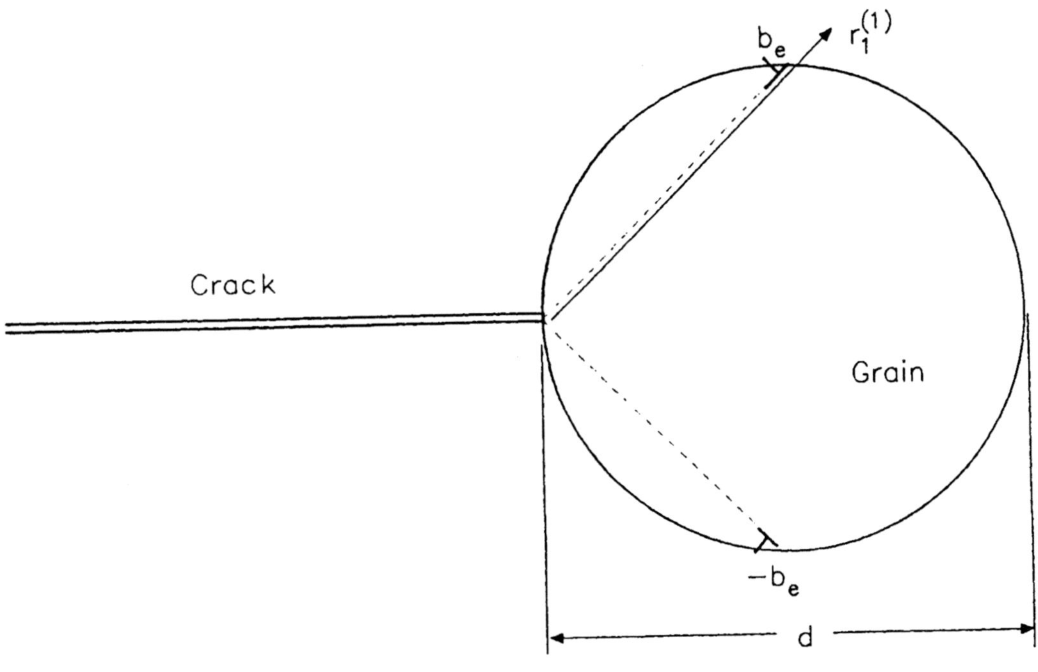

Figure 6.14. Amorphous state of polycrystalline materials

So, there may be an amorphous state in any polycrystalline material, if the size of clusters / grains is sufficiently small.

If the microstructure of a polycrystalline material satisfies the condition (6.92), no free edge dislocations can be born even by the most powerful source of dislocations, that is, the atomically sharp crack.

It is noteworthy that the superplastic and amorphous states of solid matter cannot be understood in the framework of classical continuum theories of elasticity, plasticity or fracture mechanics.

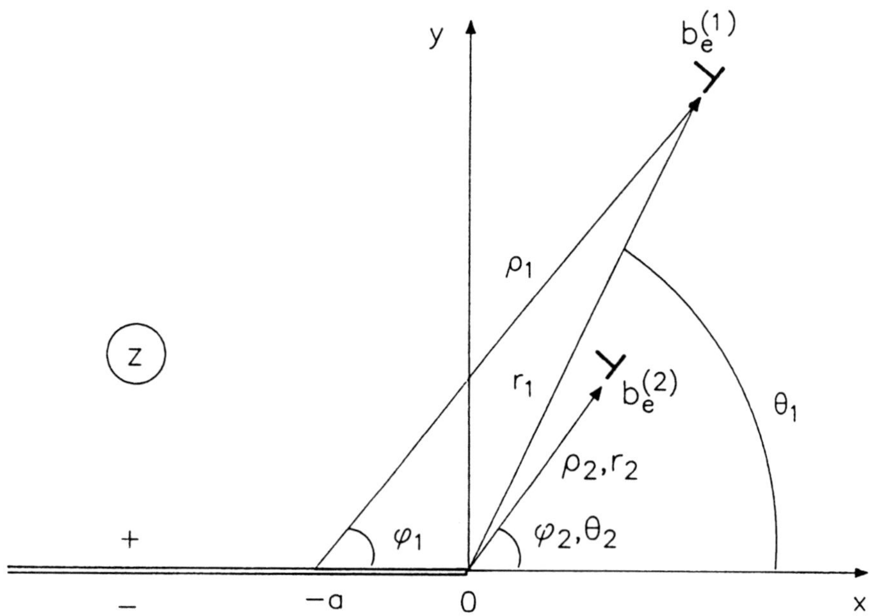

Figure 6.15. Two pairs of elementary edge dislocations in the neighborhood of a crack tip

6.9. Generation of the Second Pair of Edge Dislocations

Consider the process of monotonic loading of the cracked plane in a cubic lattice of atoms after the emission of the first pair of elementary edge dislocation (Figures 6.1 and 6.10). We introduce the second pair of (virtual) dislocations at the crack tip, and study the elastic field of the crack with one pair of actual dislocations and one pair of virtual dislocations (Figure 6.15). The coordinates of the first pair of dislocations are $z_1, \bar{z}_1$ and of the second pair $z_2, \bar{z}_2$, which are given by

$$z_1 = r_1 e^{i\theta_1} = -\alpha + \rho_1 e^{i\pi/4} \quad , \quad z_2 = r_2 e^{i\theta_2} = \rho_2 e^{i\pi/4} \tag{6.93}$$

Table 6.2 shows that $r_1^{(1)} >> a$ in ductile crystals, so that it can be assumed in this case that

$$r_1 \approx \rho_1 \, , \quad \theta_1 \approx \pi / 4 \tag{6.94}$$

while $r_2 \approx \rho_2 \, , \quad \theta_2 \approx \pi / 4 \, .$

The elastic field of these two pairs of dislocations at the point z is defined by the complex potentials $\Phi(z)$ and $\Psi(z)$,

$$\begin{aligned} \Phi(z) &= \Phi(z, z_1) + \Phi(z, z_2) \\ \Psi(z) &= \Psi(z, z_1) + \Psi(z, z_2) \end{aligned} \tag{6.95}$$

where $\Phi(z, z_k)$ and $\Psi(z, z_k)$ for $k = 1$ and 2, are given by (6.66) and (6.68), respectively.

The stress intensity factor induced by two pairs of dislocations can be calculated from (6.73) and (6.69) using the assumption (6.94),

$$k_{Id} = -1.96 \mu_0 a \left(\frac{1}{\pi r_1} + \frac{1}{\pi r_2} \right) \tag{6.96}$$

The regular component $\tau_{\rho\varphi}$ of the shear stress at the singular points on the slip plane can be calculated from (6.51), (6.66), (6.68), and (6.95),

$$\begin{aligned} \tau^0_{\rho\varphi} &= -\frac{\mu_0 b_e}{4\pi r_1} + \frac{\mu_0 b_e}{4\pi} f_1(r_1, r_2) \qquad as\ z \to z_1 \, , \\ \tau^0_{\rho\varphi} &= -\frac{\mu_0 b_e}{4\pi r_2} + \frac{\mu_0 b_e}{4\pi} f_1(r_2, r_1) \qquad as\ z \to z_2 \, , \end{aligned} \tag{6.97}$$

where the function $f_1(r_1, r_2)$ is given by

$$\begin{aligned} f_1(r_1, r_2) = \frac{1}{r_1 - r_2} &\left[1 + \frac{3}{4}\sqrt{\frac{r_2}{r_1}} + \frac{r_1 r_2 \left(1 - 2\sqrt{r_2 / r_1}\right) + r_2^2}{(r_1 - r_2)^2} \right] + \\ + \frac{1}{r_1^2 + r_2^2} &\left[\frac{3}{4}(r_1 - r_2)\sqrt{\frac{r_2}{2r_1}} - r_1 - r_2 \right] + \frac{r_2}{\left(r_1^2 + r_1^2\right)^2}\left[3r_1^2 + 2r_1 r_2 - r_2^2\right] \\ + \frac{1}{(r_1^2 + r_2^2)^3} &\left[2r_1 r_2^2\left(-3r_1^2 + r_2^2\right) + r_1 r_2 \sqrt{\frac{2r_2}{r_1}} \left(r_1^3 + 3r_1^2 r_2 - 3r_1 r_2^2 - r_1^3\right) \right] \end{aligned} \tag{6.98}$$

The $f_1(r_1,r_2)$ determines the interaction force of the dislocation at point $z_2 = r_2 e^{i\theta}$ on the dislocation at point $z_1 = r_1 e^{i\theta}$; it is shown in Figure 6.16 in the dimensionless coordinates,

$$\bar{r}_1 = r_1/\lambda_1 \quad \text{and} \quad \bar{r}_2 = r_2/\lambda_1, \text{ where } f_1(r_1,r_2) = \lambda_1 f_1(\bar{r}_1,\bar{r}_2).$$

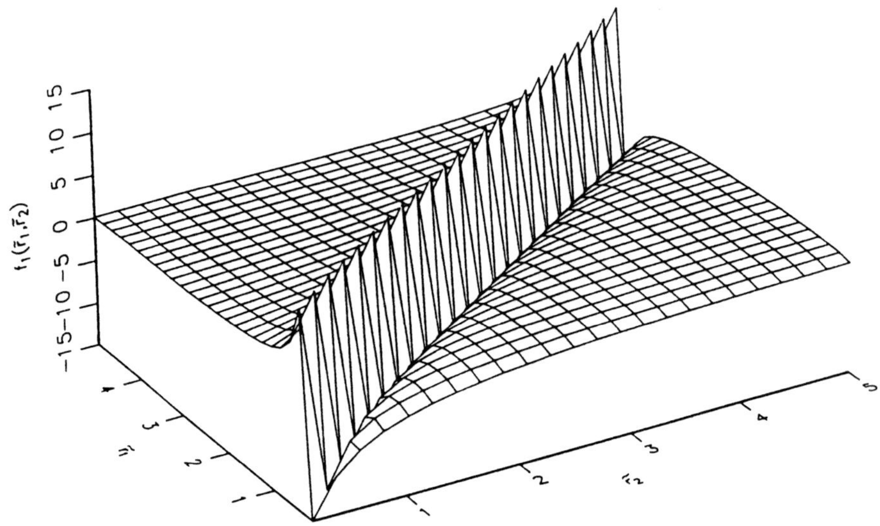

Figure 6.16. The interaction force of two pairs of edge dislocations near a crack tip

The function $f_1(r_1, r_2)$ can be written in the following form

$$f_1(r_1,r_2) = \frac{1}{r_1}\Lambda(\lambda) \quad \left(\lambda = \frac{r_2}{r_1}\right), \tag{6.99}$$

where

$$\Lambda(\lambda) = \frac{1}{1-\lambda}\left[1+\frac{3}{4}\sqrt{\lambda}+\frac{\lambda^2+\lambda\left(1-2\sqrt{\lambda}\right)}{(1-\lambda)^2}\right]+\frac{\lambda}{(1+\lambda^2)^2}(3+2\lambda-\lambda^2) \tag{6.100}$$

$$+\frac{1}{1+\lambda^2}\left[\frac{3}{4}(1-\lambda)\sqrt{\frac{\lambda}{2}}-1-\lambda\right]+\frac{1}{(1+\lambda^2)^3}\left[2\lambda^2(\lambda^2-3)+\lambda\sqrt{2\lambda}(1+3\lambda-3\lambda^2-\lambda^3)\right].$$

The first term in the right side of equation (6.97) for $\tau^0_{\rho\varphi}$ corresponds to the self-induction force, defined in Section 6.4, and the second term to the force of interaction between the first and second pairs of dislocations. The interaction force is the force of repulsion between the dislocation pairs of the same sign.

The equilibrium of a dislocation depends also on the external load K_I, and friction τ_0 on the slip plane. Taking into account the latter forces and Equation (6.97) we find the balance equations for two dislocation pairs:

$$\begin{aligned}\frac{0.327}{\sqrt{2\pi r_1}}K_I &= \tau_0 + \frac{\mu_0 b_e}{4\pi r_1} - \frac{\mu_0 b_e}{4\pi} f_1(r_1, r_2)\\ \frac{0.327}{\sqrt{2\pi r_2}}K_I &= \tau_0 + \frac{\mu_0 b_e}{4\pi r_2} - \frac{\mu_0 b_e}{4\pi} f_1(r_2, r_1)\end{aligned} \qquad (6.101)$$

The system of equations (6.101) determines a line $K_I\ (r_1,r_2)$ in the three-dimensional space (r_1, r_2, K_I). Figure (6.17) shows this line in the dimensionless coordinates:

$$\bar{r}_1 = \frac{4\pi\tau_0}{\mu_0 b_e} r_1, \qquad \bar{r}_2 = \frac{4\pi\tau_0}{\mu_0 b_e} r_2, \qquad \bar{K}_I = \frac{K_I}{\sqrt{2a\mu_0\tau_0}} \quad . \qquad (6.102)$$

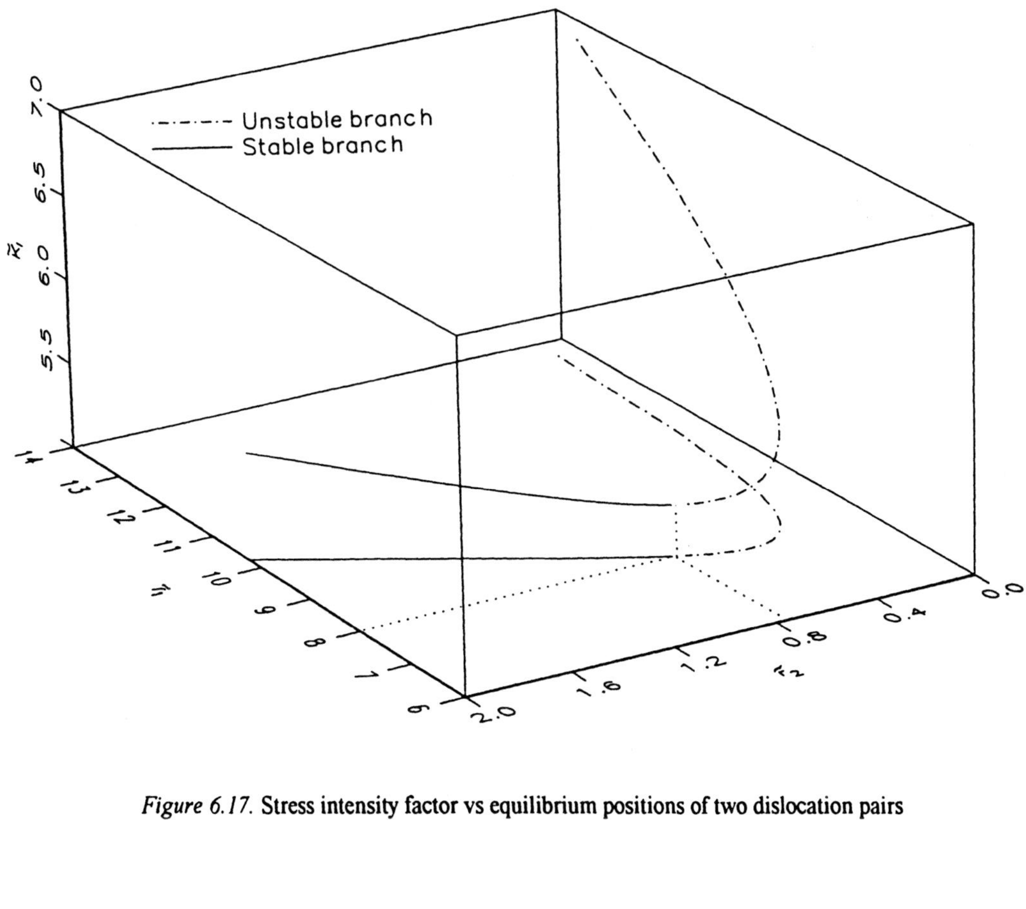

Figure 6.17. Stress intensity factor vs equilibrium positions of two dislocation pairs

Solution of the equation system (6.101) can be written as

$$\bar{r}_2 = \sqrt{\lambda} + \frac{\lambda}{1-\sqrt{\lambda}}\left[A(\lambda) - \frac{1}{\sqrt{\lambda}}A\left(\frac{1}{\lambda}\right)\right] \equiv A(\lambda) \tag{6.103}$$

$$\bar{K}_I = 1.82\sqrt{\frac{\lambda}{A(\lambda)}}\left[1 + \frac{1}{\lambda}A(\lambda) - A(\lambda)\right] \equiv B(\lambda) \tag{6.104}$$

The function $B(\lambda)$ has the minimum at point $\lambda = 0.095$, which corresponds to the values of the dimensionless stress intensity factor $\bar{K}_I^{(2)} = 5.297$ and dislocations positions, $\bar{r}_1^{(2)}=8.013$ and $\bar{r}_2^{(2)} = 0.758$, at this point. As $\lambda \to 0$ or $\lambda \to 1$, the $B(\lambda)$ tends to plus infinity. The minimum point separates the unstable and stable branches of the curve $B(\lambda)$.

Thus, the second pair of elementary edge dislocations is born at the external load

$$K_I = K_I^{(2)} = 5.297\sqrt{2a\mu_0\tau_0} \quad , \tag{6.105}$$

and the pairs of dislocations settle at the positions

$$r_1^{(2)} = 0.638\frac{\mu_0 b_e}{\tau_0}, \quad r_2^{(2)} = 0.06\frac{\mu_0 b_e}{\tau_0} \quad . \tag{6.106}$$

This solution complies with the assumption (6.94) that $r_1 \gg a$ and $r_2 \gg a$.

It is noted that the value of the stress intensity factor (6.105) corresponding to the emission of the second pair of dislocations is about 1.5 times greater than that for the first pair (6.85). The first pair of dislocations moves along its slip plane from the first stable position given by (6.86) to the second stable position (6.106). The second newborn pair of dislocations is closer to the crack tip than the first one was when it was born.

6.10. Emission of the *N*th Pair of Edge Dislocations

For three and more pairs of emitted dislocations, numerical methods are necessary which are treated in this section for any number of pairs.

Exact solution

The monotonic loading of a crack in the cubic lattice of atoms is accompanied by the emission of dislocations at certain critical values of K_I, and by crack jumps equal to the interatomic spacing. Consider the Nth step of loading, when $N - 1$ dislocation pairs have already been generated, and the crack has advanced a distance of $l_C = (N - 1)a$ (Figure 6.18). The problem is to determine the value of $K_I = K_I^{(N)}$ at which the crack advances by a further distance a.

According to the method of nanofracture mechanics, a virtual dislocation is introduced at the crack tip. Consider the elastic field created by N dislocation pairs, of which N - 1 already exist, and Nth is virtual (Figure 6.18). The coordinates of the dislocations in the complex plane are:

$$z_k = a_k + \rho_k e^{i\pi/4} = r_k e^{i\theta_k}, \quad a_k = -(k-1)a, \tag{6.107}$$

where a_k is the distance from the crack tip to the kth slip plane along the x axis, and ρ_k is the radius-vector of the kth dislocation in the polar coordinate system (ρ, φ) associated with the kth slip plane; r_k and θ_k are the coordinates of the kth dislocation in the polar coordinate system (r, θ) associated with the crack tip (Figure 6.18).

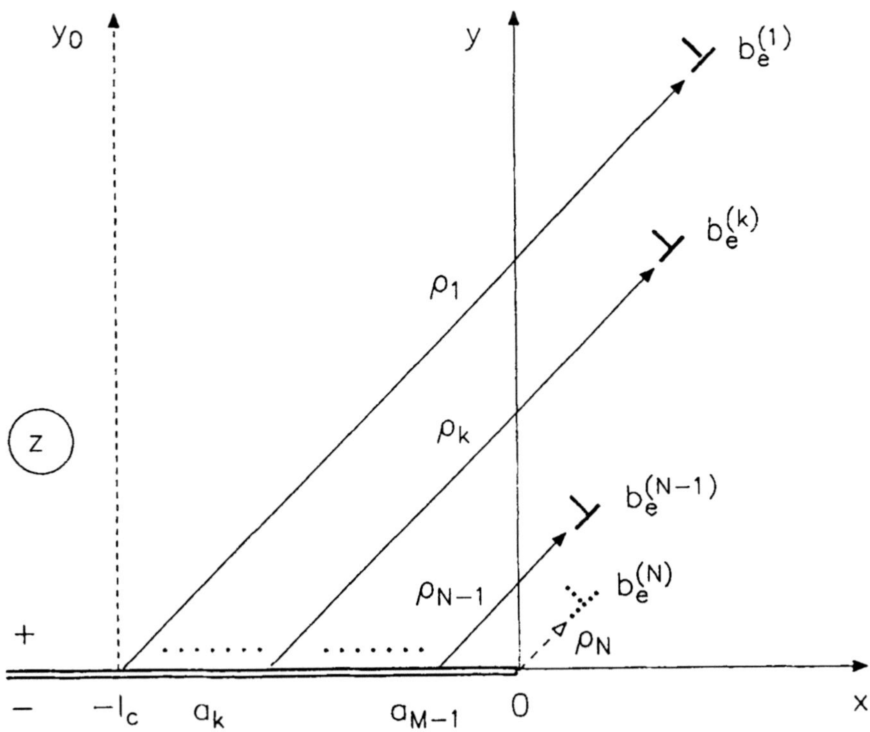

Figure 6.18. Emission of the Nth pair of edge dislocations

The potentials delivering the solution of the linear theory of elasticity for this problem are:

$$\Phi(z) = \sum_{k=1}^{N} \Phi(z, z_k), \quad \Psi(z) = \sum_{k=1}^{N} \Psi(z, z_k) \quad , \tag{6.108}$$

where $\Phi(z,z_k)$ and $\Psi(z,z_k)$ given by (6.66) and (6.68) yield the elastic field of a crack with single dislocation source at point $z = z_k$.

Consider the equilibrium of the kth dislocation on the slip plane $y = (k - 1)\,a + x$. This equilibrium depends on the external load K_I, friction on the slip plane τ_0, self-induction force and the force of interaction with other dislocations. The shear stress on the slip plane produced by the external field K_I is given by

$$\tau^e_{\rho\varphi}(z_k) = \frac{K_I}{2\sqrt{2\pi}} Im\left\{e^{2i\varphi_k}\,[\bar{z}_k \Phi'_e(z_k) + \Psi_e(z_k)]\right\}, \tag{6.109}$$

where the complex potentials $\Phi_e(z)$ and $\Psi_e(z)$ have the following form:

$$\Phi_e(z) = \frac{1}{\sqrt{z}} \quad , \quad \Psi_e(z) = \frac{1}{2\sqrt{z}}. \tag{6.110}$$

The shear stress component on the kth slip plane caused by self-induction force is equal to $\tau^0_{\rho\varphi}(z_k)$ as given by Equation (6.72).

The force of interaction with other dislocations is equal to $b_e \tau^i_{\rho\varphi}(z_k)$, where $\tau^i_{\rho\varphi}(z_k)$ is the component of shear stress induced by dislocations other than the kth dislocation on the kth slip plane. The value of the component $\tau^i_{\rho\varphi}(z_k)$ is given by

$$\tau^i_{\rho\varphi}(z_k) = \sum_{m=1,\,m\neq k}^{N} \tau^i_{\rho\varphi}(z_k, z_m), \tag{6.111}$$

where

$$\tau^i_{\rho\varphi}(z_k, z_m) = \mathrm{Im}\left\{e^{2i\varphi_k}\left[\bar{z}_k \frac{\partial\Phi(z_k, z_m)}{\partial z_k} + \Psi(z_k, z_m)\right]\right\}. \tag{6.112}$$

is the shear stress component produced by the mth dislocation on the kth dislocation position.

The equilibrium equation for the kth dislocation can be written as

$$\tau^e_{\rho\varphi}(z_k) + \tau^0_{\rho\varphi}(z_k) + \sum_{m=1,\,m\neq k}^{N} \tau^i_{\rho\varphi}(z_k, z_m) = \tau_0\,. \tag{6.113}$$

In general, the equilibrium of N dislocations in the cubic lattice of atoms with crack under plane strain conditions is governed by the following system of equations:

$$f_{ext}(z_k) + f_{self}(z_k) + \sum_{m=1,\,m\neq k}^{N} f_{\mathrm{int}}(z_k, z_m) = \tau_0\,, \tag{6.114}$$

where

$$f_{ext}(z_k)=\tau^{e}_{\rho\varphi}(z_k)=\frac{K_I}{2\sqrt{2\pi}}f^{\bullet}_{ext}(z_k),\quad \kappa=1,\ldots,N,$$

$$f_{self}(z_k)=\tau^{0}_{\rho\varphi}(z_k)=\frac{\mu_0 b_e}{4\pi\sqrt{2}}f^{\bullet}_{self}(z_k), \tag{6.115}$$

$$f_{int}(z_k,z_m)=\tau^{'}_{\rho\varphi}(z_k,z_m)=\frac{\mu_0 b_e}{4\pi\sqrt{2}}f^{\bullet}_{int}(z_k,z_m).$$

To determine the value of external load $K_I = K_I^{(N)}$ at which the Nth pair of dislocations is emitted and the crack tip advances the Nth interatomic spacing, it is necessary to find the minimum value of K_I for which the solution of the system of equilibrium equations (6.114) exists.

Let us formulate the problem in terms of dimensionless quantities. The dimensionless complex coordinate is introduced,

$$\widetilde{z}=\frac{z}{L}, \tag{6.116}$$

where the length dimension parameter L is given by

$$L=\frac{\mu_0 b_e}{4\pi\tau_0\sqrt{2}}. \tag{6.117}$$

The $f^{\bullet}_{ext}(z)$, $f^{\bullet}_{self}(z)$ and $f^{\bullet}_{int}(z,z_0)$ have the following properties

$$f^{\bullet}_{ext}(z)=\frac{1}{\sqrt{L}}f^{\bullet}_{ext}(\widetilde{z}),\quad f^{\bullet}_{self}(z)=\frac{1}{L}f^{\bullet}_{self}(\widetilde{z}),\quad f^{\bullet}_{int}(z,z_0)=\frac{1}{L}f^{\bullet}_{int}(\widetilde{z},\widetilde{z}_0). \tag{6.118}$$

The system of equilibrium equations (6.114) in dimensionless form is given by

$$\widetilde{K}_I f^{\bullet}_{ext}(\widetilde{z}_k)+f^{\bullet}_{self}(\widetilde{z}_k)+\sum_{m=1,m\neq k}^{N}f^{\bullet}_{int}(\widetilde{z}_k,\widetilde{z}_m)=1$$
$$k=1,\ldots,N \tag{6.119}$$

The system of N non-linear equations (6.119) is used in numerical experiments on dislocation emission and crack growth under monotonic loading. The solution of (6.119) depends on $\widetilde{K}_I$ and is denoted by

$$\widetilde{z}_k^{(N)}=\widetilde{z}_k^{(N)}(\widetilde{K}_I^{(N)}) \tag{6.120}$$

It should be noted that

$$\tilde{z}_k^{(N-1)}(\tilde{K}_I^{(N)}) \neq \tilde{z}_k^{(N)}(\tilde{K}_I^{(N)}), \quad 1 \le k \le N-1 \tag{6.121}$$

i.e. the positions of dislocations experience a discontinuous growth each time a dislocation pair is emitted. Some results of accurate solution of equation system (6.119) up to $N=150$ are given in Tables 6.3 and 6.4 in Section 6.12 below.

Approximate solution

Exact analysis of the Nth step of monotonic loading of a crack requires the solution of system of N non-linear equations (6.119). Experiments and theoretical estimates[22,23] indicate that the number of generated dislocation pairs can reach 10^5 for such common crystals as aluminum, copper or nickel. This means that a system of several thousands non-linear equations must be solved several times to find the minimum K_I.

To reduce the order of equation system (6.119) the following approximate method of self-consistent field was suggested in Reference[23]. Suppose the exact solution is known for N_0 dislocation pairs. The method of self-consistent field can be employed for any $N>N_0$.

Consider each of N_0 dislocations not as a single dislocation but as a cluster of dislocations or dislocation band with effective Burgers vector $b_e^{(k)}, k = 1,\dots,N_0$. The system of equilibrium equations (6.119) transforms to

$$\tilde{K}_I f_{ext}^{\bullet}(\tilde{z}_k) + b_e^{(k)} f_{self}^{\bullet}(\tilde{z}_k) + \sum_{m=1, m\neq k}^{N} b_e^{(m)} f_{int}^{\bullet}(\tilde{z}_k, \tilde{z}_m) = 1, \quad k = 1,\dots,N_0 \tag{6.122}$$

Different distributions of dislocations over N_0 clusters can be studied. One possible distribution of dislocations over clusters involves combining all existing dislocations into identical clusters and considering a virtual dislocation separately. At the $(N+1)$th step of loading there will be N_0 clusters, each with the Burgers vector $b_e^{(k)} = b_e N / N_0$, for $k = 1,\dots,N_0$ and a virtual dislocation with the Burgers vector $b_e^{(N+1)} = b_e$. In this case there are N_0 equilibrium equations in the equation system (6.122). This system should be solved on each step to find the stable positions of clusters and the virtual dislocation. The first approximation corresponds to the case when all existing dislocations are combined into a single dislocation cluster[23]. The distribution of dislocations described above is used in the numerical experiments that follow. Some results of independent numerical experiments by two different methods are given in Section 6.12 below.

Brittle vs. ductile mechanism of crack growth

The crack tip can advance not only by the ductile mechanism with emission of dislocations, but also by a brittle mechanism. If the brittle mechanism is realized the atomic bonding ahead of the crack tip is ruptured without emission of dislocations. It occurs when the superfine scale stress intensity factor k_I at the crack tip exceeds the critical value k_{Ic}, corresponding to the reversible surface energy of crack growth (see Chapter 1).

The superfine scale stress intensity factor k_I is the total stress intensity factor at the crack tip, which includes the stress intensity factors due to the external load K_I and due to the elastic field of dislocation sources k_{Id} as given by (6.73). Equation (6.73) shows that the generation of dislocations leads to relaxation of the stresses near the crack tip, so that $k_I < K_I$. The value of the superfine scale stress intensity factor can be calculated as

$$k_I = K_I + k_{Id}. \tag{6.123}$$

The brittle crack grows if:

$$k_I = K_I + k_{Id} \geq k_{Ic}. \tag{6.124}$$

Once the positions of all dislocation pairs are known the dimensionless superfine scale stress intensity factor, $\widetilde{k}_I$, can be calculated as follows

$$\widetilde{k}_I = \widetilde{K}_I + \sum_{k=1}^{N} \widetilde{k}_{Id}(\widetilde{z}_k) \tag{6.125}$$

where the dimensionless stress intensity factor of the kth dislocation source, $\widetilde{k}_{Id}(\widetilde{z}_k)$ is :

$$\widetilde{k}_{Id}(\widetilde{z}_k) = \frac{k_{Id}(z_k)}{\sqrt{2a\mu_0\tau_0}}, \tag{6.126}$$

and $k_{Id}(z_k)$ is given by (6.73).

When the (N-1)th dislocation is generated, the value of $\widetilde{k}_I$ grows with increasing K_I until the next dislocation pair is born. At this moment the dislocations jump to new stable positions as shown in Figure 6.19, and the value of $\widetilde{k}_I$ drops significantly, so that $\widetilde{k}_I^{(N-1)}(K_I^{(N)}) > \widetilde{k}_I^{(N)}(K_I^{(N)})$. Denote the upper bound of $\widetilde{k}_I$ at $\widetilde{K}_I = \widetilde{K}_I^{(N)}$ as $\widetilde{k}_{I\,\mathrm{sup}}$:

$$\widetilde{k}_{I\,\mathrm{sup}}(N) = \widetilde{k}_I^{(N-1)}(\widetilde{K}_I^{(N)}) \tag{6.127}$$

and the lower bound of $\widetilde{k}_I$ at $\widetilde{K}_I = \widetilde{K}_I^{(N)}$ as $\widetilde{k}_{I\mathrm{inf}}$:

$$\widetilde{k}_{I\,\mathrm{inf}}(N) = \widetilde{k}_I^{(N)}(\widetilde{K}_I^{(N)}) \tag{6.128}$$

The value of the superfine scale stress intensity factor $\widetilde{k}_I$ oscillates between two bounds $\widetilde{k}_{I\,\mathrm{inf}}$ and $\widetilde{k}_{I\mathrm{sup}}$ during monotonic loading when the ductile mechanism is realized as shown in Figure 6.20.

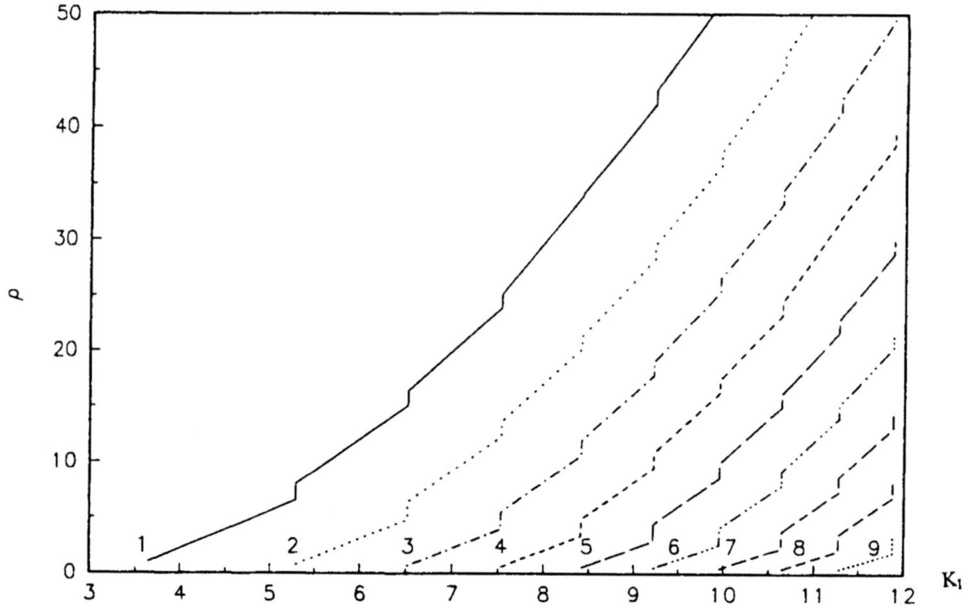

Figure 6.19. Dislocations coordinates vs K_I during monotonic loading

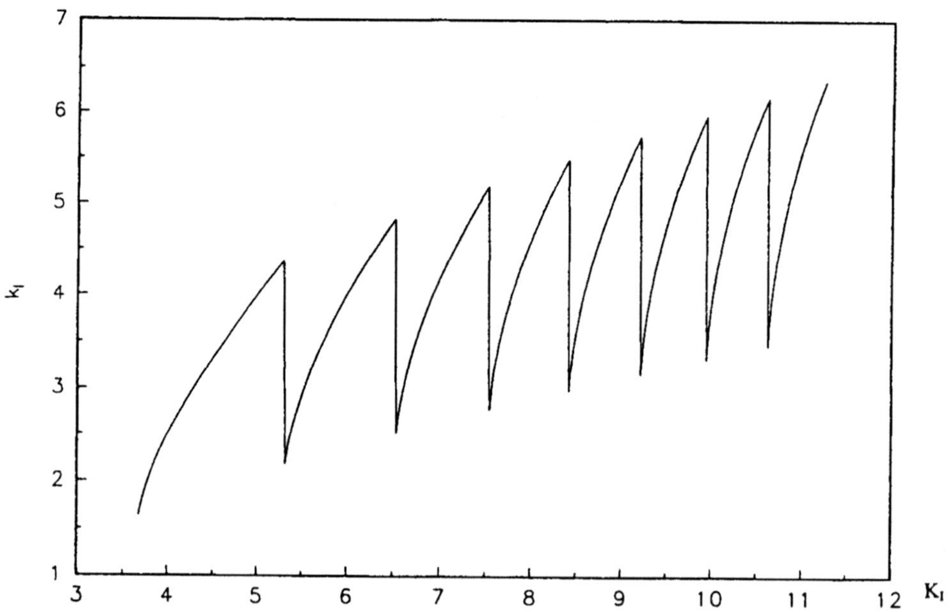

Figure 6.20. Superfine scale stress intensity factor, k_I, vs external field stress intensity factor, K_I

The values of the lower and upper bounds of the superfine scale stress intensity factor increase monotonically with increase in the external load.

At a certain value of $\widetilde{K}_I$ the superfine scale stress intensity factor $\widetilde{k}_I$ reaches the critical value $\widetilde{k}_{Ic}$, which is the dimensionless critical stress intensity factor at the crack tip

$$\widetilde{k}_{Ic} = \frac{k_{Ic}}{\sqrt{2a\mu_0\tau_0}} \tag{6.129}$$

and the crack tip advances by the brittle mechanism, i.e. an atomic bonding ahead of the crack tip is ruptured but no dislocation is generated. In this case the radius-vectors of dislocation do not change but the slip planes shift backwards relative to the crack tip, and the complex coordinates of dislocations $\widetilde{z}_k$ must be recalculated.

As a result of brittle crack tip advancement, the superfine scale stress intensity factor can either decrease or increase depending on the distribution of dislocation sources. In the beginning of the brittle crack growth the value of $\widetilde{k}_I$ will decrease because the radius-vectors of dislocation $\rho_k >> l_C$ and the crack tip will move closer to the dislocation sources. The subsequent increase in external load $\widetilde{K}_I$ is necessary to make the crack tip advance.

At a certain stage the crack tip advances a distance which suffices for the crack tip to be driven forward by the external elastic field and the brittle crack grows unstable, i.e. the external load decreases with crack growth.

6.11. Numerical Experiments

Numerical modeling involves analysis of the crack tip advancement under monotonic loading. Both ductile and brittle mechanisms of crack growth must be taken into account. In nanofracture mechanics, it is assumed that a pair of dislocations is emitted at the crack tip at the minimum external load K_I for which this pair of dislocations can find a stable position. Two numerical methods are examined independently to calculate the minimum value of K_I, namely, the iterative and minimization methods. The iterative method computes the minimum value using a step-by-step searching algorithm, while the minimization method makes use of a minimization technique, which is popular in optimal design.

Iterative method

The iterative method divides the process of crack growth into two stages. First the ductile crack growth mechanism is considered, and dislocation pairs are emitted simultaneously with the crack tip advancement. When the superfine scale stress intensity factor $\widetilde{k}_I$ reaches the critical value $\widetilde{k}_{Ic}$ only the brittle mechanism is taken into account and the positions of dislocation pairs are assumed to be "frozen" or fixed.

On the first stage the system of equilibrium equations (6.119) is solved using the iterative algorithm described below. The value of $\widetilde{K}_I$ is taken as a parameter. A virtual dislocation introduced at the crack tip may or may not have an equilibrium position depending on the value of the parameter $\widetilde{K}_I$ and the positions of other dislocations.

The main procedure of this algorithm calculates the equilibrium position of each dislocation, one by one, until all equations (6.119) are satisfied with prescribed accuracy. One cycle of the iteration process can be described as follows.

Suppose the solution is known for a certain (N -1) number of dislocations at a certain level of external load $\widetilde{K}_I$. The value of $\widetilde{K}_I$ is incremented by a small value $\Delta\widetilde{K}_I$. The system (6.119) is analyzed for N dislocations including the virtual dislocation in the dislocation-free zone. If the solution does not exist, the positions of the existing N -1 dislocations are adjusted to the new, greater load by solving system (6.119) for N-1 dislocations with the new $\widetilde{K}_I = \widetilde{K}_I + \Delta\widetilde{K}_I$. The process of incrementing the external load $\widetilde{K}_I$ is repeated until the equilibrium position of the virtual dislocation is found.

The superfine scale stress intensity factor $\widetilde{k}_I$ is calculated on each step of iteration. When $\widetilde{k}_I$ reaches the critical value $\widetilde{k}_{Ic}$ the brittle mechanism of crack growth is considered. The crack tip is assumed to advance without emanating dislocation pairs, and the positions of dislocations are considered to be "frozen" on their slip planes, i.e. they do not move with change in the external load. But the coordinates of dislocations should be recalculated with the crack tip advancement. Accordingly, the superfine scale stress intensity factor changes its value, and the corresponding value of the external load $\widetilde{K}_I$ is calculated which provides further crack growth by brittle mechanism.

This algorithm simulates the monotonic loading directly and allows to observe the change in the value of $\widetilde{k}_I$ at the time between the moments of dislocation emission. However, it is less effective for determining the values of $\widetilde{K}_I$ at which the emission of a dislocation pair occurs, since the increment in the external load $\widetilde{K}_I$ is of the same order as error in the algorithm.

Minimization method

The minimization method provides the solution of the system of equilibrium equations (6.119) in a different way. In this approach both the ductile and brittle mechanisms of crack growth are taken into account, so that it is free of the assumption of "frozen" dislocations in the iterative method. At each step the condition is checked to determine which one of the two mechanisms is realized.

Suppose the positions of the existing N-1 dislocations are known. First the ductile mechanism is examined. A virtual dislocation is introduced at the crack tip and the system of N equilibrium equations (6.119) is considered. The equilibrium equations of dislocation pairs define $\widetilde{K}_I$ as a function of dislocation coordinates. This function is presented as a line in the (N + 1) - dimensional space (for N = 1 and 2 see Figures 6.11 and 6.17, respectively). The minimum point of this line delivers the solution of the problem, i.e. the level of external load at which the new dislocation is generated, and the stable position where the dislocation is settled.

Similarly, the system of equations (6.119) defines the stress intensity factor $\widetilde{K}_I$ as a function of dislocation coordinates $(\widetilde{z}_1,....,\widetilde{z}_N)$, and can be rewritten in the following form:

$$\widetilde{K}_I = \left[1 - f^{\bullet}_{self}(\widetilde{z}_N) - \sum_{m=1}^{N-1} f^{\bullet}_{int}(\widetilde{z}_N, \widetilde{z}_m)\right] \frac{1}{f^{\bullet}_{ext}(\widetilde{z}_N)} \tag{6.130}$$

with the additional constrains:

$$\begin{gathered} f^{\bullet}_{ext}(\widetilde{z}_2)\left[1 - f^{\bullet}_{self}(\widetilde{z}_1) - \sum_{m=2}^{N} f^{\bullet}_{int}(\widetilde{z}_1, \widetilde{z}_m)\right] = f^{\bullet}_{ext}(\widetilde{z}_1)\left[1 - f^{\bullet}_{self}(\widetilde{z}_2) - \sum_{\substack{m=1\\ m\neq 2}}^{N} f^{\bullet}_{int}(\widetilde{z}_2, \widetilde{z}_m)\right], \\ \ldots \\ f^{\bullet}_{ext}(\widetilde{z}_N)\left[1 - f^{\bullet}_{self}(\widetilde{z}_{N-1}) - \sum_{\substack{m=1\\ m\neq N-1}}^{N} f^{\bullet}_{int}(\widetilde{z}_{N-1}, \widetilde{z}_m)\right] = f^{\bullet}_{ext}(\widetilde{z}_{N-1})\left[1 - f^{\bullet}_{self}(\widetilde{z}_N) - \sum_{m=1}^{N-1} f^{\bullet}_{int}(\widetilde{z}_N, \widetilde{z}_m)\right] \end{gathered} \tag{6.131}$$

Equation (6.130) defines a line $\widetilde{K}_I(\widetilde{z}_1,....,\widetilde{z}_N)$ in the $(N + 1)$ - dimensional space. The minimum point of this line yields the stable positions of N dislocation pairs and the value of external load $\widetilde{K}_I$ at which the Nth dislocation pair is emitted.

The following algorithm is suggested to find the minimum of $\widetilde{K}_I(\widetilde{z}_1,...,\widetilde{z}_N)$. The position of the virtual dislocation $\widetilde{z}_N$ is taken as the variable of the minimization problem. The property that the radius-vector of each newborn dislocation is less than the radius-vector of the dislocation emitted at the previous step $\left|\widetilde{z}_N^{(N)}\right| < \left|\widetilde{z}_{N-1}^{(N-1)}\right|$ is used to define the range for the minimization variable. The minimization problem is stated as follows

$$\begin{gathered} \widetilde{K}_I^{(N)} = \min \widetilde{K}_I(\widetilde{z}_1,....,\widetilde{z}_N) \\ 0 \leq \left|\widetilde{z}_N^{(N)}\right| < \left|\widetilde{z}_{N-1}^{(N-1)}\right| \end{gathered} \tag{6.132}$$

with the constraints given by (6.131). The additional condition

$$\rho_k^{(N-1)} < \rho_k^{(N)}, \quad k = 1,...,N-1 \tag{6.133}$$

is imposed on the radius-vectors of existing dislocations to ensure that the existing dislocations move either away from the crack tip or remain "frozen" at the previous position under loading.

The reverse motion of dislocations during unloading can be described by a system of equilibrium equations (6.117) with the opposite direction of the friction force, i.e. the reverse sign of τ_0.

The problem (6.119) is solved using the Brent minimization algorithm. A value of ρ_N is chosen from the range $0 < \rho_N < \rho_{N-1}^{(N-1)}$, the system of equations (6.131) is solved using the Newton-Raphson method and the computed coordinates of dislocations are used in (6.130) to evaluate $\widetilde{K}_I$. Each time the system of non-linear equations (6.131) is solved, the previous solution is taken as the first approximation. Then for these positions of dislocations, the superfine stress intensity factor $\widetilde{k}_I$ is calculated and the condition $\widetilde{k}_I < \widetilde{k}_{Ic}$ is checked. If this condition is satisfied, the ductile solution is retained, the dislocations settle at the computed stable positions and the crack advances one interatomic spacing. Otherwise, the crack advances by the brittle mechanism, when one interatomic bond ahead of the crack tip is ruptured, no dislocation is generated and the existing dislocations remain at previous positions. This method requires fewer iterations than the iterative method to reach the same level of accuracy.

6.12. Some Results of Numerical Experiments

Numerical modeling of crack growth under monotonic loading involves determination of the critical values of external load causing dislocation emission and crack advancement. The corresponding stable positions of dislocations are also calculated. The other objective of the numerical procedure is to construct the fracture curve or R-curve, $K_I(l_c)$. i.e. to find K_I necessary to advance the crack tip to length l_C. The fracture curve characterizes the resistance of an elastoplastic material to crack growth. It can be observed experimentally. Nanofracture mechanics is employed to determine this curve theoretically.

The numerical procedure involves the solution of the system of non-linear equations (6.119) by either of the two approaches described in Section 6.11. The procedure provides $K_I^{(n)}$ and $r_k^{(n)}$ for $k = 1,2,\ldots,n$ and $n = 1,2,\ldots$ Once the coordinates of dislocations are calculated, the superfine scale stress intensity factor is evaluated. Then the criterion of brittle fracture $k_I = k_{Ic}$ is checked and the mode of fracture is determined as either ductile or brittle. The upper bound of the superfine scale stress intensity factor $k_{I\,\sup}$ must be used in this criterion. In other words, the condition (6.124) must be checked a moment before a dislocation pair is emitted. After deciding which mechanism of crack growth, ductile or brittle, is realized, we allow the crack tip to advance one interatomic spacing. Then the coordinates of dislocations are recalculated and the system of equations (6.119) is solved again.

Numerical experiments are conducted for a crack in an aluminum specimen. The following elastic and physical properties of aluminum are used in the calculations: $\mu = 25GPa$, $\nu = 0.35$, $\tau_0 = 0.75MPa$, $\gamma = 0.84Pam$, $a = 2.85 \cdot 10^{-10}\ m$, where γ is the surface energy of crystal lattice. In the numerical experiments on crack growth only two dimensionless constants are employed, viz the dimensionless interatomic spacing

$$\widetilde{a} = \frac{a}{L} = 2.45 \cdot 10^{-4}, \tag{6.134}$$

and the dimensionless critical value of the stress intensity factor

$$\tilde{k}_{Ic} = \frac{k_{Ic}}{\sqrt{2a\mu_0\tau_0}} = \sqrt{\frac{2\gamma}{a\tau_0}} = 88.655$$

corresponding to the threshold when the ductile mode of fracture changes to the brittle mode.

First the approximate solutions are studied for the first, second and higher orders of approximation using the method of self-consistent field suggested in Reference[23] and described in Section 6.10. Then the exact solution is given by using both iterative and minimization methods.

First approximation

At the first approximation all generated dislocations are combined into a single cluster and the equilibrium of a virtual dislocation is studied. In this case $N_0 = 1$ and the system of non-linear equations (6.122) consists of two equilibrium equations, one for the dislocation cluster and the other for the virtual dislocation.

Consider the Nth step of loading, when there are $N - 1$ actual dislocations. The Burgers vector of the dislocation cluster is equal to $b_e^{(1)} = (N-1)b_e$, and the Burgers vector of the virtual dislocation is $b_e^{(2)} = b_e$. Numerical simulation takes into account both ductile and brittle crack growth mechanisms.

Initially the crack grows by the ductile mechanism with emission of dislocations from the crack tip. At this stage the value of K_I providing for the crack growth increases monotonically. The ductile section of the fracture curve is shown separately in Figure 6.21. The number of dislocations emitted at this stage equals $N_d = 4582$. The brittle crack growth initiates at $K_I/k_{Ic} = 2.78$. It is observed that in this approximation the ductile crack growth does not resume once the brittle crack growth mode has started.

The fracture curve is shown in Figures 6.22 and 6.23 in two different scales. The dimensionless quantity ξ is defined as

$$\xi = \begin{cases} (l_C - aN_d)/aN_d & \text{if} \quad 0 \le l_C < aN_d\,, \\ (l_C - aN_d)/L & \text{if} \quad l_C \ge aN_d\,, \end{cases}$$

so that the ductile section of the fracture curve corresponds to the interval $-1 \le \xi < 0$, and the brittle section to the interval $\xi \ge 0$.

Figure 6.22 shows the fracture curve in the interval $-1 \le \xi \le 3$, and Figure 6.23 in the interval $-1 \le \xi \le 20$.

It is observed that the K_I providing crack growth continues to increase after the brittle fracture mode initiates for $0 \le \xi < 0.312$ from $K_I = 2.78\, k_{IC}$ to $K_I = 2.86\, k_{IC}$, so that the stable brittle crack growth is realized in the interval $0 \le \xi < 0.312$. The stable brittle crack growth can be explained by the interaction of the crack tip with the dislocation

dislocation cluster. When the brittle crack growth starts, the radius-vector of the cluster is equal to $\tilde{\rho}_1^{(1)} = 4{,}582$. Thus, $\tilde{\rho}_1^{(1)} >> \tilde{a}$ and the dislocation source is located ahead of the crack tip. The elastic field of the dislocation cluster hinders crack growth until the crack tip advances the distance $l_C = \rho_1^{(1)} \cos\varphi_1$, which corresponds to the maximum point $\xi = 0.312$.

On the other hand, for $l_C > \rho_1^{(1)} \cos\varphi_1$ the elastic field of the dislocation source drives the crack tip forward, and the unstable crack growth occurs. The value of K_I decreases and reaches its minimum $K_I = 0.88$ k_{Ic} at point $\xi = 2.276$. The elastic field of dislocation cluster becomes negligible at $\xi > 10$, and K_I tends to k_{Ic}.

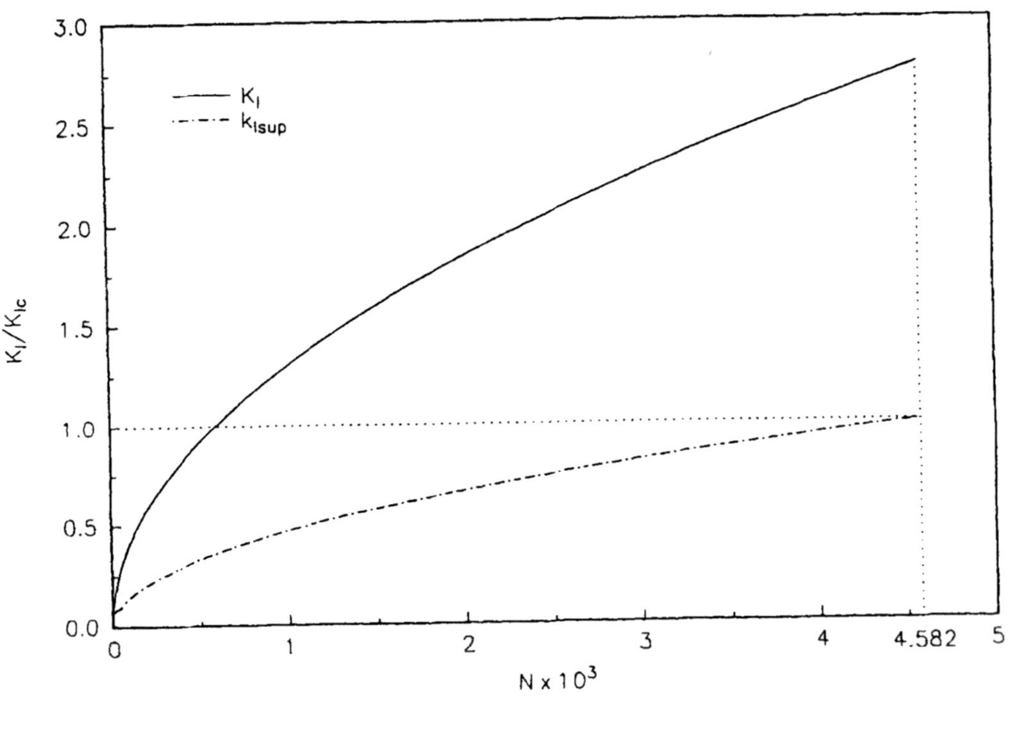

Figure 6.21. Quantum levels of K_I and the superfine scale stress intensity factor $k_{I\,sup}$ during the ductile crack growth (first approximation of the method of self-consistent field)

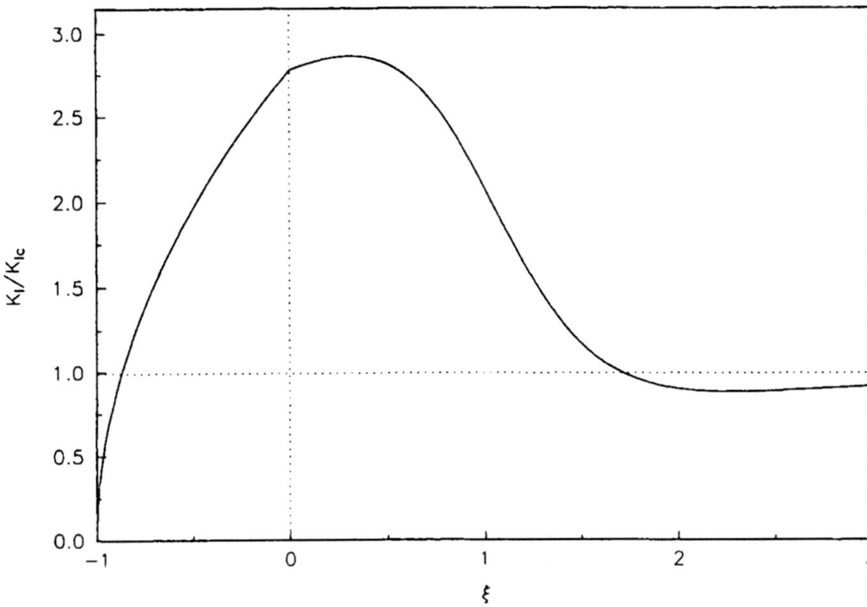

Figure 6.22. Fracture curve for $-1< \xi \leq 3$ (first approximation of the method of self-consistent field)

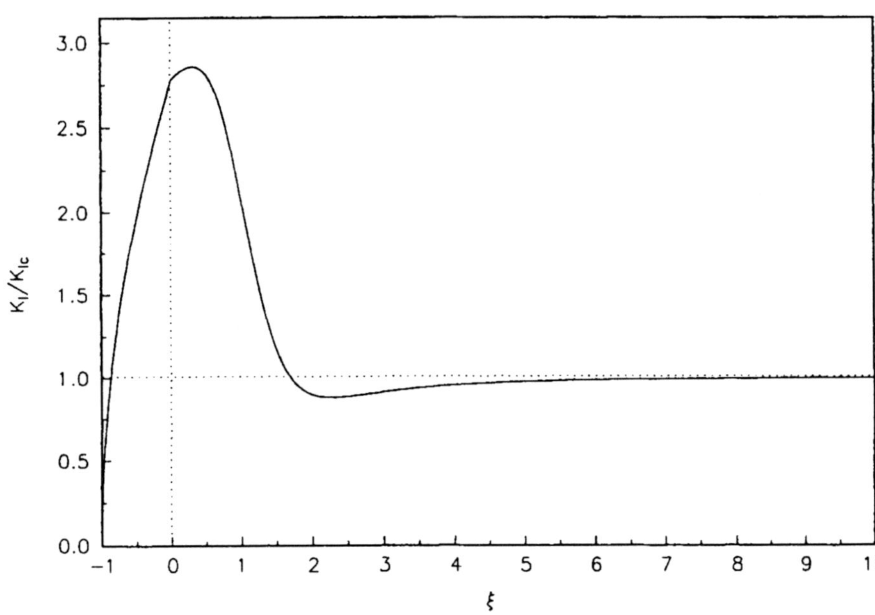

Figure 6.23. Fracture curve for $-1< \xi \leq 20$ (first approximation of the method of self-consistent field)

Second approximation

There are two dislocation clusters and a virtual dislocation in second approximation. The generated dislocations are distributed evenly over the two clusters, so that there are $N/2$ dislocations in each cluster. In this case the system of equilibrium equations consists of three equations. The system of equations (6.122) is solved by using the minimization method, taking into account both ductile and brittle crack growth mechanisms.

The ductile section of the fracture curve is shown separately in Figure 6.24. The value of K_I increases monotonically with crack growth until the brittle mode of fracture is initiated. The ductile mode does not resume after the brittle mode starts. The number of dislocations generated at this stage is $N_d = 4{,}867$, which is 6% more than in the first approximation.

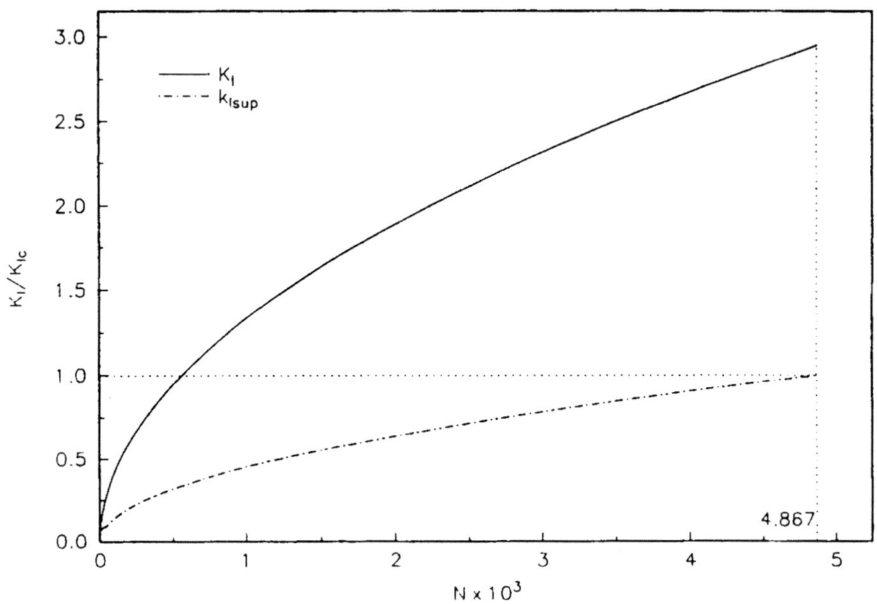

Figure 6.24. Quantum levels of K_I and the superfine scale stress intensity factor $k_{I\,sup}$ during the ductile crack growth (second approximation of the method of self- consistent field)

The fracture curve obtained in the second approximation is shown for $-1 \le \xi \le 3$ in Figure 6.25 and for $-1 \le \xi \le 20$ in Figure 6.26. The brittle crack growth starts at the external load corresponding to $K_{Ib} = 2.947\ k_{Ic}$. Then the stable crack grows by the brittle mechanism in the interval $0 \le \xi \le 0.118$. At $\xi_m = 0.118$ the value of K_I providing crack growth reaches its maximum $K_{Ic} = 3.019\ k_{Ic}$.

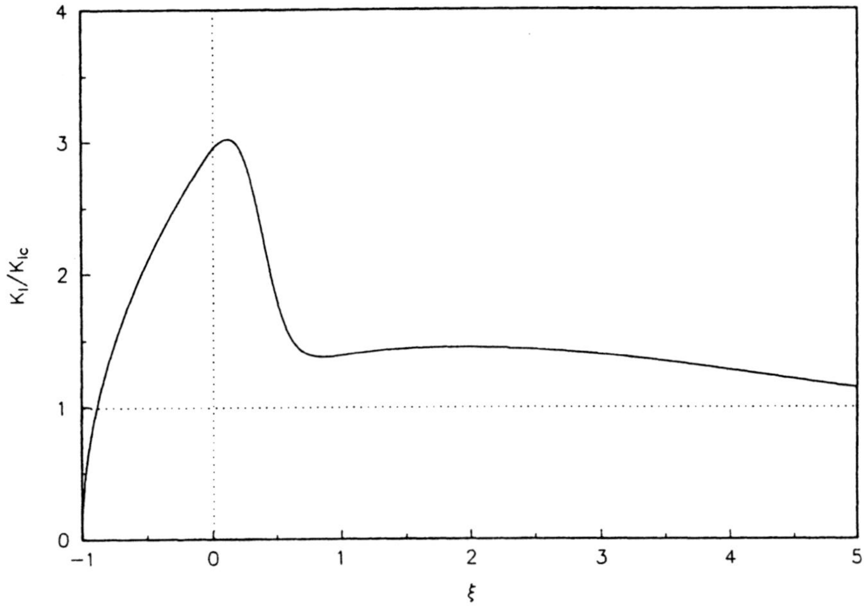

Figure 6.25. Fracture curve for $-1< \xi \leq 5$ (second approximation of the method of self-consistent field)

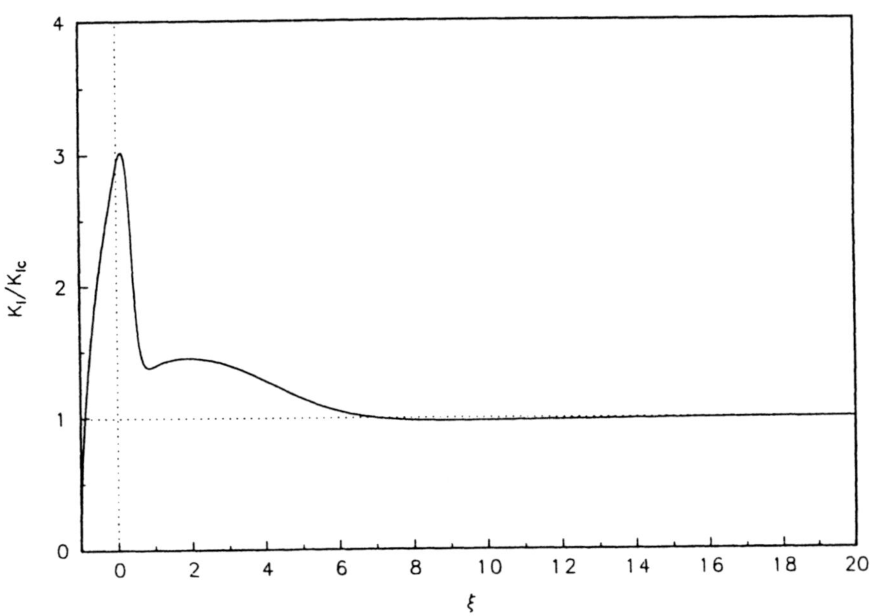

Figure 6.26. Fracture curve for $-1< \xi \leq 20$ (second approximation of the method of self-consistent field)

Between the point $\xi_m = 0.118$ and $\xi_n = 0.854$ the crack tip advances by the brittle mechanism, and the value of K_I decreases from $3.019k_{Ic}$ to $1.376k_{Ic}$. This unstable branch of the fracture curve is explained by the interaction with the second cluster of dislocations, nearest to the crack tip. The radius-vector of this dislocation cluster is equal to $\rho_2^{(2)} = 1845$. This value is less than $\rho_1^{(1)}$, so that the x coordinate of the minimum point is less than that in the first approximation.

The radius-vector of the first cluster of dislocations is equal to $\rho_1^{(2)} = 19{,}497$. The elastic field of this dislocation source provides the second stable section of the fracture curve between the points $\xi_n = 0.854$ and $\xi_{m2} = 1.90$ where K_I increases from $1.376k_{Ic}$ to $1.449\ k_{Ic}$. Then the value of K_I decreases due to the influence of the two dislocation clusters and reaches its second minimum point $K_{In2} = 0.969k_{Ic}$ at the point $\xi_{n2} = 9.104$. After that K_I increases slowly to k_{Ic} with further crack growth as the influence of dislocation clusters diminishes.

Higher order approximations

Consider the third and higher orders of approximation. Actual dislocations are combined into N_0 clusters for the N_0th order of approximation. The minimization method is employed to solve the system of $N_0 + 1$ equilibrium equations (6.122).

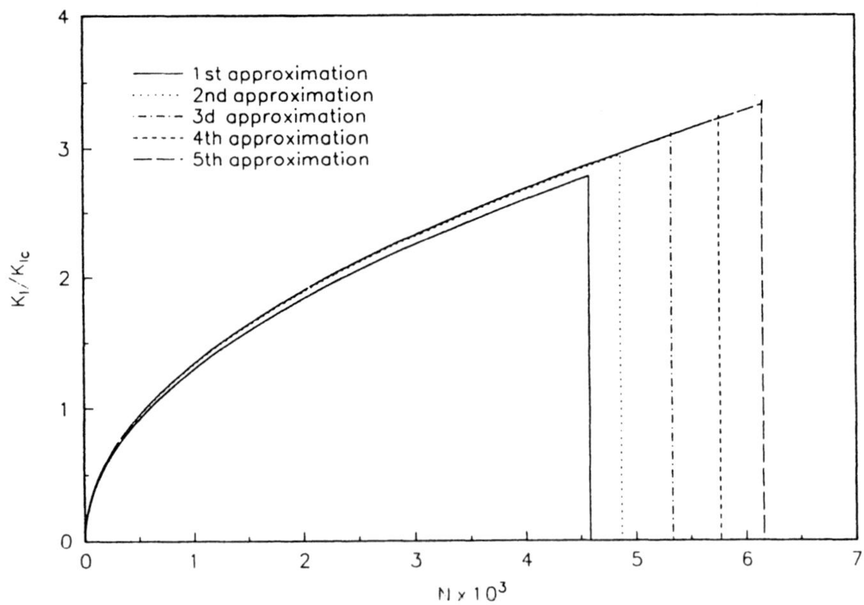

Figure 6.27. Quantum levels of K_I during the ductile crack growth (first five approximations of the method of self- consistent field)

The ductile sections of fracture curve corresponding to the first five approximations are shown in Figure 6.27. It is observed that the fracture curves at this stage for the third and higher orders of approximation are very close and differ only by the number of generated dislocations. The number of dislocations generated at the ductile crack growth stage are given in Table 6.3.

Table 6.3. Approximate calculation of the fracture curve.

N_0	N_d	K_{Ib}/K_{Ic}	K_{Im_1}/K_{Ic}	K_{In_1}/K_{Ic}	K_{Im_2}/K_{Ic}	K_{In_2}/K_{Ic}
1	4581	2.7803	2.8608	0.8799	-	-
2	4867	2.9470	3.0188	1.3760	1.4486	0.9687
3	5330	3.0965	3.1645	1.6438	1.7115	0.9770
4	5765	3.2220	3.2876	1.8347	1.8996	0.9787
5	6162	3.3298	3.3939	1.9856	2.0484	0.9792
6	6526	3.4245	3.4874	2.1116	2.1728	0.9793
7	6863	3.5097	3.5712	2.2206	2.2804	0.9792
8	7176	3.5864	3.6474	2.3167	2.3755	0.9791

The curves indicating the upper bound of the superfine scale stress intensity factor are shown in Figure 6.28.

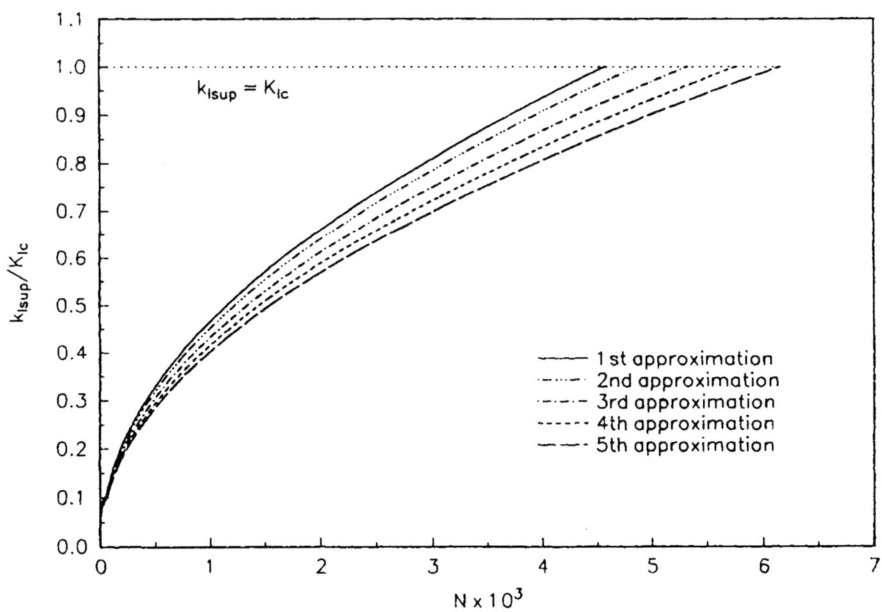

Figure 6.28. Superfine scale stress intensity factor during the ductile crack growth (first five approximations of the method of self- consistent field)

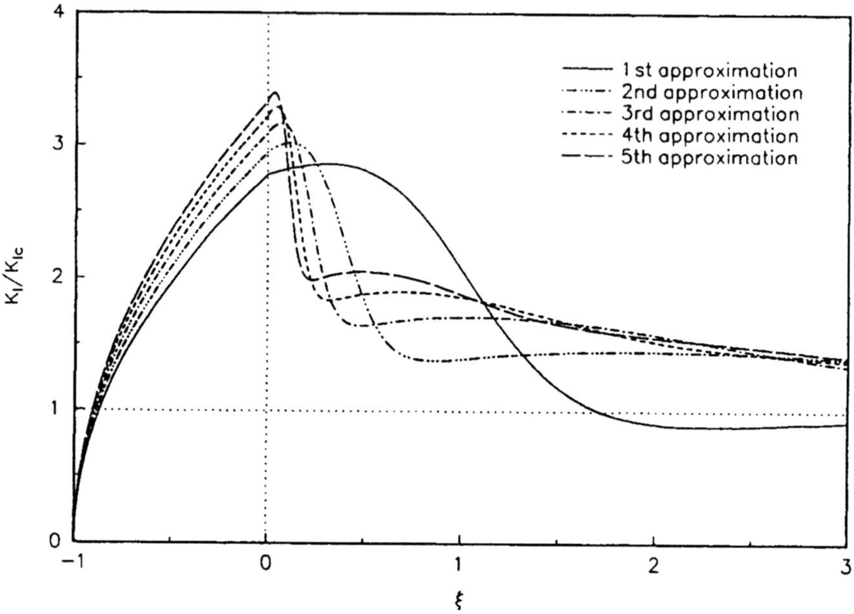

Figure 6.29. Fracture curve for -1< $\xi \leq 3$ (first five approximations of the method of self-consistent field)

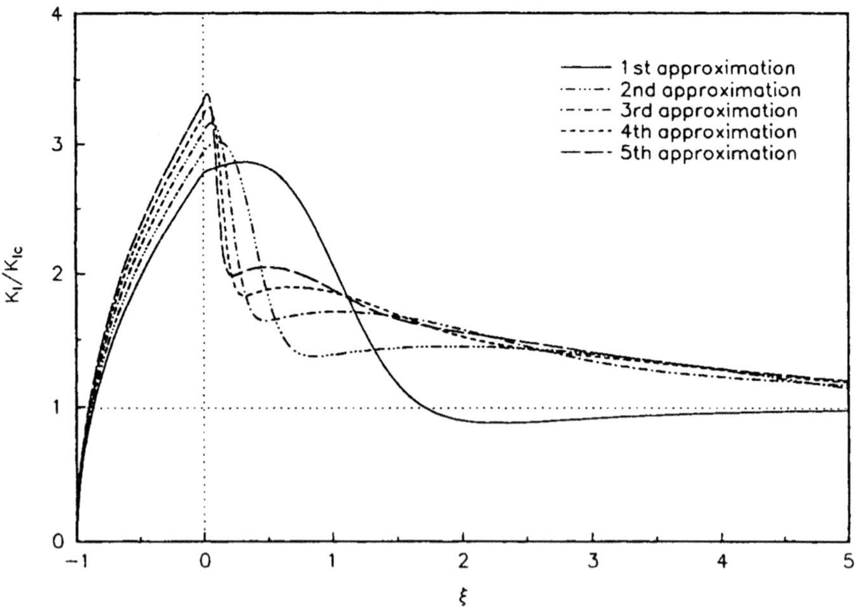

Figure 6.30. Fracture curve for -1< $\xi \leq 5$ (first five approximations of the method of self-consistent field)

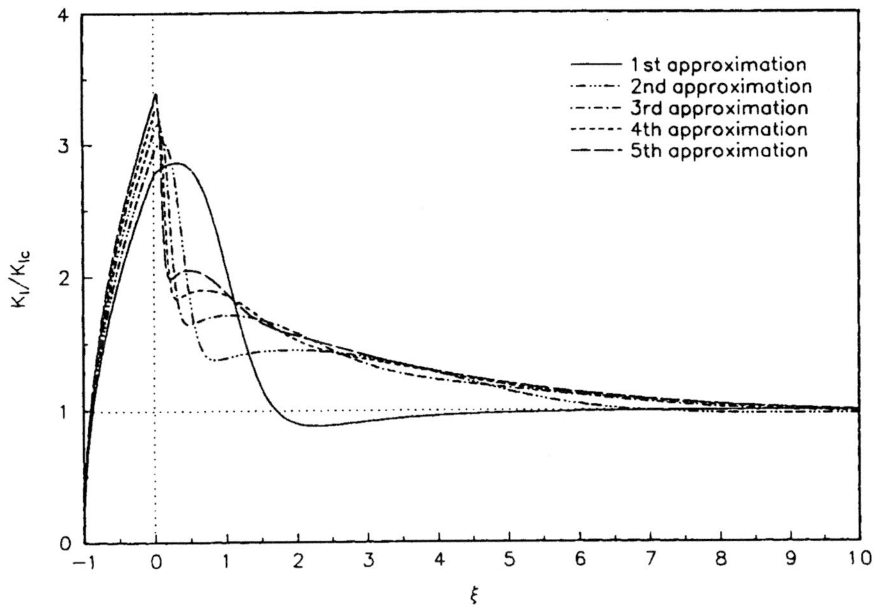

Figure 6.31. Fracture curve for -1< $\xi \leq 10$ (first five approximations of the method of self-consistent field)

The fracture curves obtained for the first five approximations are shown in Figures 6.29, 6.30 and 6.31 for $-1 \leq \xi \leq 3$ in Figure 6.29, $-1 \leq \xi \leq 5$ in Figure 6.30 and $-1 \leq \xi \leq 20$ in Figure 6.31.

It is observed that the ductile crack growth mechanism does not resume after the brittle crack growth initiation when the third approximation is used. At the fourth and higher orders of approximation only the brittle fracture is considered after the initiation because of computational restrictions.

Figures 6.29, 6.30 and 6.31 show that there are some characteristic points on the fracture curve. The first one, ξ_b , is the point of the brittle crack growth initiation. The coordinate ξ is chosen such that $\xi_b = 0$. The second characteristic point ξ_{m1} is the point of maximum $K_I = K_{Ic}$ (fracture toughness). The stable brittle crack growth occurs in the interval $\xi_b \leq \xi < \xi_{m1}$. At this stage the dislocation clusters are located ahead of the crack tip and the elastic field of the dislocation sources hinders the crack growth. Further increase in external load K_I is necessary to maintain the crack growth. The third characteristic point of the fracture curve ξ_{n1} is the point of minimum $K_I = K_{In1}$. The section $\xi_{m1} < \xi \leq \xi_{n1}$ of the fracture curve is unstable, and K_I decreases from K_{Ic} to K_{In1}.

Here the crack tip is located ahead of the nearest dislocation cluster. The elastic field of this dislocation source helps the advancement of the crack tip while the influence of other dislocation cluster is small. The interaction of the crack tip with the elastic field of the dislocation clusters other than the first one leads to the second stable section of the fracture curve. This section corresponds to the interval $\xi_{n1} \le \xi < \xi_{m2}$. At $\xi \ge \xi_{m2}$ the elastic field of all dislocation clusters helps the crack tip advancement and K_I decreases from K_{In2} to the second minimum point K_{im2}. If $\xi \ge \xi_{m2}$ the influence of the dislocation clusters diminishes and K_I increases to the value of k_{Ic}.

Figures 6.29, 6.30 and 6.31 show that the stable section of the fracture curve becomes shorter at higher orders of approximation. The maximum values of K_I and the corresponding values of ξ are given in Table 6.3. The radius-vector of the nearest cluster $\rho_{N_0}^{(N_0)}$ decreases for the higher order approximations. The values of the stress intensity factor at the characteristic points are given in Table 6.3.

The fracture curves for the third and higher orders of approximation are similar to the curve for the second approximation. There is excellent convergence in the number of dislocations generated at the ductile stage of crack growth, N_d, and in the brittle fracture initiation threshold, K_{Ib}, as the order of approximation increases.

Exact Solution

The system of equilibrium equations (6.119) was solved to determine the exact solution. The number of equations is equal to the number of generated dislocations. The fracture curve was calculated up to 102 dislocations using the iterative method and up to 150 dislocations using the minimization method.

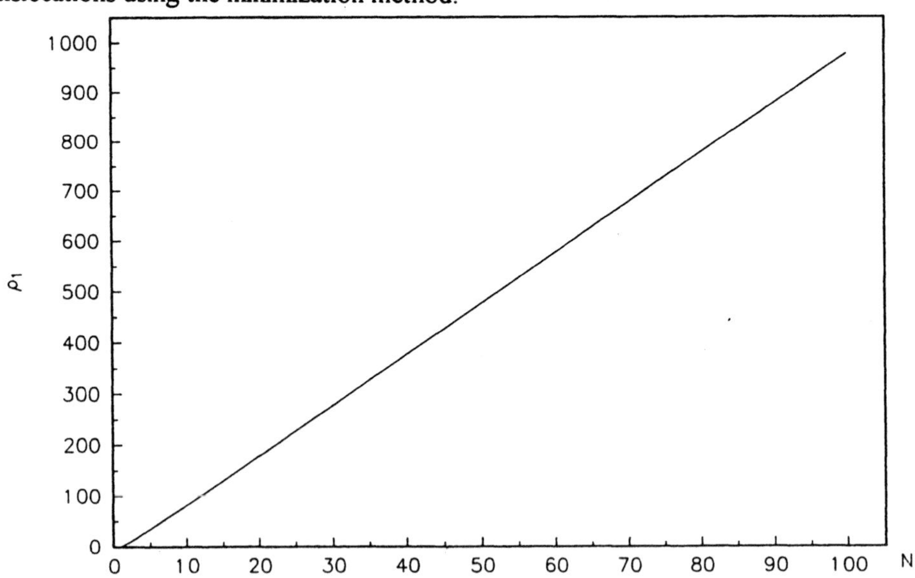

Figure 6.32. Stable positions of the first dislocation during monotonic loading (iterative method).

Figures 6.32 - 6.34 illustrate some results achieved by iterative method. Figure 6.32 shows the stable positions of the first dislocation during monotonic loading. The stable position of the first dislocation at the 100th step is equal to 1000. Figure 6.32 shows that the function $\rho_1^{(1)}(N)$ is nearly a linear function of N.

The curves of the upper and lower bounds of the superfine scale stress intensity factor are shown in Figure 6.33. The $\tilde{k}_{I\,\mathrm{sup}} = 11.6 = 0.13\tilde{k}_{Ic}$ at $N = 100$.

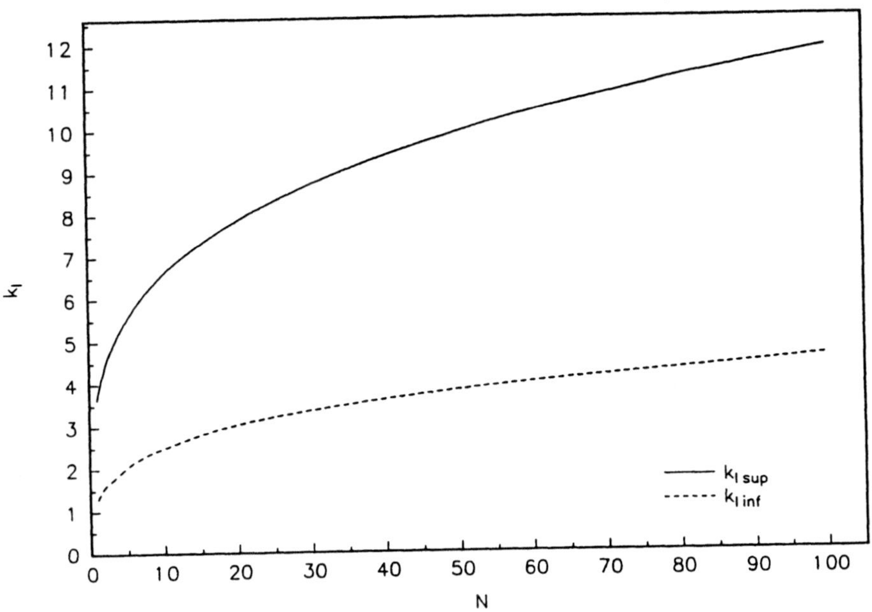

Figure 6.33. Upper and lower bounds of the superfine scale stress intensity factor during monotonic loading (iterative method)

Figure 6.34 shows the fracture curve up to 102 dislocations. It is observed that the value of $\tilde{K}_I$ maintaining the crack growth increases in proportion to $\sqrt{N}$.

The values of external load $\tilde{K}_I^{(N)}$ corresponding to the generation of the Nth dislocation, the upper bound of the superfine scale stress intensity factor $\tilde{k}_{I\,\mathrm{sup}}^{(N)}$ and the coordinates of the stable position of the first dislocation $\tilde{\rho}_1^{(1)}$ are given in Table 6.4.

Figures 6.35 - 6.37 illustrate some results achieved by minimization method. Figure 6.35 shows the stable positions of the first, 20th, 40th, 60th, 80th, 100th, 120th and 140th dislocations during monotonic loading. The stable position of the first dislocation at N = 100 is equal to 977.4 and at N = 150 to 1482.2.
The upper and lower bounds of the superfine scale stress intensity factor vs N are shown in Figure 6.36. At N = 100 the value of $\tilde{k}_{I\,\mathrm{sup}}$=11.66, i.e. 0.132 $\tilde{k}_{Ic}$ and $\tilde{k}_{I\,\mathrm{inf}}$=6.47. At N = 150 the value of $\tilde{k}_{I\,\mathrm{sup}} = 12.97$, i.e.$0.146\tilde{k}_{Ic}$ and $\tilde{k}_{I\,\mathrm{inf}} = 7.32$.

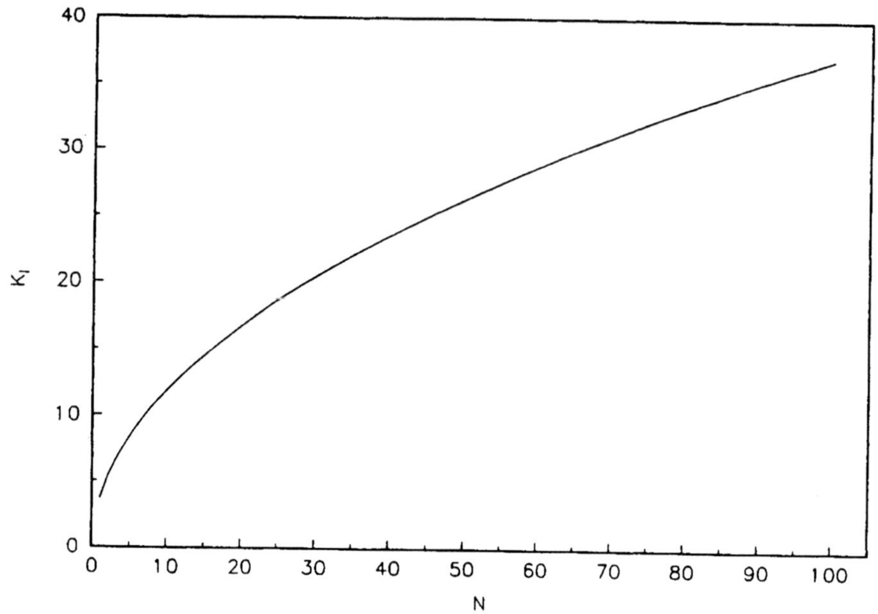

Figure 6.34. Quantum levels of K_I (iterative method)

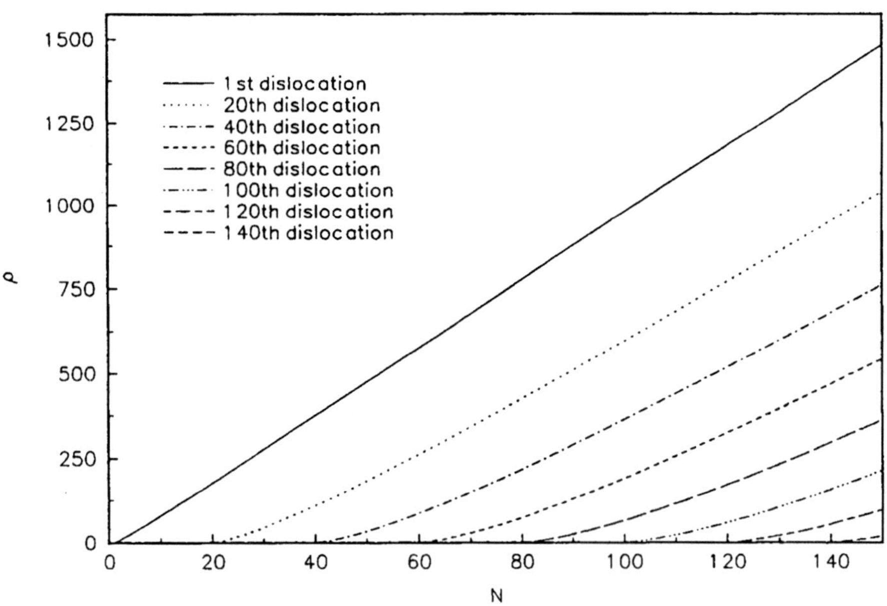

Figure 6.35. Stable positions of dislocations during monotonic loading (minimization method)

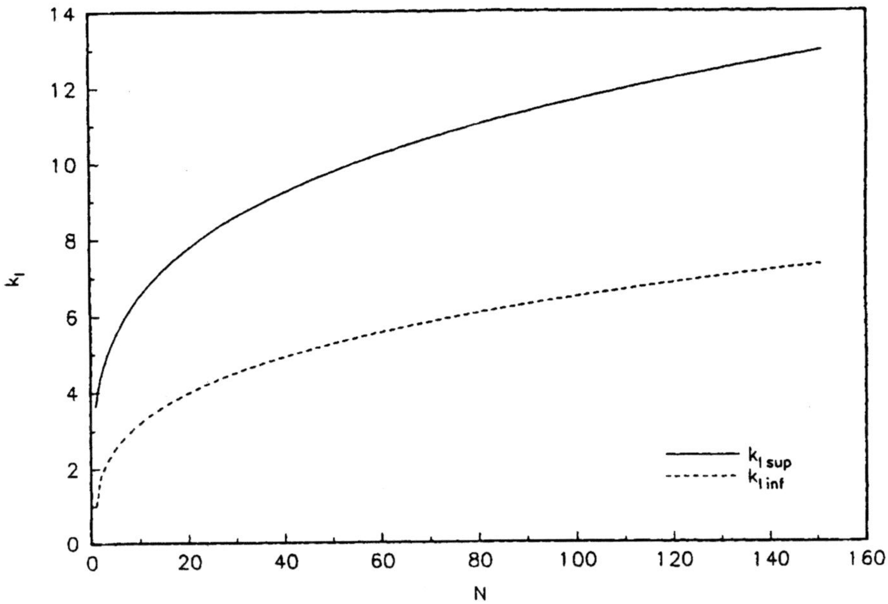

Figure 6.36. Upper and lower bounds of the superfine scale stress intensity factor during monotonic loading (minimization method)

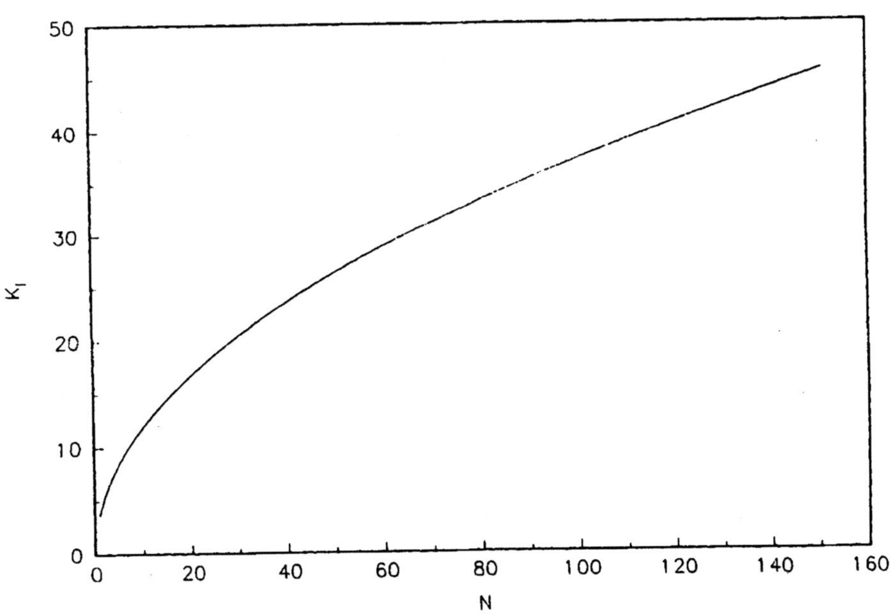

Figure 6.37. Quantum level of K_I (minimization method)

Table 6.4. Quantum levels for the generation of dislocations

	Iterative approach			Minimization approach		
N	$\tilde{\rho}_N$	$\tilde{k}_{Isup}$	$\tilde{K}_I$	$\tilde{\rho}_N$	$\tilde{k}_{Isup}$	$\tilde{K}_I$
1	1.00000	3.64055	3.64055	1.00000	3.64055	3.64055
2	0.65730	4.36570	5.29435	0.75827	4.36946	5.29727
3	0.52990	4.84169	6.51885	0.64133	4.83503	6.51398
4	0.44777	5.19365	7.52732	0.56771	5.19219	7.52533
5	0.39588	5.49348	8.41579	0.51550	5.48683	8.41019
6	0.35828	5.76223	9.21902	0.47598	5.74003	9.20692
7	0.32925	5.99900	9.95538	0.44429	5.96343	9.93765
8	0.30423	6.21190	10.6388	0.41837	6.16421	10.6166
9	0.28723	6.39371	11.2744	0.39684	6.34716	11.2534
10	0.26971	6.57431	11.8799	0.37819	6.51564	11.8550
11	0.25621	6.73436	12.4522	0.36207	6.67209	12.4269
12	0.24544	6.87768	12.9982	0.34785	6.81837	12.9729
13	0.23493	7.02781	13.5209	0.33518	6.95593	13.4964
14	0.22636	7.14594	14.0245	0.32384	7.08590	13.9999
15	0.21800	7.28042	14.5116	0.31357	7.20921	14.4856
16	0.20999	7.40214	14.9814	0.30424	7.32661	14.9552
17	0.20343	7.51397	15.4380	0.29570	7.43874	15.4102

continued on next page

continued from previous page

N	$\bar{\rho}_N$	$\bar{k}_{Isup}$	$\bar{K}_I$	$\bar{\rho}_N$	$\bar{k}_{Isup}$	$\bar{K}_I$
18	0.19697	7.63497	15.8793	0.28784	7.54612	15.8520
19	0.19133	7.76154	16.3144	0.28059	7.64920	16.2815
20	0.18645	7.83944	16.7279	0.27387	7.74837	16.6998
21	0.18129	7.93057	17.1341	0.26760	7.84397	17.1078
22	0.17665	8.03598	17.5346	0.26172	7.93628	17.5061
23	0.17256	8.12140	17.9225	0.25625	8.02556	17.8954
24	0.16853	8.21261	18.3049	0.25107	8.11204	18.2763
25	0.16495	8.30059	18.6784	0.24624	8.19592	18.6493
26	0.16157	8.38444	19.0440	0.24186	8.27737	19.0149
27	0.15816	8.45715	19.4019	0.23753	8.35655	19.3736
28	0.15513	8.54704	19.7553	0.23343	8.43361	19.7256
29	0.15223	8.61023	20.1001	0.22953	8.50868	20.0713
30	0.14938	8.69884	20.4414	0.22582	8.58187	20.4112
31	0.14684	8.78094	20.7750	0.22229	8.65330	20.7454
32	0.14434	8.83018	21.1028	0.21891	8.72306	21.0743
33	0.14195	8.89440	21.4267	0.21569	8.79124	21.3980
34	0.13958	8.98141	21.7486	0.21259	8.85792	21.7169
35	0.13748	9.04840	22.0617	0.20965	8.92318	22.0311
36	0.13541	9.12806	22.3732	0.20680	8.98708	22.3408

N	$\bar{\rho}_N$	$\bar{k}_{Isup}$	$\bar{K}_I$	$\bar{\rho}_N$	$\bar{k}_{Isup}$	$\bar{K}_I$
37	0.13334	9.19365	22.6764	0.20407	9.04970	22.6463
38	0.13145	9.26213	22.9801	0.20145	9.11109	22.9477
39	0.12963	9.30110	23.2746	0.19893	9.17131	23.2451
40	0.12783	9.36203	23.5689	0.19651	9.23041	23.5387
41	0.12620	9.42236	23.8589	0.19414	9.28843	23.8287
42	0.12473	9.47839	24.1453	0.19192	9.34543	24.1151
43	0.12318	9.55345	24.4308	0.18973	9.40144	24.3982
44	0.12146	9.61907	24.7115	0.18754	9.45651	24.6780
45	0.11997	9.64586	24.9844	0.18548	9.51067	24.9546
46	0.11884	9.70543	25.2595	0.18349	9.56395	25.2282
47	0.11722	9.76013	25.5289	0.18156	9.61639	25.4988
48	0.10905	9.81255	25.7975	0.17970	9.66802	25.7665
49	0.11456	9.86309	26.0623	0.17798	9.71887	26.0315
50	0.11330	9.91642	26.3253	0.17623	9.76896	26.2937
51	0.11210	9.96696	26.5842	0.17453	9.81833	26.5534
52	0.11094	10.1045	26.8421	0.17288	9.86699	26.8105
53	0.10977	10.0681	27.0965	0.17128	9.91497	27.0652
54	0.10868	10.1161	27.3486	0.16972	9.96229	27.3175
55	0.10763	10.1822	27.5997	0.16821	10.0090	27.5674

N	$\bar{\rho}_N$	$\bar{k}_{Isup}$	$\bar{K}_I$	$\bar{\rho}_N$	$\bar{k}_{Isup}$	$\bar{K}_I$
56	0.10018	10.2254	27.8469	0.16669	10.0550	27.8151
57	0.10550	10.2635	28.0923	0.16526	10.1005	28.0606
58	0.10455	10.3097	28.3354	0.16384	10.1454	28.3039
59	0.09720	10.3599	28.5769	0.16249	10.1897	28.5452
60	0.10263	10.4036	28.8162	0.16116	10.2335	28.7844
61	0.10173	10.4424	29.0532	0.15985	10.2768	29.0217
62	0.10081	10.4778	29.2880	0.15858	10.3195	29.2570
63	0.10001	10.5283	29.5221	0.15733	10.3617	29.4904
64	0.09991	10.5711	29.7541	0.15611	10.4035	29.7219
65	0.09823	10.6116	29.9841	0.15492	10.4447	29.9517
66	0.09127	10.6615	30.2126	0.15376	10.4855	30.1797
67	0.09668	10.6837	30.4371	0.15262	10.5259	30.4060
68	0.09609	10.7414	30.6632	0.15150	10.5658	30.6306
69	0.09515	10.7704	30.8852	0.15042	10.6053	30.8535
70	0.09441	10.8090	31.1072	0.14935	10.6444	31.0749
71	0.09373	10.8615	31.3275	0.14828	10.6830	31.2946
72	0.08691	10.9133	31.5461	0.14737	10.7213	31.5129
73	0.09227	10.9487	31.7619	0.14643	10.7592	31.7296
74	0.09157	10.9789	31.9765	0.14526	10.7967	31.9448

N	$\bar{\rho}_N$	$\bar{k}_{Isup}$	$\bar{K}_I$	$\bar{\rho}_N$	$\bar{k}_{Isup}$	$\bar{K}_I$
75	0.09093	11.0090	32.1898	0.14437	10.8338	32.1586
76	0.08443	11.0520	32.4033	0.14350	10.8706	32.3710
77	0.08976	11.0990	32.6150	0.14257	10.9070	32.5820
78	0.08911	11.1483	32.8253	0.14169	10.9431	32.7916
79	0.08847	11.1858	33.0340	0.14061	10.9788	32.9999
80	0.08183	11.2214	33.2410	0.13974	11.0142	33.2068
81	0.08723	11.2409	33.4458	0.13885	11.0493	33.4125
82	0.08674	11.2790	33.6509	0.13803	11.0841	33.6169
83	0.08605	11.3057	33.8535	0.13719	11.1185	33.8201
84	0.08552	11.3451	34.0552	0.13642	11.1527	34.0220
85	0.07931	11.3812	34.2562	0.13563	11.1866	34.2228
86	0.08441	11.4099	34.4552	0.13483	11.2201	34.4224
87	0.08404	11.4477	34.6545	0.13404	11.2534	34.6208
88	0.08337	11.4679	34.8503	0.13311	11.2864	34.8181
89	0.08286	11.5251	35.0469	0.13257	11.3192	35.0142
90	0.08243	11.5568	35.2425	0.13177	11.3516	35.2093

Figure 6.37 shows the fracture curve up to 150 dislocations. The value of $\widetilde{K}_I^{(N)}$ corresponding to the generation of the Nth dislocation, the upper bound of the superfine scale stress intensity factor $\widetilde{k}_{I\,\mathrm{sup}}^{(N)}$, and the coordinates of the stable position of the first dislocation $\widetilde{\rho}_1^{(1)}$ are given in Table 6.4. It is observed that the value of $K_I^{(N)}$ calculated by the minimization method is only 0.1% lower than the corresponding values calculated by the iterative method. The difference in the values of the superfine scale stress intensity factor $\widetilde{k}_{I\,\mathrm{sup}}^{(N)}$ is around 1.5%.

6.13. Conclusions

Nanofracture mechanics has been applied to the dislocation emission and crack growth in a cubic lattice of atoms under monotonic loading using the plane model. The linear theory of elasticity is employed to study the cubic lattice. A mathematical method is treated to solve the elastic problem for a crack with any number of dislocations under screw shear and plane strain conditions. The complex potentials are derived which yield the efficient solution of the problem.

For a screw crack with any number of screw dislocations along one line, an analytical solution is derived which appears to hold from one emitted dislocation to any large number of dislocations. The Dugdale solution of macroscopic plasticity was derived from nanofracture mechanics based on the linear theory of elasticity.

The crack tip is proven to initiate by ductile mechanism with the emission of the first dislocation pair at the external load $K_1^{(1)} = 3.641\sqrt{2a\mu_0\tau_0}$ settled at the stable position $r_1^{(1)} = \mu_0 b_e / (4\pi\tau_0)$.

The second pair of elementary edge dislocations is proven to get emitted at $K_I^{(2)} = 5.297\sqrt{2a\mu_0\tau_0}$ and settled at $r_1^{(2)} = 0.638\mu_0 b_e / \tau_0$ and $r_1^{(2)} = 0.060\mu_0 b_e / \tau_0$.

The equilibrium of an arbitrary number of dislocations is analyzed, and two algorithms for the numerical analysis of the dislocation emission and crack growth under monotonic loading are suggested. It is shown that the mathematical problem is reduced to a parametric system of non-linear equations.

The approximate method of self-consistent field is treated which allows one to solve the parametric system for any number N of emitted dislocations. The dislocations are combined into a certain number of dislocation clusters, so that a system of N_0 equilibrium equations needs to be solved for N_0 dislocation clusters.

The approximate method is used to conduct numerical experiments for a cracked aluminum crystal with many thousands of emitted dislocations. It is proven that the crack growth by brittle mechanism is always stable initially. The practical convergence of the procedure is observed. Identification analysis for the ductile or brittle crack growth mechanism is performed in the first three approximations. It turned out that the ductile crack growth does not resume once the brittle crack growth has started. Some characteristic points of the fracture curve are observed and explained.

The exact solution is obtained by using both iterative and minimization methods. The critical values of external load $K_I^{(N)}$, the superfine scale stress intensity factor $k_{I\,\mathrm{sup}}$

and the stable positions of dislocations are calculated up to N = 102 using iterative method and up to $N = 150$ using the minimization method. The results found by the two methods are observed to be close.

6.14. Problems

6.1. Consider the elementary screw dislocation at the coordinate origin, $x_1 = 0$, $x_2 = 0$, with slip plane along $x_2 = 0$,

$$\mu u_{3,1} = \sigma_{31} = \sigma_{13}^0 + \frac{\mu b_s \cos\theta}{2\pi r} \quad , \quad \sigma_{32} = \sigma_{23}^0 + \frac{\mu b_s \sin\theta}{2\pi r} \quad . \tag{6.135}$$

Here, σ_{13}^0 and σ_{23}^0 are the regular components of the stress tensor and $r\theta$ are the polar coordinates.

Using Equation (1.75) calculate the invariant integral over the rectangular contour, $x_2 = \pm\delta$, $-L_1 \le x_1 \le L_2$ when $\delta/L_1 \to 0$ and $\delta/L_2 \to 0$, and prove that

$$\begin{aligned} \Gamma = 2 \lim_{\substack{\delta/L_1 \to 0 \\ \delta/L_2 \to 0}} \int_{-L_1}^{+L_2} \sigma_{32} u_{3,1}\, dx_1 = b_s \sigma_{13}^0 + \frac{2}{\mu} \sigma_{13}^0 \sigma_{23}^0 (L_1 + L_2) + \\ + \frac{1}{\pi} b_s \sigma_{23}^0 \int_0^{\pi} cotan\varphi\, d\varphi + \frac{\mu b_s^2}{2\pi^2} \lim_{\substack{\varepsilon_1 \to 0 \\ \varepsilon_2 \to 0 \\ r \to 0}} \frac{\varepsilon_1 - \varepsilon_2}{r} . \end{aligned} \tag{6.136}$$

Here, the first term corresponds to the Peach - Koehler equation, the second term is indefinite and path-dependent, and the third and fourth term are divergent. When $L_1 + L_2 \to 0$, the second term vanishes, but the third and fourth remain divergent.

This result shows that the "rule of Γ - integration" must be applied that states: *Only the finite part of the integral should be taken, while integrating over a closed contour shrinking to a singularity. All other parts should be omitted.*

Because the invariant integral expresses the energy conservation law, try to substantiate this seemingly illogical rule.

6.2. Consider the elementary edge dislocation at the coordinate origin, with slip plane along $x_2 = 0$, and study the problem which is analogous to Problem 6.1. Show that to rescue the energy conservation law the rule of Γ - integration should be used in this case.

6.3. Consider the small scale yielding of a dihedral angle with angle $\alpha < \pi$ under screw shear (Figure 6.38). The faces of the dihedral angle are free of tractions, and the external stresses at infinity are directly proportional to $K\, r^{-\lambda(\alpha)}$ where K is the stress intensity factor (loading parameter) with dimension Newton × $(\text{Meter})^{\lambda-2}$, and $\lambda(\alpha)$ is a certain positive function. Suppose at a certain $K = K^{(1)}$ by loading the dihedral angle vertex emits a screw dislocation along the symmetry line.

Find λ, $K^{(1)}$ and the stable position $x^{(1)}$ of the generated dislocation.

Hint: Use the conformal mapping onto the half-plane.

6.4. Under the conditions of Problem 6.3, consider the emission of the first, second, third, ..., and Nth screw dislocation. Find the analytical solution for the second quantum level, $K^{(2)}$ and $x^{(2)}$, characterizing the emission of the second dislocation.

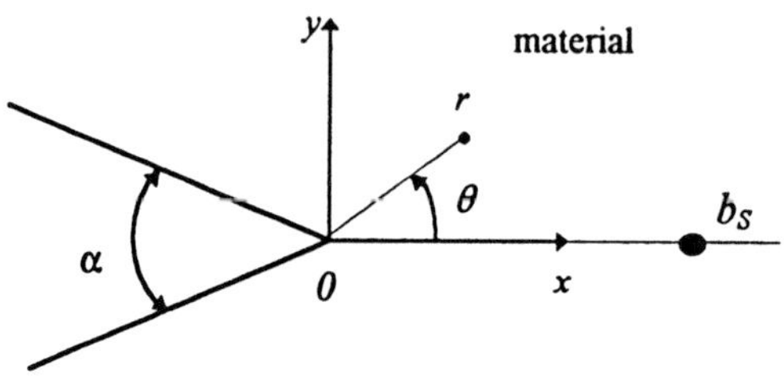

Figure 6.38. An edge dislocation near the corner point

6.5. Under the conditions of Problem 6.4, derive the non-linear equation characterizing the emission of the Nth dislocation. Consider the continuum approximation for large N and find the corresponding singular integral equation. Derive the closed-form analytical solution to the latter equation and prove that it coincides with the solution of the corresponding elastic-plastic problem (with a Dugdale-type plastic line segment).

6.6. Under the conditions of Problem 6.4 conduct numerical experiments on the emission of the first, second and so on dislocations using method of Section 6.3. Prove that, practically, for N>2 one can use the continuum approximation of Problem 6.5 to determine any quantum levels.

6.7. Suppose an interface between two crystal grains has a corner point that can be represented by the dihedral angle in the neighborhood of the corner point (Figure 6.39). Consider screw shear and find the self-similar solution which has a stress singularity $Kr^{-\lambda}$ as $r \to 0$ as in stress, where $1 > \lambda > 0$, and K is the stress intensity factor. This solution describes the external elastic field near the corner point. Find λ in terms of α and μ_1 / μ_2.

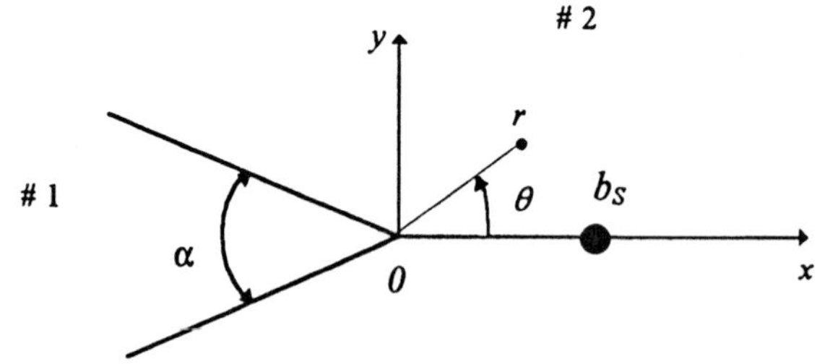

Figure 6.39. A screw dislocation emitted by the vertex of the interface dihedral angle

6.8. Under the conditions of Problem 6.7, consider the emission of a screw dislocation along the symmetry line from the vertex. Find the corresponding $K^{(1)}$ and stable position of the dislocation.

Hint: Use polar coordinates, Mellin's transform and the superposition principle to derive the solution.

6.9. Under the conditions of Problem 6.7, consider the emission of a pair of the misfit screw dislocations along the interface (from the vertex). Find the $K^{(1)}$ and stable position of the dislocations.

6.10. Under the conditions of Problem 6.7, consider the emission of the second, third, ..., and Nth pair of misfit screw dislocations along the interface (from the vertex). Derive non-linear equation for the quantum levels of the emission and conduct numerical experiments. Study the continuum approximation for a large number of dislocation pairs and compare it with the numerical experiment.

6.11. Consider the emission of a pair of elementary screw dislocations from the tip of the screw crack, $y = 0 \quad x < 0$, along the directions $\theta = \pm\alpha$. Find the first quantum level, $K_{\mathrm{III}}^{(1)}$ and $r^{(1)}$, for arbitrary α. Define α using the principle of maximum energy dissipation from Section 3.2.

Hint: The energy of external field spent to generate the pair of the dislocations (energy dissipation) is equal to $2(2/\pi)^{1/2} b_s K_{\mathrm{III}}^{(1)}(\alpha)\,[r^{(1)}(\alpha)]^{1/2} \cos(\alpha/2)$.

6.12. Consider a finite screw crack along $y = 0 \quad 0 < x < l$ in half-plane $x > 0$ with traction-free boundaries. The stress $\tau_{yz} = \tau$ is applied at infinity. Study the emission of elementary screw dislocations: (i) along the crack continuation, and (ii) under an angle, $\theta = \pm\alpha$, to the crack continuation (dislocation pairs). Find the critical values of τ and stable positions of the dislocations.

6.13. Consider the screw shear plane with a traction-free elliptical cavity. At infinity, the uniform shear stress is applied. Consider the emission of the first screw dislocation by loading. Find the critical load and stable position of the dislocation using methods of Section 6.2 and the principle of maximum energy dissipation (Section 3.2).

6.14. Consider Problems 6.3 - 6.13 for edge dislocations under the corresponding plane strain conditions.

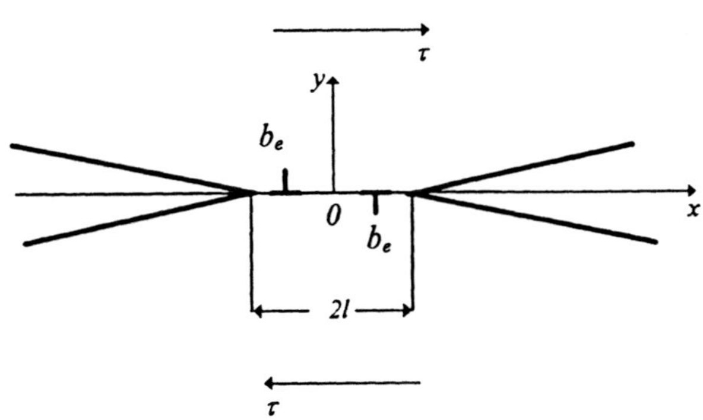

Figure 6.40. Shear of a cohesion junction of two half-spaces

6.15. Consider a cohesion junction neck between two half-spaces along $y = 0$ $|x| < l$ under plane strain (Figure 6.40). The sides of the crack along $y = 0$ $|x| > l$ are free of tractions. Shear stress $\tau_{xy} = \tau$ is applied at infinity. Consider the emission of the first pair of elementary edge dislocations moving one towards the other along $y = 0$. Find the critical $\tau = \tau^{(1)}$ and stable positions of the dislocations. Show that the annihilation of one pair of dislocations at the coordinate origin means the mutual shift of the half-planes on b_e. Find the condition of annihilation characterizing the continuous slip of the half-planes. Consider the emission of the second, third, ... and Nth pairs of dislocations and derive the non-linear equation of emission. Find the condition of annihilation for any number of generated dislocations. Show that the annihilation condition (continuous slip) does not depend on stress $\sigma_y = \sigma_\infty$ at infinity.

6.16. Consider the whole plane of a complex variable z cut along some closed and/or unclosed lines L. Find functions $F_1(z)$, $F_2(z)$,..., $F_N(z)$, which are analytic in z outside L, on the following boundary value condition equations:

$$F_n^+(z) = a_{mn}F_n^-(z) + f_n(z), \quad \text{where} \quad m,n = 1,2,\ldots,N \qquad (z \in L) \tag{6.137}$$

where a_{mn} are some given constants, and $f_n(z)$ are some functions given on L. Superscripts + and – denote the values of the corresponding function on the opposite sides of cut L.

Hint: Reduce the Riemann problem (6.137) for N functions to the following N Riemann problems for one function,

$$G_n^+(z) = b_n G_n^-(z) + g_n(z) \quad \text{where} \quad n = 1,2,\ldots,N \tag{6.138}$$

(do not sum over n)

Here, $G_n(z)$ is a linear combination of F_1, F_2, ..., and F_N to be found; b_n are some constants expressed through a_{mn} and g_n are some functions to be found in terms of a_{mn} and f_n.

Introduce the function, which is analytic in z outside L,

$$X_n(z) = \prod_{m=1}^{M}(z - z_m)^\lambda \quad \text{where} \quad \lambda = \frac{\log b_n}{2\pi i} \tag{6.139}$$

Here, $z = z_m$ is an end point of an unclosed cut, and M is the number of the ends.

Write Equation (6.138) in the following form,

$$\left(\frac{G_n}{X_n}\right)^+ = \left(\frac{G_n}{X_n}\right)^- + \frac{g_n(z)}{X_n^+(z)} \tag{6.140}$$

and apply the Sokhotsky-Plemelj formula

$$\frac{G_n(z)}{X_n(z)} = \frac{1}{2\pi i}\int_L \frac{g_n(t)dt}{X_n^+(t)(t-z)} \tag{6.141}$$

that was derived by Sokhotsky in 1874 and re-derived by Plemelj in 1908.

References

1. E. Smid and W. Boas (1935), *Kristallplastizitat*, Springer Verlag, Berlin.
2. J. P. Hirth and J. Lothe (1982), *Theory of Dislocations*, New York, McGraw Hill.
3. A. M. Kosevich (1981), *Physical Mechanics of Real Crystals* [in Russian], Naukova Dumka, Kiev.
4. R. W. K. Honeycomb (1968), *Plastic Deformation of Metals*, Edward Arnold Publ, London.
5. A. H. Cottrell (1964), *The Mechanical Properties of Matter*, Wiley, New York.
6. G. P. Cherepanov (1974), *Mekhanika Khrupkogo Razrushenia*, Nauka Press, Moscow: English edition (1979), *Mechanics of Brittle Fracture* ed. by R. de Wit and W. C. Cooley, Mc Graw Hill, New York.
7. R. Hill (1950), *The Mathematical Theory of Plasticity*, Oxford, Clarendon Press, London and New York.
8. B. D. Annin and G. P. Cherepanov (1988), *Elastic-Plastic Problems*, ASME Press, New York.
9. R. Bullough and J. R. Matthews (1991), Deformation of crystalline materials, in *Encyclopedia of Physics*, ed. by R. G. Lerner and G. L. Trigg, pp. 243-244, VCH Publishers, New York.
10. J. R. Rice and R. Thomson (1974), Ductile versus brittle behavior of crystals, *Philosophical Magazine*, **29**, pp. 73-97.
11. R.Thomson (1986), Physics of fracture, in *Solid State Physics*, ed. by H. Ehrenreich and D. Turnbull, **39**, pp. 1-129.
12. I.- H. Lin and R. Thomson (1986), Cleavage, dislocation emission, and shielding for cracks under general loading, *Acta Metallurgica*, **34**, pp. 176-190.
13. J. R. Rice, G. E. Beltz, and Y. Sun (1992), Peierls framework for dislocation nucleation from a crack-tip, in *Topics in Fracture and Fatigue*, ed. by A. S. Argon, Springer Verlag, pp. 1-58.
14. J. R. Rice (1992), Dislocation nucleation from a crack tip: an analysis based on the Peierls concept, *J. Mechanics and Physics of Solids* **40**, pp. 239-271.
15. J. C. M. Li, (1981) Dislocation sources in *Dislocation Modeling of Physical Systems*, ed. by M. Ashby, R. Bullough, B. Hartley, and J. Hirth, pp. 498-518, Pergamon Press, New York.
16. J. C. M. Li (1986), Computer simulation of dislocations emitted from a crack, *Scripta Metallurgica*, **20**, pp. 1477.
17. J. S. Wang and P. M. Anderson (1991), *Acta Metallurgica Materialia*, **39**, pp. 779.
18. G. E. Beltz and J. S. Wang (1992), *Acta Metallurgica Materialia*, **40**, pp. 1675.
19. G. E. Beltz and J. R. Rice (1992), *Acta Metallurgica Materialia*, **40**, S321.
20. G. E. Beltz and J. R. Rice (1991), Dislocation nucleation versus cleavage decohesion at crack-tips, in *Modeling the Deformation of Crystalline Solids*, ed. by T. C. Lowe, A. D. Rollett, P. S. Follansbee, and G. S. Daehn, The Minerals, Metals and Materials Society.
21. J. R. Rice and G. E. Beltz (1994), The activation energy for dislocation nucleation at a crack, *J. Mechanics and Physics of Solids*, **42**, pp. 333.
22. G. P. Cherepanov (1988), Initiation of microcracks and dislocations, *Soviet Applied Mechanics*, **23**(12), pp. 1165-1176, American edition.
23. G. P. Cherepanov (1988), Growth of a microcrack under monotonic loading, *Soviet Applied Mechanics*, **24**(4), pp. 396-409, American edition.
24. G. P. Cherepanov (1989), Closure of microcracks during unloading and the formation of reverse dislocations, *Soviet Applied Mechanics*, **24**(7), pp. 635-648, American edition.

25. G. P. Cherepanov (1990), Construction of fracture mechanics, *Soviet Applied Mechanics*, **26**(6), pp. 515-523, American edition.
26. G. P. Cherepanov (1988), Quantum fracture mechanics. *Proceedings of the 7th Int. Conference on Fracture*, Houston. Invited paper delivered by J. Rice at author's request. Published also in *Problemy Prochnosti (Strength of Materials)*, No. 2, pp. 3-9 (Translation by Plenum Press, pp. 155-163, 1990).
27. G. P. Cherepanov (1994), Nanofracture mechanics approach to dislocation generation and fracturing. Invited paper at the 12th U.S. National Congress of Applied Mechanics published in *Applied Mechanics Review*, **47**(6), part 2., S326, (1994).
28. G. P. Cherepanov (1993), Some novel approaches in mechanics of composites, Invited paper at the 1993 Winter ASME Meeting published in *Composite Materials and Structures*, ed. by C. W. Bert , V. Birman, and D. Hui, ASME Press, pp. 1-12.
29. G. P. Cherepanov, A. Richter, V. E. Verijenko, S. Adali, and V. Sutyrin (1995), Dislocation generation and crack growth under monotonic loading, *J. Applied Physics*, **78**(10), pp. 6249-64.
30. R. W. Armstrong (1966), *Material Science and Engineering*., **1**, pp. 251.
31. A. Kelly, W. R. Tyson, and A. H. Cottrell (1967), *Philosophical Magazine*, **15**, pp. 567.
32 C. Atkinson (1966), Dislocations and cracks in the theory of anisotropic elasticy, *Int. J. Fracture*, **2**(2), pp. 567-578.
33. B. S. Majumdar and S. J. Burns (1981), Crack-tip shielding - an elastic theory of dislocations and dislocation arrays near a sharp crack, *Acta Metallurgica*, **29**(2), pp. 579-588.
34. Shiue Sham-Tsong and Lee Sanboh (1985), A thermodynamic approach to the interaction between dislocation and crack and its applications, *Engineering Fracture Mechanics*, **22**(6), pp. 1105-1115.
35. S. N. G. Chu (1982), Elastic interaction between a screw dislocation and surface crack, *J. Applied Physics*, **53**(12), pp. 8678-8685.
36. S. J. Burns (1986), *Scripta Metallurgica*, **20**, pp. 1489.
37. G. Michot and A. George (1986), *Scripta Metallurgica*., **20**, pp. 1495.
38. Y. H. Chiao and D. R. Clarke (1989), Direct observation of dislocation emission from crack tip in silicon at high temperature, *Acta Metallurgica*, **37**, pp. 203-219.
39. S. M. Ohr (1985), An electron microscope study of crack tip deformation and its impact on the dislocation theory of fracture, *Materials Science Engineering*, **72**, pp. 1-35.
40. S. Kobayashi and S. M. Ohr (1984) , *J. Materials Science*, **19**, pp. 2273.
41. J. J. Gilman and H. C. Tong (1971), Quantum tunneling as an elementary fracture process, *J. Applied Physics*, **42**(9), pp. 3479-3486.
42. S. N. Zhurkov (1965), Kinetic concept of the strength of solids, *Int. J. Fracture Mechanics*, **1**(4), pp. 311-323.
43. G. P. Cherepanov (1993), Introduction to singular integral equations in aerodynamics, Chapter in *Method of Discrete Vortices* by S. M. Belotserkovsky and I. M. Lifanov, CRC Press, London-Boca Raton.
44. G. P. Cherepanov (1986), On quasibrittle fracture, *Applied Mathematics and Mechanics (PMM)*, **32**(6), pp. 1320-1337.
45. D. S. Dugdale (1960), Yielding of steel sheets containing slits, *J. Mechanics Physics Solids*., **8**, pp. 100-108.

CHAPTER 7

RELATIVISTIC ELECTRON BEAMS IN A SOLID

Since the 1960s the military have considered high-intensity electron beams as the most advanced space weapon of the future. Later some uncommon relativistic effects of the beams on solids were discovered experimentally and explained theoretically. These effects are treated in the present Chapter based on fracture mechanics methods including the method of invariant Γ integrals, the method of singular and generalized solutions, and the method of driving forces.

Other points of interest are the modeling methods based on comprehensive analysis of test data, the interpretation of theoretical solutions and the application of Mach's principle of the "economy of thought " encouraging us to invent the simplest model for the description of the observed properties. Solving partial differential equations analytically in terms of time and three spatial coordinates, the methods of characteristic lines and self-similar solutions may also be of interest for those who study these methods.

7.1. Introduction

Superluminal electrons which appeared to violate Einstein's theory of relativity, were discovered by Cherenkov and Vavilov in the 1930s. Since then, Cherenkov irradiation has become one of the main methods for the detection of relativistic particles.

In the mid 1960's, high-power pulse electron-beam accelerators having voltages of some millions of volts were invented and later used to fracture various materials. Experimental data analysis led to the discovery of a new mode of fracture in several ductile crystals, caused by a specific energy supply to the crack tip. The mode differs from the well-known thermomechanical modes of fracture caused by the "heat-thermostress-crack" mechanism. We call this new mode the electron-beam-induced fracture mode (EFM). It is characterized by the following three special features:

(i) Initial macrocracks in a specimen do not affect the threshold of fracture, i.e. the value of the beam intensity at which the specimen breaks;

(ii) fracture of different materials, which can be very ductile under the usual mechanical loads, occurs in a brittle manner; that is, the specimen usually cracks without any residual deformations;

(iii) the splitting cracks propagate with supersonic speeds.

These data are in contradiction with fracture mechanics, and cannot be understood or explained by traditional theories.

Our basic viewpoint can be briefly summarized as follows: *During irradiation of a solid by a high-intensity relativistic electron beam, some electron clusters are formed and act as "blades" or "wedges," cutting the crystalline specimen*[1].

In this section, we analyse and discuss experimental data on the EFM, and specify the peculiarities of the EFM. We will find that the processes caused by the EFM are unusual from the point of view of ordinary fracture mechanics.

In Section 7.2 the invariant Γ integrals of an electromagnetic deformable medium are modified for superluminal and supersonic singularities, and basic model and some problems explaining and describing the EFM are formulated.

In Section 7.3 we consider relativistic electron interactions in beams. Using Γ integrals, we derive the law of interaction of two moving relativistic charges, that is, the generalized Coulomb's law for relativistic charges. In particular, when two relativistic electrons e, move with the same speed υ, one behind the other along a rectilinear trajectory, the force F acting upon the rear electron is equal to:

$$F = \frac{e^2}{2\pi\varepsilon' R^2}\,\frac{\upsilon^2/a^2 - 1}{1-\upsilon^2/c^2}\,, \quad \text{where} \quad a = \frac{c}{\sqrt{\mu\varepsilon}}\,. \tag{7.1}$$

Here R is the distance between the electrons, c is the speed of light in vacuum, and a is the phase speed of light in a medium having electromagnetic constants, μ, ε, and ε'. It appears that two electrons moving faster than the phase speed of light attract each other, contrary to the common Coulomb's law; hence, beams of such relativistic electrons tend to self-pack and coalesce into a solid cluster. The latter problem is studied using a periodic chain model of the electron beam.

In Section 7.4 the dynamic elastic problem of supersonic cutting by a thin wedge is formulated and solved, and the drag force is calculated. In Section 7.5 the problem of deceleration of a moving wedge is solved in a quasi-steady approximation. The length of a resulting cut, that is, the final crack, is determined. Some applications of the analytical solutions are given. In Section 7.6 the theoretical results are analyzed and compared with experimental results. The role of relativistic electrons is estimated and some parameters of solid-state electron clusters are defined. The necessity of further study of this mysterious phenomenon is emphasized in Section 7.7.

Principal characteristics of the means of fracture and experimental conditions

In the 1960's, investigation of the fracture of solids by beams of electrons and other particles was connected with the creation of powerful sources of such radiation, for example, accelerators using shock autoemission from a pointed cathode in strong electric fields. The first results of the fracture of Si, Ge, InSb semiconductor crystals by high-energy electron beams were published[2] by Oswald in 1966. In the works that followed, one can also find some investigations of the fracture of certain metals,[3-10] certain dielectrics,[11-13] ionic crystals,[14-23] inorganic glass,[24,25] and rocks.[26-29]

Irradiation of the materials investigated in fracture experiments was done by pulses of homogeneous electron beams with a pulse span 1 *ns* to 1 μs. The average kinetic energy of a single electron in the beam was equal to $E_e \sim 0.1$ to 15 *MeV*. For reference, the chemical bond energy in solids is of the order of 1 *eV*. The beam intensity was measured within the limits from 10^{22} to 10^{27} $s^{-1}m^{-2}$ (current density, 10^3–10^8 A/m^2) so that the average particle density equals 10^{14}–10^{19} particles/m^3. The pulse radiation frequency reached 360 cycles per second. It is significant that for these energies electrons in the beam are relativistic. The relationship between energy E and the speed of a particle υ in relativistic mechanics has the form

$$E = \frac{m_0 c^2}{\sqrt{1 - \upsilon^2 / c^2}} - m_0 c^2 , \qquad (7.2)$$

where m_0 is the mass of the particle at rest, and c is the speed of light in a vacuum. Thus,

$$\frac{\upsilon}{c} = \sqrt{1 - (1 + E_*)^{-2}} , \qquad E_* = \frac{E}{m_0 c^2} . \qquad (7.3)$$

For instance, for an electron with energy E of 0.5 MeV, we get υ / c =0.85, c/υ =1.18, and υ = 0.25 Gm/s. This speed exceeds the phase speed of light for many dielectrics. And the scattering of particle velocities in the beam ensures the presence of superluminal electrons for even smaller energies of a beam.[30] The dimensionless speed of a particle υ / c vs. its dimensionless energy E_* is shown in Figure 7.1. Superluminal electrons are electrons moving in a medium with a speed exceeding the phase speed of light.

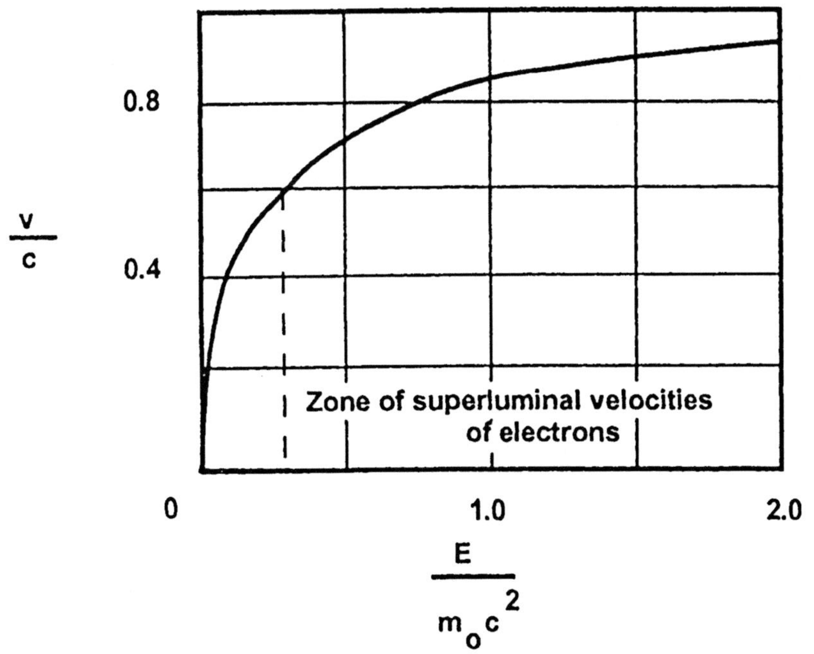

Figure 7.1. The speed of a particle versus its energy

For energies of about 1 *MeV*, relativistic effects are essential for light particles only (because for electrons, $m_0 c^2 = 0.5\ MeV$, and for neutrons and protons, $m_0 c^2 = 0.94\ GeV$). A further increase of energy changes the particle speed only insignificantly, bringing it closer to the speed of light in the vacuum.

The average distance between the particles in the beam can be estimated, according to the data mentioned above, to be about 1−10 *μm* (in the laboratory system of coordinates of the observer, since with the speed approaching that of light, relativistic effects begin to play a role). Substituting these values into Heisenberg's uncertainty principle, namely 1 *MeV* for ΔE and 10^{-3} *ps* for Δt, we obtain:

$$\Delta E \Delta t = 10^{-27}\ Js >> \hbar . \tag{7.4}$$

Hence, interactions of particles in the beam are of a nonquantum character, and the electrons can be considered material particles having no dimensions. The quantum character of the interaction between the electrons of such energies arises at distances of the order of 10^{-1} *pm* (the dimension of the electron localization region, according to the uncertainty principle).

Let us also consider the character of the interaction between radiation and the material exposed to radiation. The totality of atomic nuclei forming the structure of the substance, and the electrons, take up an insignificantly small volume in comparison with the volume of the solid they form. Interatomic distances have the order of 10^{-1} *nm* (e.g., for NaCl, the lattice spacing, $a = 0.5$ *nm*), the radius of the nucleus, 10^{-14} *m*, and the classic radius of the electron is 2.8×10^{-15} *m*. Evaluating the character of the interaction of the beam electrons with the nucleons of the solid, with the help of Equation (7.4), we find that this interaction can also be considered nonquantum.

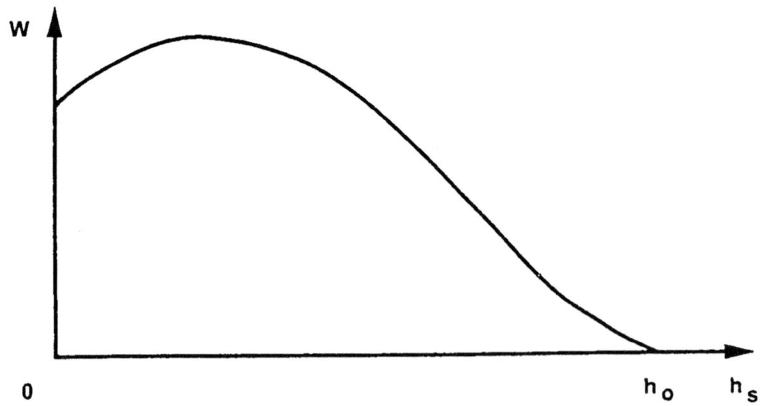

Figure 7.2. Qualitative dependence of the specific absorbed energy, *W*, on the sample thickness, h_s . Here, h_0 is the maximum penetration depth of an electron in a substance (e.g., for NaCl $h_0 = 0.45mm$).

In order to eliminate additional mechanical effects, the ticknesses of the irradiated samples in the fracture experiments were chosen so that the absorbed energy distribution was homogeneous in the sample volume. The qualitative pattern of distribution of specific absorbed energy within sample thickness is presented[31] in Figure 7. 2.

Table 7.1. Thresholds of irradiation power Φ_c and fracture energy W_c for some materials

Materials	Φ_c $(Watt/nm^2)$	W_c (eV/nm^3)	Reference
Rb Cl	0.09	0.12	16
Zn S	0.36	0.14	16
In Sb	0.17	0.50	2
Si	0.50	2.05	2
Li F	0.17	0.24	16
	0.17	0.30	19
Na F	0.13	0.18	16
	0.13	0.23	19
Na Cl	0.11	0.14	16
	0.11	0.20	19
Ge	0.3	1.20	16
	0.3	1.50	19
	0.3	1.17	2
Al	0.67	...	41
K Br	0.09	0.12	16
Whisker crystal KBr	0.42	0.57	16
K Cl	0.07	0.11	19
	0.08	0.09	20
	0.07	0.09	16
Whisker crystal KCl	0.42	0.47	16
Rocks[a]	...	450-750[b]	26,27

[a]Basalt, granite, sandstone, limestone, greenstone, marble, and clayey schist were studied.
[b]Specific absorbed energy of irradiation that ensures fracture.

From Figure 7.2, one can see that the requirement of homogeneity is approximately fulfilled with the sample thickness, $h_S \sim h_0/2$. In the experiments the thickness of the samples varied from 0.05 *mm* to some millimeters (for metals); while investigating the influence of the inhomogeneity of the absorbed energy, the sample thickness greatly exceeded the values mentioned. Other sample dimensions were usually restricted by the cross-section sizes of the homogeneous electron beam; in high-voltage accelerators, the beam radius was several millimeters. Under laboratory conditions the accelerator and the samples investigated were placed in a vacuum chamber to reduce energy losses. The materials investigated also included strong, defect-free whisker crystals and radiation-hardened materials whose mechanical properties differ greatly from those of common materials.[17]

The average absorbed specific energy during irradiation, W, was usually from 10^{-1} to 0 eV/nm^3 (10^{-2} –0 GJ/m^3), while threshold fracture energies were of the order of 10^{-1} eV/nm^3 for many crystals. The critical values of irradiation power, Φ_c, and $W = W_c$ for the fracturing of some materials are presented in Table 7.I. The quantities, W_c, proved to be two or three orders less than the binding energy of ideal crystals (for NaCl, according to reference[32], the binding energy U_c = 400 eV/nm^3 = 7.5 GJ/m^3), that is the theoretical strength of the material. Until recently, the cause of the discrepancy has been poorly understood. However, it has been clear that the classic theory of cracks has been of no help because the test data demonstrated no effect of initial cracks (See below).

Irradiation by high-power electron beams on solids (experimental data)

When solids are irradiated by powerful pulse beams of relativistic electrons, a number of specific mechanical and physical phenomena appear. Let us mention the main effects:

1. Fracture of various materials, even those that are very ductile under usual conditions, occurs in a "brittle" way[2,5,8,15,16];

2. Fracture is of a threshold nature characterized by the absence of accumulation effects and determined by the value of the critical irradiation density or by the corresponding density of absorbed energy[2,16,25,33-37];

3. Cracks fracturing the sample propagate with super-Rayleigh and supersonic speeds[16,20,27,38,39];

4. Macrodefects and preliminary man-made cracks have negligible influence on the fracture of dielectrics[16,22,36,40];

5. Thresholds of fracture depend slightly on the temperature and purity of crystals and also on the energy of relativistic particles[16,24,37];

6. In an ultrashort range (10^{-1} -10 ps), the conductance of dielectrics grows many times[12,22];

7. Dielectrics glow in ultraviolet and visible spectral regions[20];

8. Powerful electron emissions turn into a vacuum discharge between the surface of the dielectric and the anode[17].

We will consider these peculiarities in more detail below. First, we shall consider the fracture of samples with small thicknesses for which the absorbed energy distribution can be regarded as close to homogeneous. All of the materials are characterized by brittle fracture of the samples along one or two main cracks; in fact, any effects of ductility, even for metals, are not observed at all.[8,41,42] The second characteristic feature comprises low fracture thresholds in terms of specific absorbed energy W_c (or the density of irradiation Φ_c). The values of W_c and Φ_c for some materials are shown in Table 7.1. If the beam intensity exceeds the threshold value ($\Phi > \Phi_c$), a single pulse can produce fracture. Fracture accumulation at precritical irradiation is very weak; when $\Phi = 0.7\ \Phi_c$, the samples can endure a considerable number of pulses without fracture. The latter can be explained by the change of the material properties connected with rebuilding the structure under the action of irradiation.[4,16,20] Fracture thresholds of thin samples do not depend, in fact, on the average electron energy E_e when energies are in the interval of 0.1–15 *MeV*, pulse durations equal 1–10 *ns*, and sample temperatures, 80-500 K. For irradiation at considerably smaller energies and densities, fracture thresholds depend substantially on temperature.[9]

Experiments on the fracture of glass[20,24] show that the mechanical strength of glass is scattered within wide limits, but the irradiation strength of glass is evaluated with great precision.

In anisotropic crystals, several sharply expressed fracture thresholds correspond to the different directions of fracture; the typical histogram of distribution has two maxima,[13] see Figure 7.3. The maxima are caused by the possible development of the splitting crack on different cleavage planes under similar conditions of irradiation. For the fracturing of thin samples, the fracture thresholds do not depend on the macrodefects of the samples; for fracturing dielectrics, splitting of the sample takes place in the undisturbed part despite the presence of microcracks or even previously created cracks[22,36,40]. (In some materials, splitting cracks can also grow from macrodefects.)

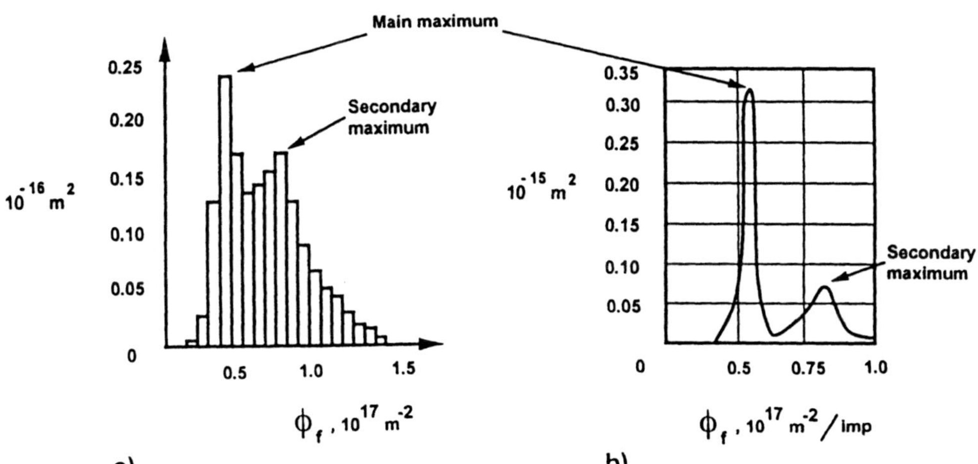

Figure 7.3. Typical histogram of fracture threshold irradiation power rate for thin KCl-specimens:
a) irradiation ensuring the fracture, b) irradiation of individual impulse.

The microdefects of materials with a dislocation density of 10^6–10^{12} mm^{-3} exercise only weak influence on their irradiation strength. Preliminary irradiation and so-called radiation hardening, change irradiation thresholds of fracture insignificantly[23,29], despite the fact that the hardening increases dislocation density 10^3–10^4 times, with considerable increase of mechanical strength. Some data on the influence of preliminary irradiation on the fracture threshold of KCl crystals for plates with a thickness of h_S=300μm are shown in Table 7.2. It should be mentioned that for preliminary hardening, the greatest effect is achieved by irradiation with light charged particles (electrons), while the slightest effect occurs with slow neutrons; however, the fracture thresholds of "defectless" whisker crystals can drop several times. (See Table 7.1.)

Table 7.2. Fracture threshold of KCl crystals in terms of the preliminary irradiation power

Type of preliminary irradiation	Energy of particles *(MeV)*	Total irradiation energy *(J/nm²)*	Fracture threshold power rate of irradiation *(W/ nm²)*
No preliminary irradiation	...	0	0.08
Protons	6	3	0.096
Thermal neutrons	< 1	100	0.083
	< 1	1000	0.087
High-speed neutrons	> 1	100	0.091
	> 1	1000	0.095
Electrons	2	10	0.097

Increasing the sample thickness does not change the general character of fracture; it remains brittle, and there is a threshold irradiation energy; however, other factors, unnoticed before, begin to influence fracture properties. In addition to the main cracks splitting the sample, one observes splitting out, swelling, and loosening of the surface on both sides of the sample[16,28]. Some materials carry traces of blast radiolysis: gas bubbles and surface erosion.[3] The density of the beam required for fracture Φ_f grows as the sample thickness increases; dependence of Φ_f on h_S for a single pulse is presented[13] in Figure 7.4. Fracture of thick samples is influenced by already existing defects. With intermediate beam density, $\Phi_c < \Phi < \Phi_f$, a crack appears in the specimen, but it does not propagate through; cyclic irradiation results in further propagation of the crack and the fracture. With subcritical density $\Phi < \Phi_c$, after several cycles of irradiation of thick samples, one can observe subcritical growth of already existing cracks or microcracks of the material, i.e., a shift of the fracture threshold occurs under prolonged exposure, and the fracture of samples is accompanied by the appearance of a great number of cracks.[3]

The speed of crack growth in the material under normal conditions is determined by the rate of energy fed to the end of the moving crack. Under usual mechanical loading in isotropic elastic media, the speed of crack growth V_c is limited by the value c_* at which branching takes place, that is, roughly $c_* = c_S/2$, where c_S is the speed of shear waves in the medium. If the rectilinear direction crack growth is predetermined, e.g., by anisotropy of the material, the maximum speed of crack growth under the influence of external forces will be[35,43] the speed of Rayleigh waves c_R. If the growth of the fracturing crack begins inside the specimen, the speed of fracture $V_f = h_S / t_f$, where h_S is the sample thickness and τ_f is the time to sample fracture, will exceed the speed of crack growth in one direction, V_c. In the limiting case of splitting from the center of the sample, the speed of fracture can reach the value of $2V_c$. Thus, the maximum value of the fracture speed for an isotropic plate under normal conditions is $2c_*$, and for an anisotropic one, $2c_R$.

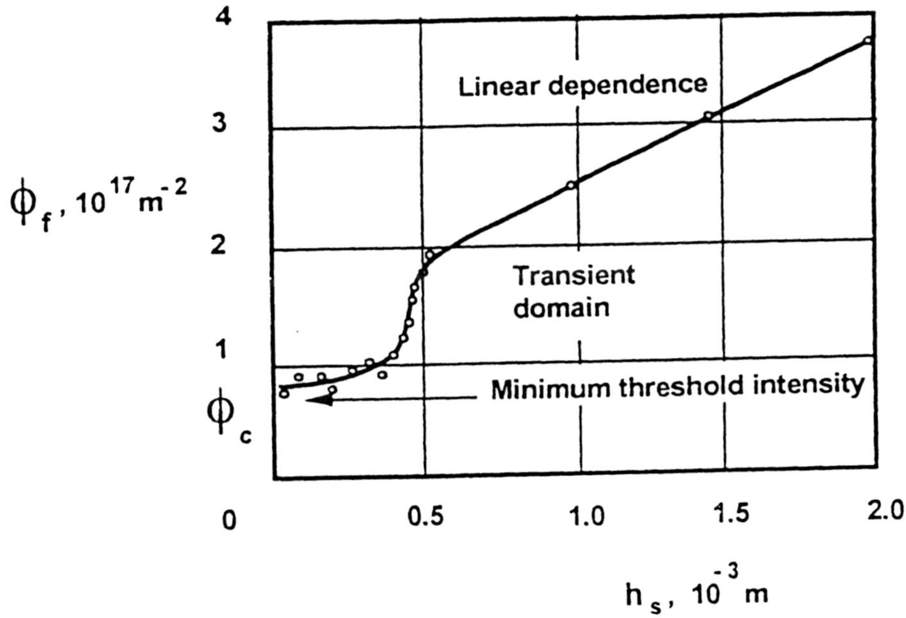

Figure 7.4. The fracturing beam density Φ_f versus sample thickness h_s for single pulse

Many authors[12,13,16,20,27] observed high speeds of fracture considerably exceeding both the speed of Rayleigh waves and that of transverse and longitudinal elastic waves while fracturing solids by powerful relativistic electron beams. Although exact experimental measurements of fracture speeds are still scarce, the data are reliable and significant: average fracture speeds mentioned in these articles exceeded the maximum speed of brittle fracture, c_R. For instance, while fracturing KCl and NaCl crystals, the average fracture speed V_f reached values of more than $3c_R$ (see Figure 7.5). In this case, the speed of longitudinal elastic waves is: $c_p = 4500$, $c_S = 2.4$, $c_R = 2.3$, and $c_* = 1.5$ *Km/s*. These data show that the fracture of solids by powerful electron beams differs considerably from usual brittle fracture and other types of quasibrittle, tough, or ductile fracture.

It should be noted that the methods of measuring fracture speeds by the secondary current pulse give only a crude estimate of the lower bound of crack growth speed for the following reasons.[13] First, the speed of fracture can exceed the average value to a great extent. Second, the fracture process lasts considerably longer ($\sim 10^{-7}$ *s*) than the irradiation pulse that causes it ($\sim 10^{-8}$ *s*); thus, the average fracture speed can arise from considerably different speeds of crack growth at different stages of propagation. Third, the exact place of crack emergence is not known, and even the average crack speed V_c determined from data on fracture speed V_f can differ within the interval $0.5\ V_f < V_c < V_f$. Hence, the speed of crack growth can exceed averaged experimental values to a very great extent.

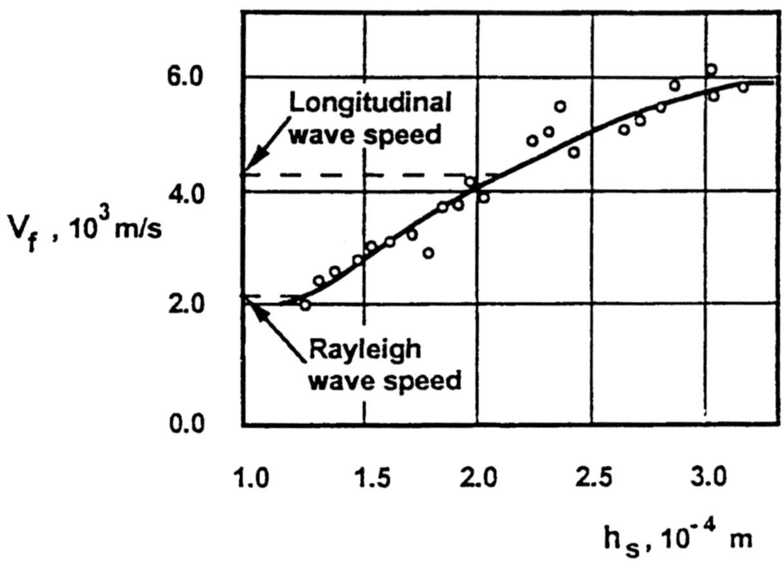

Figure 7.5. Fracture velocity, V_f, for KCl and NaCl crystals versus thickness of specimen, h_s

It is worth noting that for optical fracture caused by laser rays, the speed of crack growth can exceed the speed of elastic waves several times.[44–46] Even crack speeds of the order of $100c_S$ were registered.[46] Probably some resemblance must exist between the mechanisms of electron and optical fracture.

Another specific feature of fracture by powerful electron irradiation is connected with the small number of the cracks fracturing the sample. As already mentioned above, even for beam parameters that considerably exceed threshold values, usually only one crack appears. The number of splitting cracks increases only to two or three for a beam density two to three orders of magnitude higher than the threshold density.

All of the preceding phenomena connected with the fracture of solids by powerful relativistic electron beams manifest themselves very clearly for dielectrics (for alkalihaloid crystals, in particular). We should stress the unexpected peculiarity of the fracture of dielectrics: *Macrodefects in the samples do not affect fracture.* Even with the weakening of samples by means of cracks made beforehand, irradiation fracture takes place in some virgin region of the sample. The mechanism of the fracture of dielectrics by relativistic electron beams of high intensity appears to be not connected with the initial cracks.[3–7]

Other effects of the irradiation of solids by high energy electron beams are not connected directly with the fracture of materials. However, these are observed only while irradiating materials by relativistic electron beams and they do not depend upon the temperature and purity of the crystals.

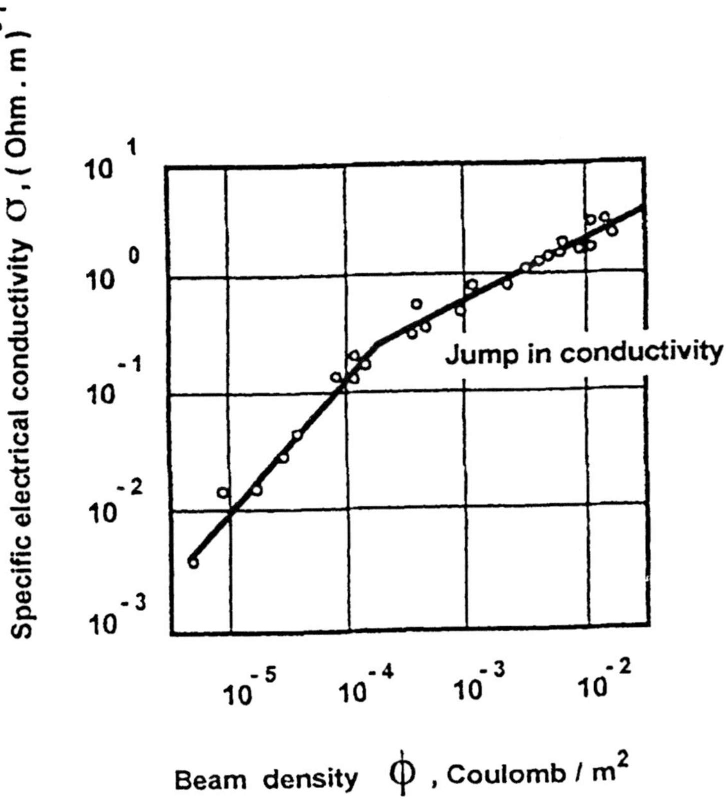

Figure 7.6. Specific electrical conductivity σ versus the electron beam density Φ for the KBr sample

The very large increase in electric conductivity of dielectrics with electron irradiation is not a threshold phenomenon and is connected only with electron beams; it is not observed with neutron, proton, or photon irradiation.[14,22,47] The electrical conductivity of the material σ can increase considerably, by three to four orders, even if irradiation intensity Φ is smaller than the threshold intensity Φ_c. Near the fracture threshold, σ reaches values of the order of 10 $\Omega^{-1} m^{-1}$, which is $10^{13}-10^{15}$ times larger than the initial value (e.g., for a threshold beam density, the resistance of the KCl sample corresponds to the resistance of a similar metal sample under usual conditions). The dependence of the pulse electrical conductivity σ on the density of electron irradiation for the KBr samples is shown[24] in Figure 7.6. Relaxation time for the intensive component of conductivity is very short. In the same work, it is noted that even for a high-speed oscilloscope with a resolving power of 10^{-1} *ns*, it is impossible to determine the shape of the current pulse signal; the authors evaluated the time for the existence of the effect with values of the order of 1 *ps*. (This value is comparable to the time, $1-10^{-1}$ *ps*, for the electromagnetic disturbance to pass through such a sample.)

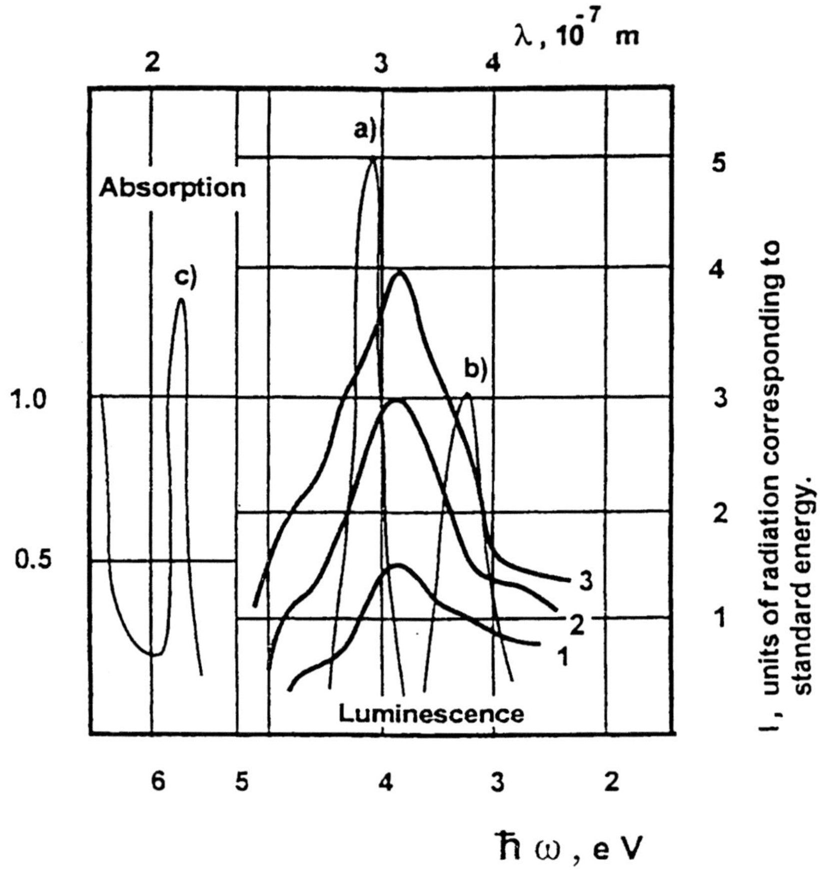

Figure 7.7. Spectra of light pulse radiation of KI crystals:
1) for $\Phi = 2.5\times10^{-4}$; 2) for $\Phi = 1.1\times10^{-3}$; 3) for $\Phi = 2.4\times10^{-3}$ Coulomb/m^2
The dotted lines are:a) standard singlet luminescence; b) triplet luminescence;
c) fundamental absorption spectrum

As external voltage increases, the pulse electrical conductivity grows weakly; for a voltage of -600 to +600 *V* (corresponding to fields of 10 *MV/m* with the intensity of the internal field of the solid being[32] of the order of 1–100 *GV/m*), the increase of σ does not exceed 30%. The immense increase in the conductivity of dielectrics in the interval 80–400 K does not depend on temperature. This fact, as well as the ultrashort time of relaxation of the effect, proves that charge carriers are fast electrons. The kinetic energy of these carriers must considerably exceed the energy of the thermal electrons, which is about $0.1eV$.

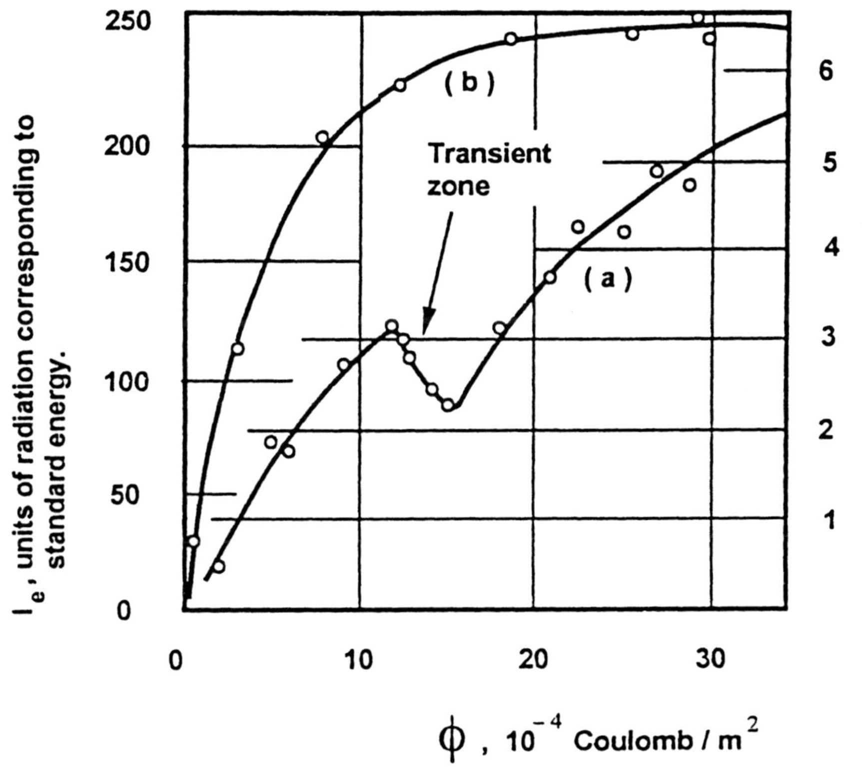

Figure 7.8. Glow intensity in (a) subnanosecond and (b) exciton range versus irradiation beam density Φ for KI crystal.

The typical dependence of σ on flux Φ (see Figure 7.6) shows a sharp change of the rate of increase with the irradiation density ($\Phi \sim 10^{-4}\ C/m^2$, where $C/m^2 = 6.2\times10^{20}$ electrons/m^2 in the case of an electron beam). Beginning from such values of Φ, of the order of $10^{-4}\ C/m^2$ for dielectrics, one can observe the characteristic glow with a subnanosecond time de-excitation, considerably shorter than for fundamental singlet luminescence of excitons[20,32]; the entire irradiated sample glows evenly. It is worth noting that the purity of the crystals has a very weak effect on glowing, and the glow intensity within the interval of 80-500 K does not depend on temperature. Characteristic features of the glow are as follows (see Figure 7.7): (a) The half-width of the radiation spectrum is 1.5–2 eV; (b) radiation occurs in the ultraviolet and visible spectral region with a frequency threshold $\hbar\omega = 5\ eV$ (corresponding to a wavelength of the order of 2500Å); (c) for a number of crystals, for KI in particular, the large wavelength tail of radiation obeys the power law of -3/2; (d) some structure appears that superimposes a wide band of spectrum characteristics.

Figure 7.8 shows[20,24] the glow intensity of KI crystals in terms of Φ for a subnanosecond glow at 80 K and exciton glow at 400 K. Note that the glow intensity grows as Φ increases, but a section of the curve after $\Phi = 10^{-3}$ C/m^2 can probably indicate the presence of some different glow mechanisms.

From a practical point of view, the most interesting prospect is the application of high-voltage accelerators for fracturing rocks. Here are the most important conclusions deduced from the first experimental data[2,26,48] concerning the fracture of a number of rocks: (a) The fracture is brittle and results in the formation of a few large fragments and many tiny particles whose total volume is very small; (b) a low-power intensity of fracture, of the order of 10^8 J/m^3, is characteristic of rocks with different mechanical strengths–up to the strongest of them, such as basalts, gneisses, etc.

Some comments

Experimental data show that electron fracture is beyond the limits of conventional notions of viscous, ductile, brittle, or mixed fracture of solids[35,48–51]. At the same time, this fracture proceeds in a "brittle" way, i.e., at the expense of one to two considerable cracks propagating in a solid, which indicates the mechanical nature of the splitting of the material. However, as one can judge from the peculiarities of the fracture of cracked samples and supersonic crack growth registered in the experiments, such splitting (or cutting) of a solid cannot be produced by means of energy flow to the end of the crack through the nuclei of the solid. It is natural to assume the existence of a physical mechanism which, beginning from a certain critical value of the beam density of relativistic electrons (or the absorbed energy), causes the formation of macroscopic electron clusters in a solid. The contradiction between the mechanical character of brittle cleavage and its peculiarities is then removed entirely. The electron clusters of solid plasma become the mechanical object whose high-speed progress causes material cleavage. (The mean-square-law speed of fluctuations in solid plasma reaches values[13,52] of the order of 1 Mm/s, which considerably exceeds the speed of elastic waves in a solid.) Furthermore, supersonic progress of the plasma cluster, like a knife or a wedge cutting a body, will be accompanied by shock waves in a solid. The existence of shock waves in material fractured by irradiation with powerful electron beams was observed in many works.[5,8,33,53,54]

The corresponding fracture pattern is called electron-beam-induced fracture mode (EFM). The problems of defining the physical mechanism causing the formation of electron clusters in a solid are also to be studied, as well as the problem of determining the laws of the motion of clusters cutting the body. To solve these problems, the present chapter considers the interaction taking place in the powerful relativistic electron beam in a continuum medium, in which the effect of the coalescence of superluminal electrons is discovered. The problem of cutting a solid by a "knife" moving with supersonic speeds is used to find the length of a final crack.

This approach can be deemed only as a first approximation to reality. The full picture of these phenomena must also take into account the effects of the usual types of viscous, brittle, or ductile fracture and effects of excitation in the electron subsystem of a solid. Some authors reported the existence of several different mechanisms for fracturing solids by means of powerful relativistic electron beams, including electrical breakdown,[3] the mechanical effect of the electron beam,[13] and the thermomechanical effect.[27] On the other hand, some experimenters believe that these three mechanisms barely influence the observed fracture.[16,17,19] There is some mention of the relativistic effects of the excitation of the bound electrons of the material.[24] Different explanations for electron fracture, such as thermal-shock fracture,[2,8] accumulation of internal microvoltages,[3,55,56] and the waves of elastic and thermal stresses,[21,41] do not explain the main peculiarities of EFM by electron irradiation.[36,57] The remarks about the leading role of solid plasma in the process of fracture[13,40,54] are close to the mechanism of plasma "wedge" or "knife" of the initial stage of the process presented here.

The non-continuum approach to the description of supersonic fracture was suggested by Sanders[58] using an ideal lattice in which the crack growth rate is determined by the rate of energy input. Fracture on the atomic level within the framework of a one-dimensional model of a limited chain, analogous to Sanders model, has also been considered.[56]

7.2. Electromagnetic Media: Invariant Integrals and Interactions

The electron beams used for the fracture irradiation of solids are relativistic and, moreover, the particle speeds surpass the phase speed of electromagnetic waves (the speed of light) in the material medium, for example, in dielectrics. The existence of these electrons was discovered by Cherenkov[59-61] and Vavilov.[62] The theoretical description of the motion of an individual superluminal electron and the explanation of the irradiation that arose in the case of this motion were given by Tamm and Frank,[63] and Tamm[64] awarded by the Nobel Prize for this work.

Relativistic electron field

The phenomenon of the motion of Cherenkov's superluminal electrons and the Cherenkov-Vavilov irradiation are analogous to the formation of shock waves produced by the motion of a source of perturbation with speeds faster than elastic-wave speeds in a continuum.[64] The electromagnetic field of a superluminal electron differs principally from the field of a stationary charge at rest as follows: in the charge wake inside the Mach cone only, the components of the electromagnetic field change sign when the charge speed passes the phase speed of light. This difference completely changes the nature of the interaction for electron beams moving with superluminal speeds.

The energy dissipation of an electron (and its deceleration in a medium) depends on the interaction of the electron's field with external electromagnetic fields existing in this media (the bremsstrahlung and ionization), on the wave losses on the Mach cone fronts (the Vavilov-Cherenkov irradiation), and on the collective interaction with the fields of other beam electrons. The interactions of the first and second kinds are approximately equal for all beam particles at great speed (they are identical if the particles form a periodic lattice), and these interactions exert negligible influence on the mutual spatial distribution of electrons in the beam. Therefore, this distribution and hence the deformation of an electron beam is determined mainly by the collective interaction. That's why only the collective interaction of electrons is studied below.

The electromagnetic interaction of two or more charges moving with superluminal speeds in a continuous medium was studied by Borzykh and Cherepanov.[65] The law governing the interaction of two charges is necessary for investigation of collective interactions. To obtain this law we use the metod of the invariant Γ integrals[66-69].

The collective interactions are considered here on the assumption that all particles have the same constant speeds. This statement of the problem is correct for times much smaller than the time for existence of superluminal electron beams in a medium, and it permits simple estimations of process parameters. We note, too, that mutual distribution and collective interaction are considered in the proper coordinate frame moving with electrons, and the relativistic effects of the shortening of the length and time are significant for these systems.[70-74]

For the study of collective interactions, we use the known solution[75] of the Maxwell equations for the proper field of a charge moving with constant speed υ, which is greater than the phase speed a of light in the medium, but less than the speed c of light in vacuum. Suppose the charge is at the coordinate origin of the proper space-time coordinate frame, x_1, x_2, x_3 and t; and the direction of its motion coincides with that of the x_3 axis. Then, the steady-state field of a superluminal negative charge e is given by

$$E_i = \frac{e M^2 x_i}{2\pi\varepsilon'(x_3^2 - M r^2)^{3/2}} \qquad (i = 1,2), \tag{7.5a}$$

$$E_3 = \frac{-eMx_3}{2\pi\varepsilon'(x_3^2 - M r^2)^{3/2}}, \tag{7.5b}$$

$$B_1 = Ax_2, \quad B_2 = -Ax_1, \qquad r^2 = x_1^2 + x_2^2, \tag{7.5c}$$

where

$$A = \frac{\mu' eV M (1 - a^2/c^2)}{2\pi(1 - \upsilon^2/c^2)(x_3^2 - M r^2)^{3/2}}, \qquad B_3 = 0. \tag{7.5d}$$

Here,

$$M = \frac{\left(\upsilon^2/a^2\right) - 1}{1 - \left(\upsilon^2/c^2\right)} = \frac{\mu\varepsilon\upsilon^2 - c^2}{c^2 - \upsilon^2} \tag{7.5e}$$

$$a=\frac{c}{\sqrt{\mu\varepsilon}},\qquad \mu'=\mu\mu_0,\qquad \varepsilon'=\varepsilon\varepsilon_0, \tag{7.5f}$$

where E_1, E_2, E_3, B_1, B_2, and B_3 are the electromagnetic field components of the $x_1\ x_2\ x_3$ system, a is the phase speed of light in the medium, ε' is the dielectric (and μ' magnetic) constant of the medium, and M is the specific relativistic Mach number (ε_0 and μ_0 are the absolute dielectric and magnetic constants of the medium, respectively). The field described by Equations (7.5) is defined inside the Mach cone, i.e., in the region $x_3^2 > Mr^2$, $x_3 < 0$; outside this cone, the field of the electron vanishes.

Mechanical model for the supersonic cutting of solids

The phenomenon of the supersonic crack can be explained by the existence of some objects of the size much greater than interatomic spacing that move in a solid with speeds faster than those of sound. In the following, we assume that an electron plasma cluster forms that cuts the solid like a "wedge" or a "blade." The cluster form at the beginning of fracture if a certain critical intensity of irradiation is achieved. This idea is analogous to that of the mechanism of fracture induced by plasma jets for high-power pulsed-laser irradiation.[44–46] The speed of plasma jet propagation is faster than elastic wave speeds, too. The large supersonic speeds of the plasma blade cause the same persistent pressure of the blade on the surface of a propagating crack at its tip. The motion of a finite-length plasma blade can form a narrow canal of fracture. The maximum probability of the crack orientation would be along a minimum-strength plane, such as a cleavage plane of a crystal. It is obvious that irreversible deformations would not have time to develop on the sides of a crack arising from supersonic cutting because the speed of dynamic dislocations responsible for irreversible strains is limited by speeds of sound.

Remember a related dynamic problem of elasticity about the supersonic motion of a point load on the boundary of a half-plane.[76] The solution to this problem is analogous to that of a similar acoustic problem,[77] complicated by the fact that in elasticity there is a system of two wave equations and two wave fronts. The supersonic motion of a finite plasma blade is analogous to the flight of a thin wing in gas dynamics.

Following these introductory notes, the problem of supersonic cutting is solved below within the framework of the model assumptions about an absolutely rigid wedge and an ideally elastic material. Electron fracture is a complicated phenomenon, and detailed information on the properties of solid plasma clusters and their behavior in a material is lacking. Therefore, the simplifications are needed for a preliminary description of the electron fracture which will permit the eventual study of the phenomenon later.

Suppose that a thin, rigid wedge of length l moves in an infinite elastic space with the supersonic speed V. The wedge angle is equal to 2α (see Figure 7.9), and the direction of movement is the x axis. The two-dimensional problem is assumed to be symmetric relative to the x axis; the z coordinate directed along the wedge edge is nonessential. The surface of the resulting cavity outside the contact zone is a free boundary. The law of wedge motion with the initial supersonic velocity V_0 must be found.

The boundary condition is

$$V_n = 0 \quad \text{on} \quad \Sigma_w \tag{7.6}$$

where V_n is the normal component of the velocity at the contact surface Σ_w. The condition for contact friction is

$$\sigma_{tn} = F(\sigma_n, V_t)\ , \tag{7.7}$$

where the directions of the tangent and the normal to the contact surface Σ_w are denoted by subscripts t and n respectively; the normal and shear stresses are σ_n and σ_{tn}; and the normal and tangential components of the material velocity are V_n and V_t. The function F is a known frictional characteristics found from an experiment, or from some model ideas; for example, we will assume $\sigma_{tn} = f\sigma_n$, where f is the Coulomb friction coefficient.

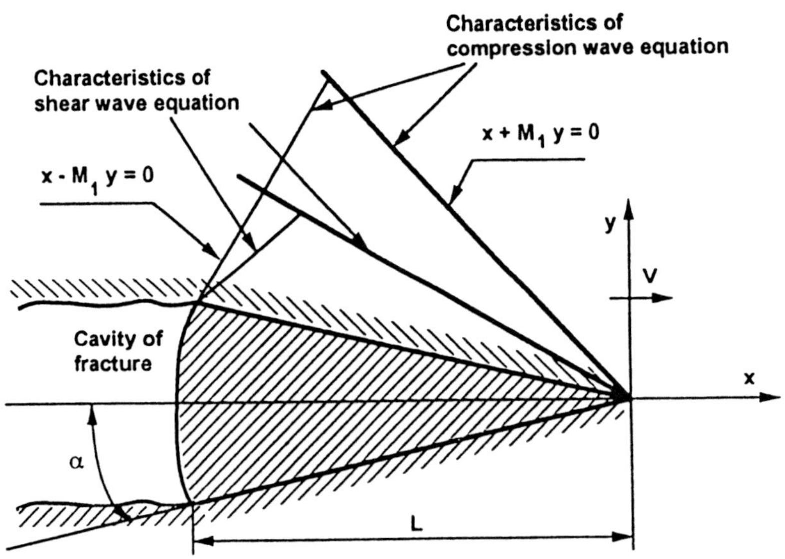

Figure 7.9. A thin wedge moving in an elastic solid with a supersonic velocity. Here M_1 is the modified Mach number, see Equation (7.43) below

The statement of the problem is closed using the conditions of mass and momentum conservation at two Mach fronts, where velocities and stresses are discontinuous.

Invariant dynamic integrals

The state of the electromagnetic deformable matter can be characterized by field vectors **E**, **B**, **D**, and **H**, the displacement vector **u**, the velocity vector **v**, the stress tensor σ_{ij} and a certain strain tensor ε_{ij}. The following equations hold:

Maxwell's equations:

$$\gamma_{ijk}\ E_{j,k}+\frac{\partial B_i}{\partial t}=0,\quad \gamma_{ijk}H_{j,k}-\frac{\partial D_i}{\partial t}=J_i\ , \tag{7.8a}$$

$$D_{i,i}=\delta,\quad B_{i,i}=0,\quad J_{i,i}+\frac{\partial\delta}{\partial t}=0\ ; \tag{7.8b}$$

Newton's equations:

$$\sigma_{ij,j}=\frac{d(\rho v_i)}{dt}-\rho F_i \quad . \tag{7.9}$$

The local law for the conservation of energy is

$$\dot{U}=q_{i,i}+E_i\dot{D}_i+H_i\dot{B}_i+\sigma_{ij}\ \dot{\varepsilon}_{ij}\,. \tag{7.10}$$

The kinematic conditions for any finite strains are

$$2\dot{\varepsilon}_{ij}=v_{i,j}+v_{j,i}\ ,\qquad v_i=\dot{u}_i \tag{7.11}$$

Here, J is the current density vector; δ is the charge density; ρ is the density of the matter; x_1, x_2, x_3 are rectangular Cartesian coordinates; t is time; u_1, u_2, and u_3 are displacement components; **q** is the vector of the net heat flux; $\dot{U}$ is the rate of change of the internal energy of the matter per unit volume (the dot over a letter refers to the total derivative with respect to time) and $\gamma_{123}=\gamma_{231}=\gamma_{312}=1$, $\gamma_{132}=\gamma_{321}=\gamma_{213}=-1$, all other γ_{ijk} being equal to 0. The strains of the medium are arbitrary ($\dot{\varepsilon}_{ij}$ is the strain rate tensor).

All functions in these equations are assumed to be continuously differentiable the necessary number of times throughout the investigated region, with the exception of singular points, singular lines, and singular surfaces, at which these equations make no sense.

The invariant Γ integrals of the first kind[67,68] are:

$$\Gamma_k = \int_\Sigma \left[\left(G + F + \frac{1}{2}\rho \dot{u}_i \dot{u}_i \right) n_k + \left(D_i E_k + B_i H_k - \sigma_{ij} u_{j,k} - q_{i,k} \right) n_i \right] d\Sigma \tag{7.12}$$

$$(i, j, k = 1,2,3)$$

Here, Σ is a surface in the coordinate system, x_1, x_2, x_3, and G and F are the following potentials:

$$G = U - E_i D_i - H_i B_i \ , \tag{7.13a}$$

$$F = -\int \left[\frac{\partial P_i}{\partial t} + \rho E_i + \gamma_{ijk} \ J_j \ B_k \right] d\,x_i \tag{7.13b}$$

$$P_i = \gamma_{ijk} B_j D_k \ . \tag{7.13c}$$

For the physical field, which is stationary with respect to Σ, the following theorems were proven.[67,68]

Theorem 7.1. Γ integrals do not change their value along any closed surface Σ encompassing a singular point, singular line, or singular surface. The surface Σ can be arbitrarily deformed without changing the value of the Γ integrals, provided that, in the deformation process, the surface Σ does not intersect a singular point, a singular line, or a singular surface.

Theorem 7.2. If an open surface Σ is limited by a contour L, the Γ integrals do not change their value with any deformation of the surface Σ, so long as (a) the contour L is fixed, and (b) the surface Σ does not intersect a singular point, a singular line, or a singular surface in the process of deformation.

In the Cartesian coordinate system x_i' moving at constant velocity V, coinciding with that of beam electrons, the Γ integrals are found from Equation (7.12) by the transformations $x_i \to x_i'$, $\dot{u}_i \to \dot{u}_i' - V_i$. Theorems 7.1 and 7.2 are true in the x_i' coordinates for a moving surface Σ fixed with beam electrons, for thermodynamically reversible media, or for steady-state field in any media.

The theory of the motion of singularities[66-68] is given here for speeds faster than the sonic or luminal speed of propagation.

To be specific, consider a singular point O_S, moving with a superluminal or supersonic velocity in the direction of the x_3 axis. This motion forms a moving, singular surface Σ_S (or a family of singular surfaces for different wave processes), joined to the singular point O_S. If the motion of O_S is a locally steady process, surface Σ_s is the conic surface, $x_3^2 = M_\bullet r^2$, $x_3 < 0$, near the point O_S where

$$M_\bullet = \upsilon^2 / c_\bullet^2 - 1 > 0 \quad \text{and} \quad r^2 = x_1^2 + x_2^2 \tag{7.14}$$

where x_1, x_2, and x_3 are coordinates in the moving reference frame fixed to O_S, and c_* is the characteristic wave speed. By considering the motion of a singularity from time $t=0$ of its creation at some inner point of a continuum, we see that in the time t a domain of the field of singularity O_S will have the form shown in Figure 7.10; outside this disturbed volume zone V_c the field will be zero. The surface that bounds V_c consists of two parts, such that the singular surface Σ_S is the surface of an axisymmetrical, curvilinear Mach front cone, that is, a shock wave front, with the components of velocities, stresses and other field parameters being discontinuous; Σ_{ct} is the rear wave front of disturbances arising at $t=0$.

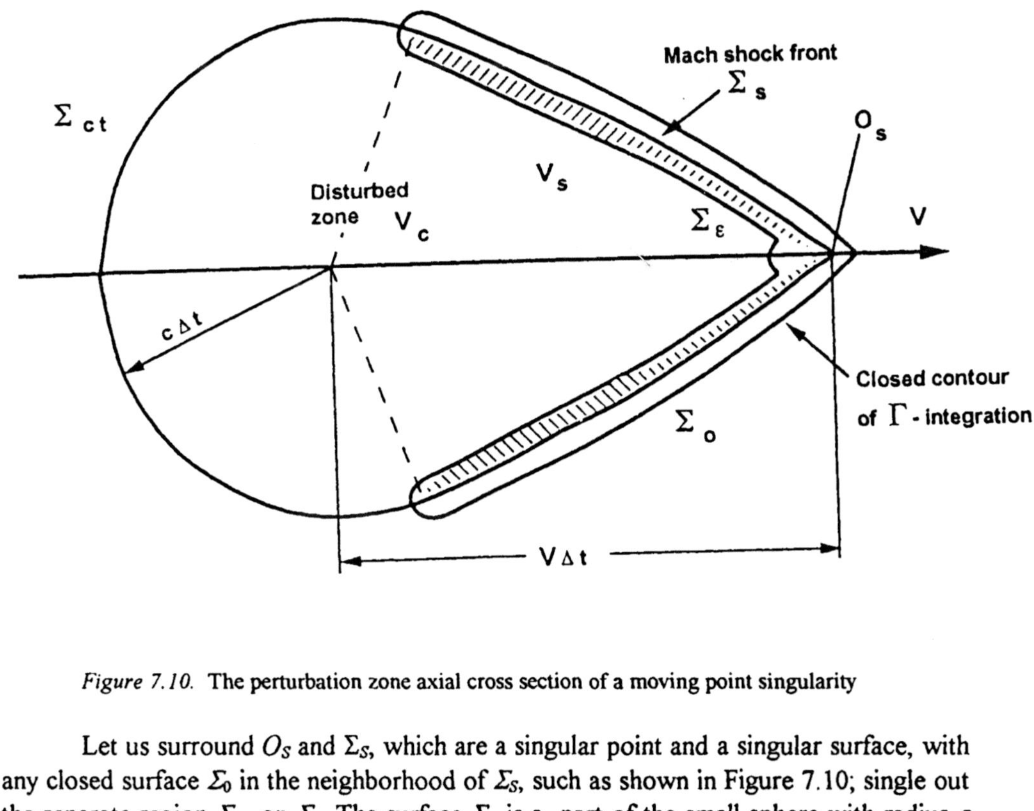

Figure 7.10. The perturbation zone axial cross section of a moving point singularity

Let us surround O_S and Σ_S, which are a singular point and a singular surface, with any closed surface Σ_0 in the neighborhood of Σ_S, such as shown in Figure 7.10; single out the separate region Σ_ε on Σ_0. The surface Σ_ε is a part of the small sphere with radius ε that tends to zero (for zero external fields the field of singularity is absent outside V_c and on the outward part of Σ_0). The field is continuous along Σ_0.

Let us consider the invariant integral Γ_3 over Σ_0. Physically, it represents the energy flux flowing into Σ_0. Surface Σ_0 can be deformed, without intersecting a singular point, a singular line, or a singular surface, to be situated wholly outside V_c. Hence, the invariant integral Γ_3 over Σ_0 is equal to zero. Surface Σ_0 can be represented as the sum of Σ_0^+, Σ_0^- and Σ_ε where Σ_0^+ and Σ_0^- are the inward and outward parts of Σ_0 respectively.

The integral Γ_3 over Σ_0^- equals zero because the external field is zero. The integral Γ_3 over Σ_0^+ represents the energy dissipation on Σ_S per unit length (of the singularity progress). Denote as Γ_ε the invariant integral Γ_3 over Σ_ε. Physically, Γ_ε represents the intensity of energy source at O_s. Denote as $\Gamma_3 = -\Gamma_\varepsilon$ the value of energy dissipation rate at the singular point O_s.

The energy dissipation on Σ_S is equal to

$$dA = \Gamma_3\, dx_3 \tag{7.15}$$

when O_S advances on $dx = \upsilon_3 dt$ during time dt. Here, υ_3 is the velocity of O_S.

Analogous equations can be derived for the singular point motion in the direction of the x_1 and x_2 axes. As a result, we have

$$\Gamma_i = \frac{\partial A}{\partial x_i}, \tag{7.16}$$

where Γ_i is the energy dissipation at O_S per unit length of progress of O_S along the x_i axis. The energy dissipation rate is equal to

$$\dot{A} = \Gamma_i\, \upsilon_i \qquad (i = 1,2,3) \tag{7.17}$$

where υ_i is the component of the singular point velocity along the x_i axis. As seen, the fact that Γ is invariant with respect to the integration surface, greatly simplifies the calculations.

For the supersonic or superluminal singularities under consideration, the singularity O_S is the source of energy flowing down the singular surface Σ_S which is the sink of energy. Some examples of this dissipation include the shock wave in gas dynamics or the Cherenkov -Vavilov irradiation of superluminal electrons in dielectrics.

In the case under study, the singularity is an electron; hence, Γ_1, Γ_2, and Γ_3 are equal to the corresponding components of the force applied to the electron (the drag).

In the general case, when an external field outside V_c is non-zero, the energy dissipation rate is given by

$$\dot{A}_0 + \dot{A}_S = \Gamma_i\, \upsilon_i\,. \tag{7.18}$$

Here $\dot{A}_0$ and $\dot{A}_S$ are dissipation rates in external and internal fields obtained by integration over the outer and inner sides of Σ_0, respectively.

In the case of the steady motion of a singularity that corresponds to the infinite time of the motion of the singularity, the disturbed zone is the infinite cone, as shown in Figure 7.11. The limits of the Γ integrals over Σ_0^+ and Σ_0^- as $t \to \infty$ provide the complete dissipation rates $\dot{A}_0$ and $\dot{A}_S$.

The sign of $\dot{A}_0$ can be positive or negative depending on whether the external field will absorb the energy of the point source or feed an internal field inside V_c. The sign of $\dot{A}_S$ is always positive because it corresponds to the dissipation of energy.

The value of Γ_i is defined as the result of the calculation of the invariant integral Γ_i over Σ_ε as $\varepsilon \to 0$ and the axis of the light cone coincides with the x_i th direction of motion of the electrons. In gas dynamics, the quantities Γ_i equal the components of wave drag experienced by a moving body, and in electrodynamics, $\Gamma_i\ \upsilon_i$ is the bremsstrahlung radiation energy rate of Cherenkov-Vavilov.

In the problem under study, according to Equations (7.5), the field is steady state, not only in the vicinity of O_S, but also all over the disturbed domain, V_c.

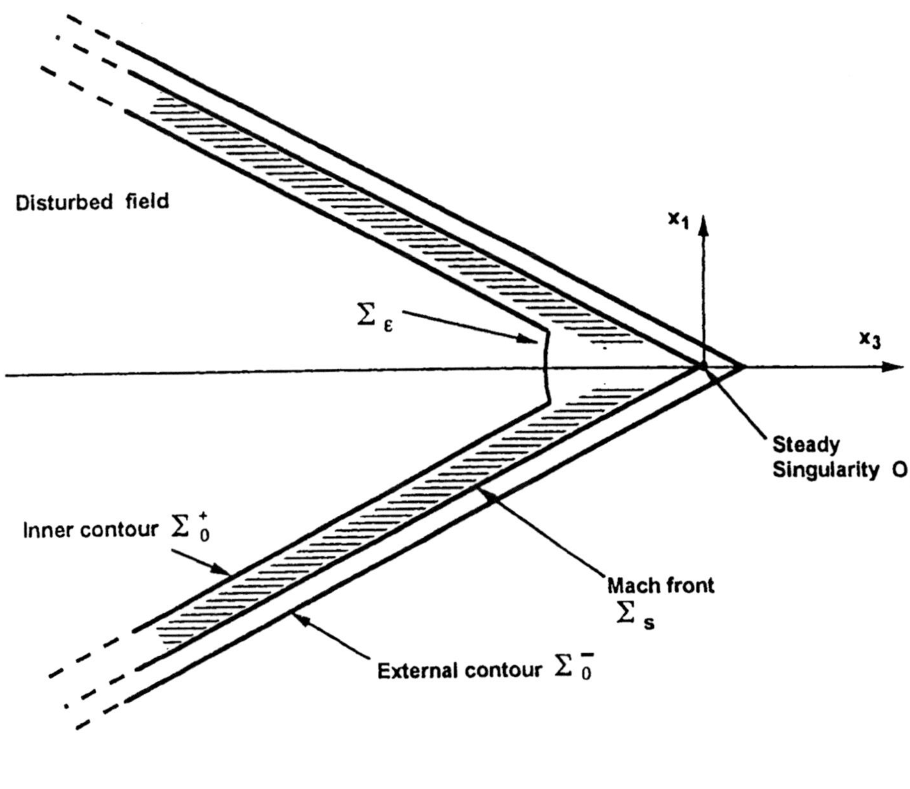

Figure 7.11. Surface of integration near a point singularity

A problem is well posed if its solution provides a finite value of Γ_i; otherwise, the problem is ill posed because an infinite energy makes no physical sence. Well-posed problems permit determination of the physical characteristics of the singularity, namely, a drag, a specific dissipation, etc. It should be noted that asymptotic fields of many singularities often yield improper Γ integrals which are usually divergent. In order to compute invariant integrals divergent at singularities, we apply a heuristic rule of Γ integration,[66-69] which requires taking into account only the finite part of the integration result. Furthermore, the integration surface in the neighborhood of a singularity should be symmetrical. As to the problem under study, symmetry is achieved provided that the integration surface Σ_ε in the neighborhood of an electron is, e.g., spherical (Figures 7.10 and 7.11).

Particular forms of invariant Γ integrals in different media used in the moving coordinate frame with respect to which physical field is steady are listed here:

(i) a compressible gas,[48,78]

$$\Gamma_i = \int_\Sigma \left(\rho v_i v_j n_j + p n_i\right) d\Sigma \qquad (i, j = 1,2,3), \tag{7.19}$$

where p is pressure, and v_i is a gas velocity component,

(ii) an electromagnetic field in a dielectric,[79]

$$\Gamma_i = \int_\Sigma \left[\left(D_j E_i + B_i H_j\right) n_j - \frac{1}{2}\left(D_j E_j + B_j H_j\right) n_i \right] d\Sigma, \tag{7.20}$$

(iii) a linearly or nonlinearly elastic medium,[66]

$$\Gamma_i = \int_\Sigma \left[\left(U + \frac{1}{2}\rho v_k v_k\right) n_i - \sigma_{kj} u_{k,i} n_j \right] d\Sigma. \tag{7.21}$$

For static problems, velocities equal zero ($v_k = 0$), and the latter equation provides an invariant J integral.[35]

7.3. The Electron Coalescence of a Relativistic Beam in a Medium

Consider interaction effects in electron beams using invariant Γ inegrals (7.20).

Individual electron with a superluminal speed in a dielectric medium

Let us consider the motion of an individual electron with negative charge e having a constant superluminal speed υ in a homogeneous, external electromagnetic field $\mathbf{E}_0 = \{E_{0i}\}$ and $\mathbf{B}_0 = 0$ in a dielectric. The proper singular field $\mathbf{E}_S$ and $\mathbf{B}_S$ of this electron is determined by Equations (7.5). According to the principle of superposition based on the linearity of Maxwell's equations, the complete field near the individual electron is

$$\mathbf{E} = \mathbf{E}_S + \mathbf{E}_0\,, \qquad \mathbf{B} = \mathbf{B}_S + \mathbf{B}_0 \tag{7.22}$$

The proper dissipation of energy of an electron in the external field per unit length which is equal to the drag of the electron, is defined according to Equation (7.20), where the integral is taken over Σ_ε. As the Σ_ε surface, we choose the small circular cross section of the Mach cone at $x_3 = -\varepsilon$ (see Figure 7.12).

Electrostriction and magnetostriction effects are ignored here. Substituting Equations (7.5) and (7.22) into Equation (7.20), and calculating the corresponding surface integrals according to the rule of Γ integration as $\varepsilon \rightarrow 0$, we obtain

$$\Gamma_i = eE_{0i} \qquad (i = 1,2,3). \tag{7.23}$$

We show only the calculation of Γ_3. In this case, we have

$$\Gamma_3 = \lim_{\varepsilon \to 0} \frac{1}{2} \int_{\Sigma_\varepsilon} (D_1 E_1 + D_2 E_2 + B_1 H_1 + B_2 H_2 - D_3 E_3 - B_3 H_3) d\Sigma . \tag{7.24}$$

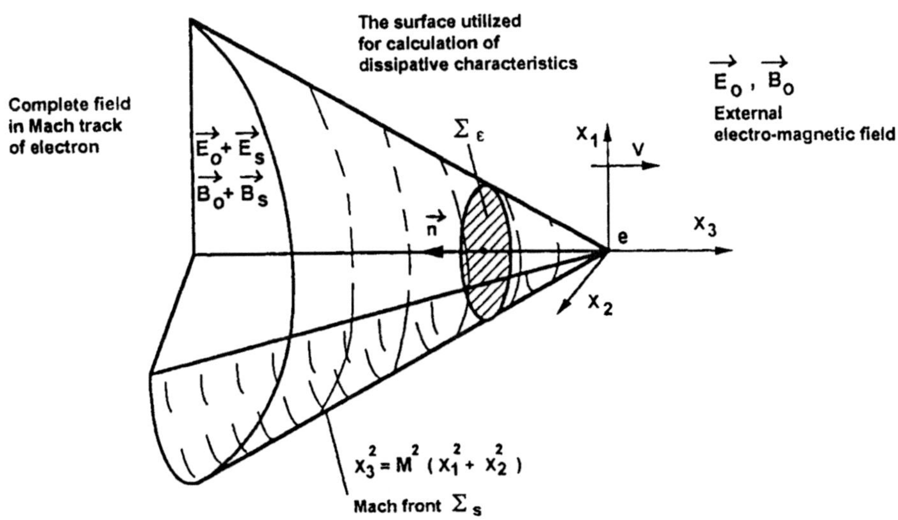

Figure 7.12. The three-dimensional Mach front picture of an electron moving in a medium with a speed exceeding the phase speed of light

Making use of the rule of Γ integration and Equations (7.5) and (7.22), we can reduce this integral to

$$\Gamma_3 = -\lim_{\varepsilon \to 0} eE_{03}\, \varepsilon \int_0^\varepsilon \frac{\tau d\tau}{(\varepsilon^2 - \tau^2)^{3/2}} = eE_{03} \tag{7.25}$$

We note that the external field is considered in the coordinate frame fixed to the relativistic electron; translation to the laboratory system is made by Lorentz's transformations. Formally, Equation (7.25) coincides with the corresponding expressions for a stationary electron.

One-dimensional, semi-infinite chain of superluminal electrons
Let another electron e_1 be situated inside the Mach cone of the first electron e_0; both move with a superluminal speed $V > a$. (The indices 0 and 1 are used to label the particles). For e_1 the external field will now be the field of the electron e_0. Therefore, if both charges move along one axis, then on the axis $\mathbf{B} = 0$, according to Equations (7.5c,d). Let z be the distance between the particles. According to Equations (7.5a,b) and (7.25), the force acting on the charge e_1 is:

$$F_1(z) = \frac{e_0 e_1 M}{2\pi\varepsilon' z^2} , \tag{7.26a}$$

where

$$M = \frac{\upsilon^2/a^2 - 1}{1 - \upsilon^2/c^2} . \tag{7.26b}$$

Equation (7.26) is the generalization of Coulomb's law. The latter is derived from (7.26) in the limit $\upsilon \to 0$. So, Coulomb's law holds only under static conditions. When the speed of electrical charges increases, the interaction force becomes less and vanishes at the phase speed of light. The interaction force of two moving electric charges versus their speed is shown in Figure 7.13a.

When $\upsilon > a$ one can see from (7.26) that the rear electron e_1 is always attracted to the frontal electron e_0. In this case, there is no reaction of rear electron on frontal electron.

Let us consider a one-dimensional, semi-infinite chain of superluminal electrons moving along the x_3 axis, the distance between any two neighboring electrons being equal to b at the initial moment. In this case, the interaction forces are directed along the axis of the chain. Let f_{mn} be the force acting on the mth electron, due to the nth electron ($n < m$); the resultant force acting on the mth electron is

$$F_m = \sum_{n=0}^{m-1} f_{mn} . \tag{7.27}$$

According to Equations (7.26) and (7.27), at the initial instant, we obtain

$$F_m(b) = \frac{e^2 M}{2\pi\varepsilon' b^2} \sum_{n=0}^{m-1} (n+1)^{-2} . \tag{7.28}$$

As is known,[80]

$$1 \le \sum_{n=0}^{m-1} (n+1)^{-2} < \frac{\pi^2}{6} . \tag{7.29}$$

Hence,

$$F_1(b) \le F_m(b) < \frac{\pi^2}{6} F_1(b) . \tag{7.30}$$

From Equation (7.30) it follows that for any m, the forces $F_m(b)$ differ little from F_1. Therefore, one can obtain a simple estimate of the deformation of a chain system by considering the motion of a single electron e_1 in the field produced by e_0.

Taking Equation (7.26a) into account, we can write the relativistic equation of motion of electron e_1 in the moving coordinate system in the form[71-74]

$$\frac{d^2 z}{dt^2} = \frac{e^2(\mu\varepsilon\upsilon^2/(c^2-1)}{2\pi\varepsilon' m_0(1-\upsilon^2/c^2)z^2} \left[1-\frac{1}{c^2}\left(\frac{dz}{dt}\right)^2\right]^{3/2} \tag{7.31}$$

(m_0 is the mass of electron at rest).

First, we limit ourselves to the case of small relative particle speeds where $(1/c^2)(dz/dt)$ may be neglected in Equation (7.31). On this assumption we solve Equation (7.31) for the following initial conditions,

$$z = -b \quad \text{and} \quad \frac{dz}{dt} = 0 \quad \text{at} \quad t = 0 , \tag{7.32}$$

and obtain the following solution

$$t K^{1/2} = b^{1/2}(-z)^{1/2} (z+b)^{1/2} + b^{3/2} \arcsin \left(1+\frac{z}{b}\right)^{1/2} , \tag{7.33}$$

where

$$K = \frac{e^2(\upsilon^2/c^2-\varepsilon^{-1})}{\pi\varepsilon_0 m_0(1-\upsilon^2/c^2)} \quad (\mu = 1 \quad \text{for dielectrics}) . \tag{7.34}$$

Let us estimate the time τ necessary for e_1 to approach e_0 to within very short distance. In this way, a dense system of two electrons is formed; quantum interactions are essential in this dense system, just as in a solid.

Substituting $z = 0$ into the solution, Equation (7.33), we obtain the time for coalescence of two electrons

$$\tau = \frac{\pi b^{3/2}}{2K^{1/2}} . \tag{7.35}$$

We note that quantities b, t, and τ are considered in the coordinate system fixed at electron e_0. Lorentz transformation provides the corresponding quantities in the laboratory frame of reference,

$$b' = b\left(1 - \upsilon^2 / c^2\right)^{1/2}, \qquad t' = t\left(1 - \upsilon^2 / c^2\right)^{-1/2}, \tag{7.36}$$

and from Equation (7. 35) it follows that

$$(\tau')^2 = \frac{\pi^3 \varepsilon_0 m_0 (b')^3}{4e^2(1 - \upsilon^2 / c^2)^{3/2}(\upsilon^2 / c^2 - \varepsilon^{-1})} . \tag{7.37}$$

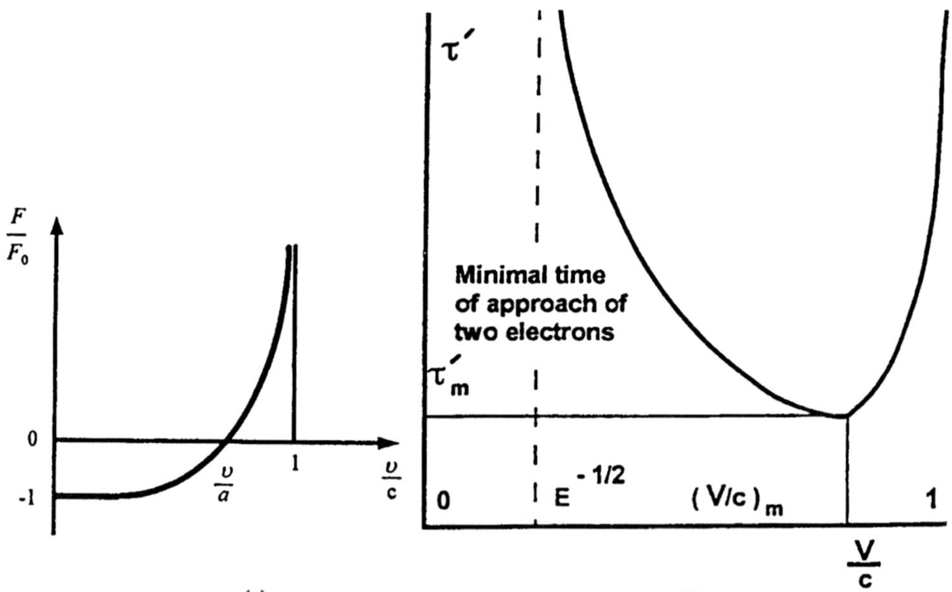

(a) (b)

Figure 7.13. Relativistic interaction of an electron pair: a) Force on the rear electron versus speed of the electron pair (F_0 is the static interaction force); b) The laboratory time τ' of the coalescence of two relativistic electrons versus their speed υ / c

The laboratory time τ' of the coalescence of two electrons in terms of their speed υ / c, for a dielectric material, is shown in Figure 7.13b. As we can see, the distance between the two electrons contracts most substantially in a narrow region of energies (speeds) of the particles, where τ' is small. For $\upsilon^2/c^2 = (3+2\varepsilon)/5\varepsilon$, the coalescence time τ' takes the minimal value

$$\tau'_m = \frac{c_m (b')^{3/2}}{(1-\varepsilon^{-1})^{5/4}}, \tag{7.38}$$

where

$$c_m = \frac{(5^{5/4}\pi)^{3/2}\varepsilon_0^{1/2} m_0^{1/2}}{12^{3/4} e}. \tag{7.39}$$

For an electron, $c_m = 6.437 \times 10^{-2} m^{-3/2} s$.

Hence, a chain of superluminal electrons condenses to a solidlike state during a time span τ' defined by Equation (7.37). For instance, for $b' = 1\mu m$ (of the order of the distance between electrons in pulsed electron beams), we obtain $\tau'_m \sim 10^{-11} s$. This means that electron plasma clusters form at a depth of the order of 1mm beneath the surface of an irradiated material.

Electron beams in a solid

As shown, relativistic electron beams exhibit the mechanism for self-compression. Since the time T of the action of the coalescence mechanism is small, this mechanism can manifest itself only for high-density beams (small b'). From Equation (7.38) the necessary particle density in the initial beam can be estimated as follows:

$$n_0 (b')^3 > c_m^2 (1-\varepsilon^{-1})^{-5/2} T^{-2}. \tag{7.40}$$

The quantity T is the time of run of the directed beam of superluminal electrons in a medium and is determined by the interplay of two factors, namely the deceleration of electrons to the phase speed of light in the medium, and the losses due to excitation and ionization of the fixed electrons of the material medium. A more precise value of critical density can be obtained from Equation (7.37).

Consideration of the spatial distribution of electrons in a beam shows that forces acting inside the system are mainly directed along the axis of motion, but there are also some forces directed towards the boundary of the Mach cone. However, the latter forces are substantial only for boundary electrons, and in the far zone these forces are balanced by the action of other superluminal electrons. Moreover, dispersion of particle energies in a beam induces a Lorentz force, which depends on the relative velocity of motion.

It follows that the effect of self-compression can be observed experimentally when electron beams pass through a material, the thickness of which is smaller than the deceleration length. The initial density can be defined by Equation (7.40), where T is the time of electron motion in the material.

Decrease of the pulse duration for self-compression of a beam can be estimated directly, if the initial duration of the pulse is smaller than T (otherwise, a series of local zones of large densities are created, and the initial pulse duration is not transformed). For example, in a dielectric 1 mm thick, an initial density on the order of $10^{22}\ m^{-3}$ is necessary in order to considerably decrease the pulse duration of a beam with a particle energy of $1MeV$.

The estimated value of the particle density is somewhat higher than the mean densities reached in electron accelerators. However, in small volumes (as compared to the dimensions of a beam) this density can be reached due to the inhomogeneities in the distribution of electrons in a beam and the avalanche of secondary ionized electrons.

7.4. Steady Supersonic Motion of an Infinite Thin Wedge

Assume that the solid plasma cluster of coalesced electrons can be considered as an absolutely rigid wedge cutting a material at the hypersonic speed. We study the latter problem in Sections 7.4 and 7.5 below.

Equations of the steady plane problem of the theory of elasticity

For the plane problem, the Lamé equations of the dynamic theory of elasticity are as follows:

$$\frac{\partial^2 \Phi_i}{\partial x_1^2} + \frac{\partial^2 \Phi_i}{\partial x_2^2} = \frac{1}{c_i^2}\frac{\partial^2 \Phi_i}{\partial t^2} \qquad (i = 1,2), \tag{7.41}$$

where Φ_1 and Φ_2 are wave potentials; c_1 and c_2 are speeds of longitudinal and transverse waves in a medium; and x_1 , x_2 is the system of coordinates fixed to the medium. In the moving system of coordinates $x = x_1 - Vt$, $y = x_2$, $t' = t$, fixed to a wedge moving with speed V, Equation (7.41) for the steady state is

$$M_i^2 \frac{\partial^2 \Phi_i}{\partial x^2} = \frac{\partial^2 \Phi_i}{\partial y^2} \qquad (i = 1, 2) \tag{7.42}$$

where

$$M_i^2 = \frac{V^2}{c_i^2} - 1 > 0 \tag{7.43}$$

Here M_1 and M_2 are modified Mach numbers.

Displacements, stresses, and velocities of the medium are expressed through the derivatives of wave potentials in the following way:

$$u_x = \frac{\partial \Phi_1}{\partial x} + \frac{\partial \Phi_2}{\partial y} - Vt$$
$$u_y = \frac{\partial \Phi_1}{\partial y} + \frac{\partial \Phi_2}{\partial x} \tag{7.44}$$

$$\sigma_{xx} = G(1 + M_2^2 - 2M_1^2)\frac{\partial^2 \Phi_1}{\partial x^2} + 2G\frac{\partial^2 \Phi_2}{\partial x \partial y}$$
$$\sigma_{yy} = G(M_2^2 - 1)\frac{\partial^2 \Phi_1}{\partial x^2} - 2G\frac{\partial^2 \Phi_2}{\partial x \partial y} \tag{7.45}$$
$$\sigma_{xy} = 2G\frac{\partial^2 \Phi_1}{\partial x \partial y} + G(M_2^2 - 1)\frac{\partial^2 \Phi_2}{\partial x^2}$$

$$\mathrm{v}_x = \frac{\partial u_x}{\partial t} = -V\left(1 + \frac{\partial^2 \Phi_1}{\partial x^2} + \frac{\partial^2 \Phi_2}{\partial x \partial y}\right)$$
$$\mathrm{v}_y = \frac{\partial u_y}{\partial t} = -V\left(\frac{\partial^2 \Phi_1}{\partial x \partial y} - \frac{\partial^2 \Phi_2}{\partial x^2}\right) \tag{7.46}$$

Here u_x and u_y are displacements, v_x and v_y are velocities, σ_{xx} , σ_{xy} , and σ_{yy} are stresses, and G is the shear modulus for the elastic medium.

Normal and tangential components of velocity, v_n and v_t, on a plane are:

$$\mathrm{v}_n = \mathrm{v}_i n_i\ , \quad \mathrm{v}_t = \mathrm{v}_i t_i \qquad (i = 1,2) \tag{7.47}$$

where n_i and t_i are the respective components of the unit vector of a normal and tangent to the plane. The general solution to Equation (7.42) is

$$\Phi_i = \varphi_i(x - M_i y) + \psi_i(x + M_i y) \qquad (i = 1,2), \tag{7.48}$$

where φ_i and ψ_i are arbitrary functions (do not sum over i!).

General solution of the problem for supersonic motion of an infinite thin wedge

First, let us consider a special case of the problem of the motion of a wedge with an angle α *as* $l \to \infty$. The boundary conditions, Equations (7.6) and (7.7), along $y + \delta x = 0,\ x < 0$ are

$$\mathrm{v}_n = 0, \qquad \sigma_{tn} = F(\sigma_n, \mathrm{v}_t), \qquad \delta = \tan\alpha \tag{7.49}$$

where the normal and tangent unit vectors are:

$$\begin{aligned} \boldsymbol{n} &= \{\delta(1+\delta^2)^{-1/2}, \quad (1+\delta^2)^{-1/2}\}, \\ \boldsymbol{t} &= \{-(1+\delta^2)^{-1/2}, \quad \delta(1+\delta^2)^{-1/2}\}, \end{aligned} \tag{7.50}$$

The wave potential Φ_i differs from zero in the corresponding area D_i situated between the surface of the wedge, $y + \delta x = 0$, and the corresponding Mach line, $x + M_i y = 0$, for the upper half-plane (Figure 7.14). The Mach lines are the shock waves on which the laws of conservation of mass and momentum should be met. The wedge is considered thin, that is, the following condition for the hypervelocity flow past a thin airfoil is satisfied:

$$\delta < \frac{1}{M_2} < \frac{1}{M_1}. \tag{7.51}$$

This means that each characteristic line of the corresponding family crosses the wedge surface only at one point and, therefore, the rectilinear shock waves are attached to the wedge vertex. Such a condition is analogous to forming a shock wave by a wedge in gas dynamics, and the mathematical problem is analogous to that of the flow of a supersonic gas around a thin wedge.[81–82]

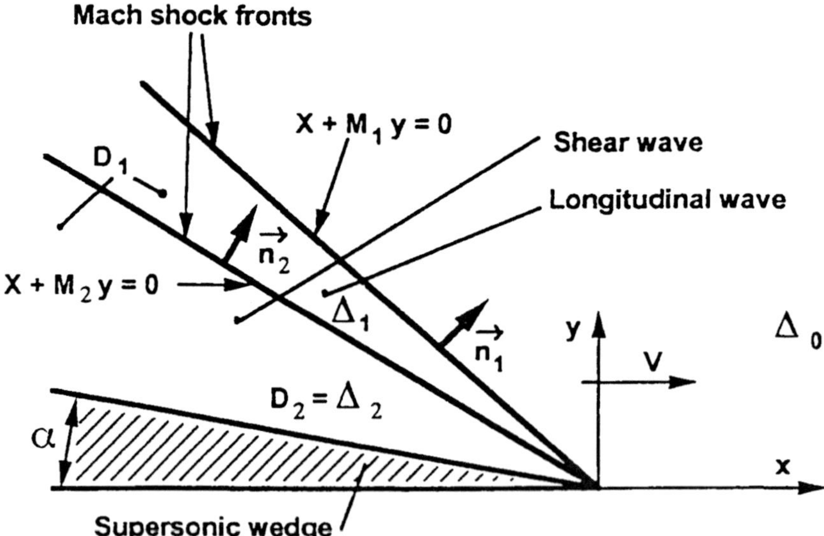

Figure 7.14. The perturbation zone of a supersonic wedge moving in an elastic solid

Let Δ_0 be the undisturbed domain of an elastic body, Δ_1 the domain in which only Φ_1 does not equal zero, and Δ_2 the domain in which both Φ_1 and Φ_2 are not zero. In each of these domains, ρ_0, ρ_1, and ρ_2, $v^{(0)}$, $v^{(1)}$, and $v^{(2)}$, and $\sigma_{ij}{}^{(0)}$, $\sigma_{ij}{}^{(1)}$, $\sigma_{ij}{}^{(2)}$ are the densities, velocities, and stresses of the medium, respectively. The laws of conservation on the Mach lines follow:

(i) Along $x + M_1 y = 0$, with normal and tangent unit vectors,

$$\mathbf{n} = \{(1 + M_1^2)^{-1/2}, M_1(1 + M_1^2)^{-1/2}\},$$
$$t = \{-M_1(1 + M_1^2)^{-1/2}, (1 + M_1^2)^{-1/2}\}, \tag{7.52}$$

we obtain

$$\rho_0 v_{n1}{}^{(0)} = \rho_1 v_{n1}{}^{(1)}, \tag{7.53}$$

$$\sigma_{n1}{}^{(0)} - \sigma_{n1}{}^{(1)} = \rho_1 v_{n1}{}^{(1)}(v_{n1}{}^{(0)} - v_{n1}{}^{(1)}), \tag{7.54}$$

$$\sigma_{tn1}{}^{(0)} - \sigma_{tn1}{}^{(1)} = \rho_1 v_{n1}{}^{(1)}(v_{t1}{}^{(0)} - v_{t1}{}^{(1)}) . \tag{7.55}$$

(ii) Along $x + M_2 y = 0$, with normal and tangent unit vectors,

$$\boldsymbol{n} = \{(1 + M_2^2)^{-1/2},\ M_2(1 + M_2^2)^{-1/2}\},$$
$$t = \{-M_2(1 + M_2^2)^{-1/2}, (1 + M_2^2)^{-1/2}\}, \tag{7.56}$$

we obtain

$$\rho_1 v_{n2}{}^{(1)} = \rho_2 v_{n2}{}^{(2)}, \tag{7.57}$$

$$\sigma_{n2}{}^{(1)} - \sigma_{n2}{}^{(2)} = \rho_2 v_{n2}{}^{(2)}(v_{n2}{}^{(1)} - v_{n2}{}^{(2)}), \tag{7.58}$$

$$\sigma_{tn2}{}^{(1)} - \sigma_{tn2}{}^{(2)} = \rho_2 v_{n2}{}^{(2)}(v_{t2}{}^{(1)} - v_{t2}{}^{(2)}). \tag{7.59}$$

Let us express the components of the velocity and stresses in Equations (7.52) - (7.59) through the derivatives of $\varphi_i(\xi)$ and $\psi_i(\xi)$ using Equations (7.44) - (7.49). We obtain

$$\varphi_1''(\xi) \equiv 0, \ \varphi_2''(\xi) \equiv 0, \tag{7.60}$$

$$\rho_2 = \rho_1 = \rho_0[1 + (1 + M_1^2)\psi_1''(0)]^{-1}, \tag{7.61}$$

$$v_x = -V[1 + \psi_1''(x + M_1 y) + M_2\psi_2''(x + M_2 y)],$$
$$v_y = -V[M_1\psi_1''(x + M_1 y) - \psi_2''(x + M_2 y)] \tag{7.62}$$

$$\begin{aligned}
\sigma_{xx}/G &= (1 + M_2^2 - 2M_1^2)\psi_1''(x + M_1 y) + 2M_2\psi_2''(x + M_2 y)\,, \\
\sigma_{yy}/G &= (M_2^2 - 1)\psi_1''(x + M_1 y) - 2M_2\,\psi_2''(x + M_2 y)\,, \\
\sigma_{xy}/G &= 2M_1\psi_1''(x + M_1 y) + (M_2^2 - 1)\,\psi_2''(x + M_2 y)\,.
\end{aligned} \tag{7.63}$$

In particular, on the wedge surface, $y + \delta x = 0$, according to Equations (7.57) and (7.58), we obtain

$$\begin{aligned}
v_n &= -V(1 + \delta^2)^{-1/2}[\delta + (\delta + M_1)\psi_1'' + (\delta M_2 - 1)\,\psi_2'']\,, \\
v_t &= -V(1 + \delta^2)^{-1/2}[(M_1\,\delta - 1)\psi_1'' - (\delta + M_2)\,\psi_2'' - 1]\,,
\end{aligned} \tag{7.64}$$

$$\begin{aligned}
\sigma_n &= G(1+\delta^2)^{-1}\{2(\delta M_2 - 1)(M_2+\delta)\psi_2'' + [(1+M_2^2-2M_1^2)\delta^2 + 4M_1\delta + M_2^2-1]\psi_1''\} \\
\sigma_{tn} &= G(1+\delta^2)^{-1}\{2(M_1+\delta)(M_1\delta-1)\psi_1'' + [(M_2^2-1)\delta^2 - 4M_2\delta - M_2^2 + 1]\,\psi_2''\}\,,
\end{aligned} \tag{7.65}$$

where the function arguments of $\psi_1(\xi_1)$ and $\psi_2(\xi_2)$, are $\xi_1 = y(M_1\delta-1)/\delta$ and $\xi_2 = y(M_2\delta-1)/\delta$.

Equations (7.62) and (7.63) constitute a general solution of the problem; the form of the functions $\psi_1(\xi_1)$ and $\psi_2(\xi_2)$ is determined by two conditions on the surface of the wedge.

Let us adduce the full account of the results of solving the problem when the condition $\sigma_{tn} = f\sigma_n$, where f is an arbitrary constant, is met on the surface of the wedge. Applying Equation (7.64), we get from Equation (7.48)

$$(1-\delta M_2)\psi_2'' - (\delta + M_1)\psi_1'' = \delta\,. \tag{7.66}$$

Here, the argument of the functions $\psi_i(\xi_i)$ on the straight line, $y + \delta x = 0$, as well as in Equations (7.64) and (7.65), is $\xi_i = y(M_i\delta-1)/\delta$. From Equations (7.50) and (7.65), we obtain

$$\begin{aligned}
&\{2(M_1+\delta)(M_1\delta-1) - f[(1+M_2^2 - 2M_1^2)\delta^2 + 4M_1\delta + M_2^2 - 1]\}\,\psi_1'' \\
&+ [(M_2^2-1)\delta^2 - 4M_2\delta + 1 - M_2^2 - 2f(\delta M_2-1)(M_2+\delta)]\,\psi_2'' = 0\,.
\end{aligned} \tag{7.67}$$

The solution of Equations (7.66) and (7.67) follows:

$$\begin{aligned}
\psi_1''(\xi) &= -\,\delta A^{-1}[(M_2^2-1)\delta^2 - 4M_2\delta + 1 - M_2^2 - 2f(\delta M_2-1)(M_2+\delta)], \\
\psi_2''(\xi) &= \delta A^{-1}\{2(M_1+\delta)(M_1\delta-1) - f[(1+M_2^2 - 2M_1^2)\delta + 4M_1\delta + M_2^2 - 1]\}\,,
\end{aligned} \tag{7.68}$$

where

$$\begin{aligned}
A = {} & (\delta+M_1)[(M_2^2 - 2M_1M_2 - 1)\delta^2 + 2(M_1 - M_2)\delta - M_2^2 - 1] \\
& - f(1-\delta M_2)[(M_2^2 - 2M_1^2 - 1)\delta^2 + 2(M_1 - M_2)\delta + M_2^2 - 2M_1M_2 - 1]\,.
\end{aligned} \tag{7.69}$$

It is also necessary to put $\psi_1'' \equiv 0$ and $\psi_2'' \equiv 0$ in Δ_0 and $\psi_2'' \equiv 0$ in Δ_1.

Superthin wedge without friction

The value of f is assumed to be known, although such information concerning the motion of plasma wedges in a solid is not available. From Equations (7.68), one can see that f has a considerable effect on the velocity and stresses only when it is close to unity, i.e., in the expansion of the solutions in Equations (7.68) into a power series of the form

$$\psi_i'' = \sum_{n=0}^{\infty} \alpha_{in} f^n \quad , \tag{7.70}$$

coefficients α_{i1} are small in comparison with α_{i0}; therefore, the simplest solution of the problem in the limiting case of zero friction, when $f = 0$, is of interest. This solution allows us to obtain effective estimates.

We shall confine ourselves to the superthin wedges, for which the following conditions are fulfilled:

$$\delta << \frac{1}{M_2} \quad \text{and} \quad \delta << \frac{1}{M_1} . \tag{7.71}$$

The solutions, Equations (7.68), for a superthin wedge, are reduced to

$$\psi_1'' = \frac{(1-M_2^2)\delta}{M_1(1+M_2^2)}, \qquad \psi_2'' = \frac{2\delta}{1+M_2^2} . \tag{7.72}$$

Velocity and stress components in this case are equal to

$$\mathrm{v}_x = -V(1+\psi_1'' + M_2\psi_2''), \quad \mathrm{v}_y = V(\psi_2'' - M_1\psi_1'') \tag{7.73}$$

$$\begin{aligned} \sigma_{xx} &= G(1+M_2^2-2M_1^2)\psi_1'' + 2GM_2\psi_2'' , \\ \sigma_{yy} &= G(M_2^2-1)\psi_1'' - 2GM_2\psi_2'' , \\ \sigma_{xy} &= 2GM_1\psi_1'' + G(M_2^2-1)\psi_2'' , \end{aligned} \tag{7.74}$$

where ψ_1'' is given by Equations (7.72) in Δ_1 , $\psi_2'' = 0$ in Δ_1 , and ψ_1'' and ψ_2'' are given by Equations (7.72) in Δ_2.

According to (7.73) and (7.74), density in the disturbed regions is determined by

$$\rho_2 = \rho_1 = \rho_0\left(1 + \frac{(M_2^2-1)(M_1^2+1)\delta}{M_1(1+M_2^2)} \right), \tag{7.75}$$

and normal and tangential stresses on the surface of the wedge in this limiting case are:

$$\sigma_n = \frac{G\delta\left[(M_2^2 - 1)^2 + 4M_1 M_2\right]}{M_1(1+M_2^2)} \quad , \quad \sigma_{tn} = 0 \tag{7.76}$$

This solution will be used in Sections 7.5 and 7.6.

Similarly, the case of the arbitrary shape of a wedge can be considered. The theory of cutting solid materials by high-power beams of photons has recently been constructed,[48,83–85] taking into account the motion and position of a target.

7.5. Deceleration of the Finite Wedge

Let us study here the supersonic motion of a finite wedge.

Drag and energy dissipation of a supersonic wedge

Solution of the plane problem of the infinite thin wedge presented in Section 7.4 holds for the domain consisting of two characteristic triangles resting on the contact surface of a wedge of length l (see Figure 7.9). Energy dissipation and forces acting upon the wedge body can be found without solving the problem as a whole, if one uses the invariant Γ integral.

Let us consider the plane problem for a cavity, at the end of which a moving supersonic wedge pierces the body. The length of the cavity is assumed to be much larger than the length of the wedge; the surface of the cavity is assumed to be free of tractions (the effect of the substance that may be present in the cavity outside the wedge is neglected). These conditions allow us to assume zero disturbances at infinity. Choose the closed surface of integration Σ consisting of Σ_∞, Σ_0, and Σ_w, as shown in Figure 7.15, where Σ_∞ is situated at an infinite distance from the wedge; Σ_0 is the free surface of the cavity; and Σ_w is the surface of the wedge contact with the medium. The Γ integral over Σ_w will provide the energy rate per unit length produced by the finite wedge; the energy flows into shock waves and dissipates on the Mach lines. The Γ integral over Σ_∞ and Σ_0 is equal zero.

Using Equation (7.21) in the moving system of coordinates fixed to the wedge we have:

$$\Gamma_x = \int_{\Sigma_w}\left[\left(U' + \frac{1}{2}\rho \mathrm{v}_i' \mathrm{v}_i'\right)n_x - \sigma_{ij}u_{i,x}n_j\right]d\Sigma \qquad (i,j=1,2)$$
$$(\mathrm{v}_1' = \mathrm{v}_1 + V\,,\quad \mathrm{v}_2' = \mathrm{v}_2\,,\quad u_1' = u_1 + Vt\,,\quad u_2' = u_2) \tag{7.77}$$

where V is the wedge velocity, and U' is the strain energy per unit volume.

It is easy to see that the second term in Equation (7.77) is the sum of the surface forces acting on the rigid wedge. Let us label this term R_x. Then Equation (7.77) becomes

$$\Gamma_x = W_x + R_x + K_x\,. \tag{7.78}$$

Here W_x and K_x is the flux of the internal and kinetic energy respectively expressed by the first two terms. The value of K_x has an order of $\rho V^2\delta^3 l$ because v_i' is of the order of $V\delta$.

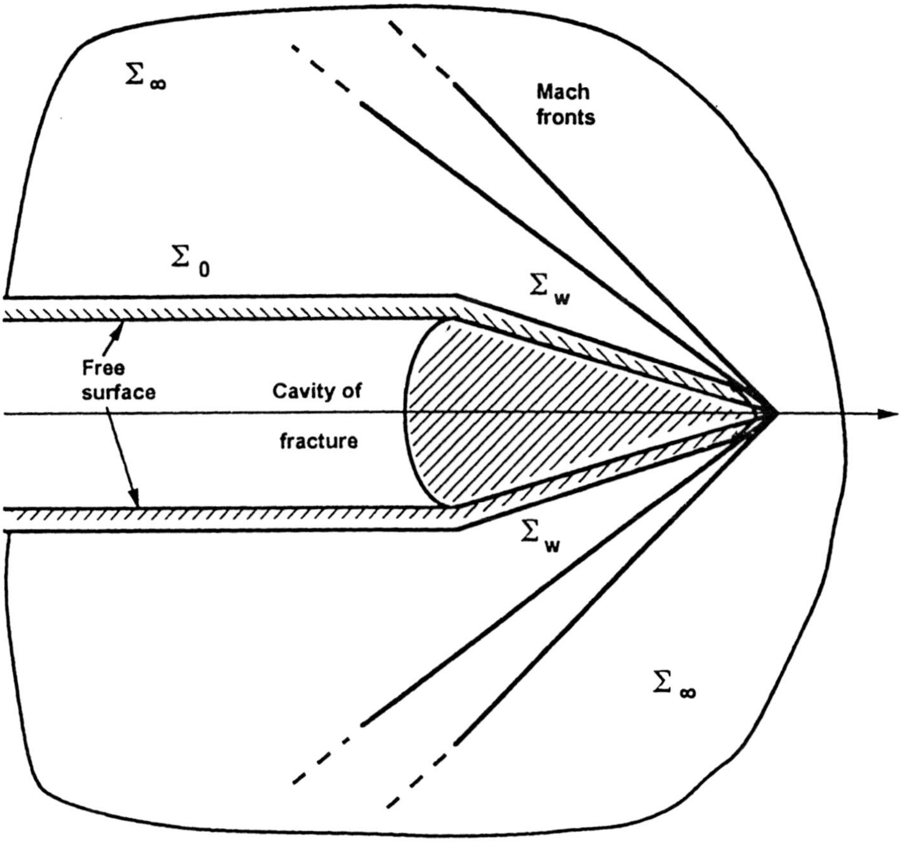

Figure 7.15. Steady-state supersonic motion of a finite wedge

Let us calculate W_x and R_x in an explicit form for the simple case of the superthin symmetric wedge without friction. Let us use Equations (7.72), (7.73), and (7.76) and the following expression:

$$U' = \frac{1}{4}\sigma_{ij}\left(u'_{i,j} + u'_{j,i}\right) = \left[(1+\nu)\sigma_{ij}\sigma_{ij} - \nu\sigma_{ii}\sigma_{jj}\right]/2E \tag{7.79}$$

for a linear-elastic medium, where ν is Poisson's ratio and E is Young's modulus. We obtain

$$W_x = \frac{\eta E \delta^3 l}{4(1+\nu)^2} \qquad (W_y = 0), \tag{7.80}$$

where

$$\eta = M_1^{-2}(1+M_2^2)^{-2}\{[4M_1M_2 + (1-M_2^2)^2]^2 + 4[M_1M_2 + (1-M_2^2)(1+M_2^2-2M_1^2]^2 + 2M_1^2(1+\nu)(1+M_2^2)^3 + 2(1+\nu)(1+M_2^2)(2M_1M_2 + 1-M_2^2)\}. \tag{7.81}$$

The value of W_x appears to be proportional to δ^3, that is, W_x has same order as K_x. Further calculations give the following results:

$$R_x = -\frac{2GR(M)\delta^2 l}{M_1(1+M_2^2)} \qquad (R_y = 0) \;, \tag{7.82}$$

where

$$R(M) = (M_2^2 - 1)^2 - 4M_1M_2 \qquad \left(M = \frac{V}{c_2}\right). \tag{7.83}$$

Comparison of Equations (7.80) and (7.82) shows that at small angles of the wedge opening, the work W_x spent to increase the internal energy of the body and the flux of kinetic energy K_x are small in comparison with the work R_x spent to overcome the drag. Thus, the dissipative characteristics for the process of motion of the thin supersonic wedge are determined by the drag of the wedge.

The formula for the drag,

$$R_x = -\int_{\Sigma_w} \sigma_{ij} u'_{i,x} n_j d\Sigma \tag{7.84}$$

is applicable for any dimensions of the cavities formed at the fracture.

If one extrapolates Equations (7.82) and (7.83) to the subsonic motion where $V < c_2$, from these equations it follows that the drag of the wedge vanishes at $V = c_R$ where c_R is the Rayleigh speed nullifying the function $R(M)$. This is a correct result which was strictly proven[86].

Finite wedge penetration

Let us consider the motion of the wedge subjected only to the forces of inertia and drag in quasisteady approximation, i.e., assuming that the drag is given by Equation (7.82). At the initial time, the wedge possessed momentum, mV_0, where m is the linear mass of the wedge. The law of motion of the wedge is

$$m\frac{dV}{dt} = R_x(V), \tag{7.85}$$

with the initial condition $V = V_0$ at $t = 0$. The solution of Equation (7.85) has the form

$$t = \frac{-m}{2G\delta^2 l}\int_{V_0}^{V}\frac{M_1(1+M_2^2)}{(M_2^2-1)^2-4M_1M_2}\,dV\ , \tag{7.86}$$

where $G = \rho_0 c_2^2$. Remind that

$$M_1 = \sqrt{(V^2/c_1^2)-1}\,,\quad M_2 = \sqrt{(V^2/c_2^2)-1}\,. \tag{7.87}$$

For the most important limiting case of hypervelocities, when $V/c_1 >> 1$, we obtain the following simple solution from Equation (7.86):

$$V = V_0\,exp\left(\frac{-2t\rho_0 c_1\delta^2 l}{m}\right). \tag{7.88}$$

The corresponding distance covered by the wedge for the period t equals

$$L(t) = \int_0^t V dt = \frac{mV_0}{2\rho_0\delta^2 c_1 l}\left[1-exp\left(\frac{-2\rho_0 c_1 t\delta^2 l}{m}\right)\right]. \tag{7.89}$$

Because $m = \rho_w \delta l^2$, where ρ_w is the density of the wedge substance, from Equation (7.89), it follows that

$$\frac{L_\infty}{l} = \frac{\rho_w V_0}{2\delta\rho_0 c_1}\ . \tag{7.90}$$

Equation (7.90) gives the length of the crack formed due to supersonic motion of the wedge, the subsonic phase of motion being neglected.

Taking into account Equation (7.71), we can also derive the following estimate:

$$\frac{L_\infty}{l} >> \frac{\rho_w V_0^2}{2\rho_0 c_1^2}\ . \tag{7.91}$$

Equations (7.90) and (7.91) will be used in the next section.

7.6. A Comparative Analysis of the Theoretical and Experimental Results

In this section we provide some estimates and compare theoretical predictions with available test data.

Preliminary remarks

The mechanical model of the supersonic cutting given in Sections 7.4 and 7.5 was not associated directly with the quantitative characteristics of electron radiation. Let us evaluate the basic parameters of the model using the known properties of the coalesced electron plasma clusters formed in the materials under study.

The orders of magnitude used for further calculation are: the coalescence time of superluminal electrons in a beam $\tau = 1\ ns$; the irradiation intensity $\Phi_0 = 10^{18}\ m^{-2}$; the electron energy $E = 10^{-13}\ J$ (1 *MeV*); the cross-sectional area of the specimen $S = 5\times10^{-5}\ m^2$; the specimen thickness $h_S = 5\times10^{-4}\ m$; the relativistic momentum of an electron,

$$p = \sqrt{(E_0 / c)^2 - m_0^2 c^2} = 5 \times 10^{-22}\, N \times s\,; \tag{7.92}$$

material density $\rho_0 = 10^3\ kg/m^3$; the speed of the longitudinal waves $c_1 = 10^3\ m/s$; the atomic spacing $a = 10^{-9}\ m$; the ideal theoretical strength of the crystal $U = 10^{11}\ J/m^3$; the specific surface energy of the crystal $\gamma = 1\ J/m^2$; the electron mass at rest $m_0 = 10^{-30}\ kg$. The total momentum of a beam for a pulse is equal to 10^{-9} *Ns*; the mean value of the beam energy absorbed in a material at the fracturing threshold equals $5\times10^7\ J/m^3$; and the energy absorbed in a specimen for a pulse equals 1 J/m^3. These numerical values are typical for the experiments for fracturing ionic crystals.

These data show that the specific energy of the fracturing is three orders less than the theoretical strength of crystals. For comparison, if one assumes that all energy of the beam would be spent to form one or two cracks observed in experiments, the surface energy would be equal to $10^4\ J/m^2$; this magnitude is four orders greater than the surface energy. A physical mechanism of cluster formation using the electron beam coalescence effect needs to be estimated.

Ionization electrons

As mentioned above, the electron density necessary for significant self-compression of primary electrons should be substantially greater than the initial density of a beam.

However, irradiation induces considerable ionization of fixed electrons of the crystal. According to experimental data,[31] the energy dissipation of fast electrons by ionization is characterized by the following parameter, $-\partial E/\partial(\rho x) = 2\times10^5\ eVm^2/kg$, for electron energy E from 0.1 to 100 *MeV*. Specific loss of energy per unit thickness is equal approximately to $k=5\times10^{-11}\ J/m$ for typical material density ρ_0. Using the Bethe-Bloch formula[31], we obtain the number η' of electrons liberated per unit-thickness by one fast electron,

$$\eta' = \frac{k}{I + E_i}\,, \tag{7.93}$$

where I is the ionization energy of a fixed electron (for many materials, it is approximately equal to 10 *eV*), and E_i is the liberated electron energy. The value of E_i may approach 10 *eV*, too; the corresponding speeds of electrons equal $10^5 - 10^6\ m/s$ which are typical speeds of electron clusters of solid plasma in irradiated materials. Hence, the mean density of the liberated electrons n_i is equal to:

$$n_i = \eta' \Phi_0 = \frac{k\Phi_0}{I + E_i} = 5 \times 10^{25} m^{-3} . \tag{7.94}$$

This value is by seven orders greater than the initial density n_0 of a relativistic beam.

Even much greater densities of electron clusters are possible. For example, in anisotropic crystals having cleavage planes, there are observed prevailing directions of motion of electron plasma clusters along the cleavage planes.[13] A considerable current of secondary electrons may be created on the cleavage planes of the material. Directed secondary currents can possibly be formed by the transference of a superluminal electron momentum to liberated electrons and the redistribution of the density of secondary electrons. Probably, the formation of clusters of secondary electrons is possible along all of the cleavage planes. An electron current passing through the weakest cleavage plane will have the form of a thin blade. The thickness of the blade d will be much less than the two other dimensions, l.

An upper bound for the density of a blade-shaped cluster of secondary electrons ρ_w can be estimated, assuming that all ionized electrons are collected in a blade of dimensions $d \times l \times l$. Using (7.94) we obtain

$$\rho_w < \rho_w^* = \frac{m_0 k h_s S \Phi_0}{dl^2 (I + E_i)} . \tag{7.95}$$

Note that the specimen lengths are usually smaller than the distance that electron clusters move, at $V_0=10^5 m/s$, during the time of fracturing τ_f. (The latter is of the order of 0.1 ms.)

We use Equation (7.95) to estimate the possibility of mechanical cutting by the liberated electrons. Substituting ρ_w^* into Equation (7.91), assuming $L_\infty = h_s$, and using the relationship $d = 2l\delta$, we find

$$d^2 < \frac{k m_0 V_0 S \Phi}{2\rho_0 c_1 (I + E_i)} = 5\times10^{-11} m^2 \tag{7.96}$$

Taking into consideration that the destruction of material begins approximately at scale 10^{-10} m, we obtain the following estimate of the wedge thickness in the mechanical model

$$10^{-10} < d < 7\times10^{-6} m . \tag{7.97}$$

(The avalanche formation of secondary liberated electrons by relativistic electrons provides another possible mechanism for the self-condensation of electrons and the nucleation of solid plasma clusters.)

Electron plasma clusters: comparison of the theory with experimental data

Let us estimate other parameters of the condensed electron clusters of solid plasma using the following scheme for fracturing:

(i) The electron fracture of solids is performed by clusters that are formed by a αth part of liberated electrons in a specimen domain with characteristic volume $h_s S = l h_s^2$;

(ii) fracturing occurs because of a breakage of the atomic bonds of the nearest lattice atoms, and therefore we can assume further that $d = 10^{-9}\ m$ in accordance with Equation (7.97).

According to Equation (7.95) the mass density of a cluster of liberated electrons (and the density of the model wedge) is:

$$\rho_w = \frac{\alpha m_0 k h_s^2 \Phi_0}{l d (1 + E_i)} . \tag{7.98}$$

The unknown quantities l, δ, ρ_w and α are connected by Equations (7.49) and (7.90) for the mechanical model,

$$\frac{L_\infty}{l} = \frac{p_w V_0}{2 p_0 c_1 \delta} \tag{7.99}$$

$$d = 2l\delta, \tag{7.100}$$

and by the obvious inequalities for the same model,

$$\delta << \frac{c_1}{V_0} \sim 10^{-3} , \tag{7.101}$$

$$\frac{L_\infty}{l} >> 1 . \tag{7.102}$$

From Equations (7.98) - (7.102) we find the following estimates for basic parameters of electron clusters of solid plasma (wedges of the mechanical model);

$$1\ \mu m < l < 50\ \mu m \tag{7.103}$$

$$10^{-5} < \alpha < 10^{-3} \tag{7.104}$$

$$5 \times 10^{-4}\ kg/m^3 < \rho_w < 10^{-1}\ kg/m^3, \tag{7.105}$$

$$5 \times 10^{26}\ m^{-3} < n_w < 10^{29}\ m^{-3}, \tag{7.106}$$

$$10 < L_\infty / l < 10^3, \tag{7.107}$$

where n_w is the electron density in the cluster.

From this analysis, it follows that the crack length L_∞ formed by the electron cluster (wedge) is proportional to the beam intensity Φ if the specimen thickness, h_s much greater than h_0 (the electron cluster is created in the layer of thickness h_0). From Equations (7.98) and (7.99), we obtain

$$L_\infty = \lambda\Phi, \qquad \lambda = \frac{\alpha m_0 k l h_0^2 V_0}{d^2 \rho_0 c_1 (I + E_i)}. \tag{7.108}$$

According to Equations(7.103) - (7.107) and the values of other parameters estimated above, the coefficient λ is of the order of $10^{-20}\ m^3$, which agrees with the experimental data (see Figure 7.4). Hence, the theoretically derived linear function L_∞ vs. Φ predicts the experimental data.

The estimate for l in Equation (7.103) provides a sufficiently narrow range for the dimension of the cluster (wedge). The dimensions of the material inhomogeneities, which may form a secondary electron cluster, are of the same order. This value of l is comparable with the lengths of material structural defects as dislocations, microcracks, and grains (in polycrystalline metals), but it is smaller than the macrocrack dimensions.

7.7. Conclusion

The model suggested allows us to predict the main experimental results specific to the electron-beam-induceed fracture mode (EFM): namely, the absolutely brittle type of the fracture, the independence of the fracturing beam intensity from the primary cracks, and the supersonic speed of the crack propagation. The model is quite simple and rather effective. Therefore, we anticipate that it will stimulate further research.

The basic result of the theory presented is that the supersonic fracture process is accompanied by a very small amount of work to render a fracture. The invention of extremely thin ultrarelativistic beams of very great density will allow one to cut solids with minimal damage and energy spent to fracture. Achieving such an invention is physically possible according to this theory.

The coalescence of electrons into a solid wedge cutting a material was considered. The most important problem to be done in the near future within the framework of this approach is the problem of explosion of the condensed cluster of electrons (solid plasma wedge) under subluminal speeds of motion in the deceleration stage. In the subluminal range of speeds, the electrons repel one another which creates a great pressure of solid electron plasma and the explosion effect. Quantitatively, the latter effect is probably much more significant for the fracturing than the cutting effect under study.

Besides relativistic electron beams of great density, beams of other elementary particles, e.g., protons and intense photon rays, can probably induce the same mode of fracture. Probably, this is valid also for giant laser beams in the nanosecond range; however, a great deal of experimental and theoretical work is required to achieve a full understanding and control of the phenomenon.

Electrons moving in an ideal vacuum always repel, which creates instability of an electron beam in the vacuum. However, from the theory treated in this Chapter it follows that, for ultrarelativistic electron beams of very high energy in the space vacuum, the coalescence effect will appear, which will allow one to create superstable and superdense relativistic electron clusters of unprecedented destroing power.

7.8. Problems

7.1. Find the solution to Maxwell's equations (7.7) for charge e moving steadily at a constant velocity **V** in a medium described by the dielectric ε', and magnetic μ', constants. Consider the cases of subluminal and superluminal speeds.

7.2. Solve the problem of three electrons moving steadily one by one in one and the same direction along one line. Study interaction forces and mutual movement for both subluminal and superluminal motion.

7.3. Consider two electrons moving steadily in the same direction, but along different lines. Study interaction forces and mutual movements for different positions of the electrons in both subluminal and superluminal motion.

7.4. Construct the generalized solution to Maxwell's equation for the electron "arrow" created by a condensed semi-infinite chain of superluminal electrons moving along one line.

7.5. Consider the cubic lattice of electrons forming a half-space and moving steadily in the direction perpendicular to the flat boundary of the half-space. Study subluminal and superluminal cases.

7.6. Consider Problem 7.5 for the case when the cubic lattice of electrons is shaped as a round semi-infinite cylinder with a flat end, and all electrons move parallelly to the axis of the cylinder with a constant speed.

7.7. Consider the simple cubic, f.c.c. and b.c.c. packing of the entire space by electrons moving with a constant velocity. Study interaction and mutual movement of the electrons in the lattice.

7.8. Study Problems 7.1 - 7.7 for point acoustical sources moving steadily in an acoustic medium (one wave equation) with a constant speed.

7.9. Study Problems 7.1 - 7.7 for point acoustical sources moving steadily in an elastic medium (two wave equations) with a constant speed.

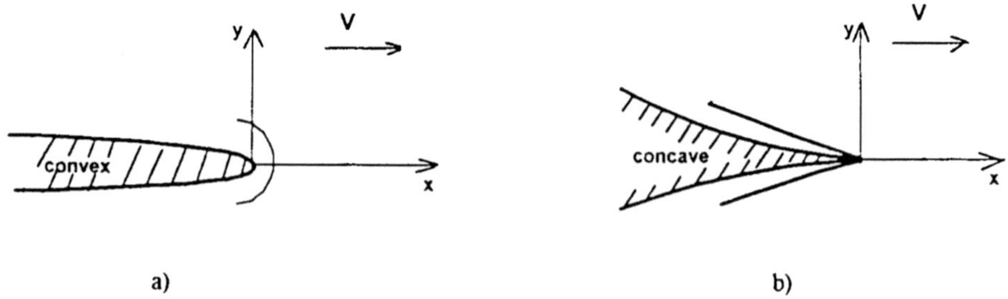

Figure 7.16. Motion of a convex and concave thin wedges in an elastic medium

7.10. Study the supersonic steady-state motion of a thin curvilinear wedge in an elastic medium. (Figure 7.16). Show that a shock wave in front of a convex wedge end is formed. Prove that the continuous solution for a concave wedge is not stable and the shock wave forms in this case, too (an analogue of the Nikolsky - Taganov theorem in gas dynamics).

7.11. Consider a photon as a condensed electron-positron pair and derive the photon field using the superposition of solutions (7.4) under the assumption that $\lim_{\Delta_e \to \infty} (e\Delta_e) = m_p$ is the singularity strength characterizing the photon (Δ_e is the distance between the electron and positron in the pair).

7.12. Study high-intensity photon beams using the methods of this Chapter and the solution to Problem 7.11.

References

1. G . P. Cherepanov and A. A. Borzykh (1993), Theory of the electron fracture mode in solids, *J. Applied Physics*, **74**(12), pp. 7134-7153
2. R. B. Oswald (1966), Fracture of silicon and germanium induced by pulsed electron irradiation, *IEEE Trans. Nuclear Sci.*, **NS-13**(6), p. 63.
3. N. K. Atamanova, A. I. Melker, and I. L. Tokmakov (1976), The fracture of aluminum alloy by pulse electron beams, *Fizika i Chimia Obrabotky Materialov (Physics and Chemistry of Material Treatment (PCMT))*, **4**, p. 29.
4. M. Hirai, Y. Konde, T . Yoshinari, and M. Ueta (1971), *J. Physical Society*, Japan, **30**, p. 1071.
5. J. I. Perkin, E. Norrin, and D. W. Large (1971), The spalling of aluminium with a pulsed electron beam, *J. Physics D: Applied Physics*, **4** (7), p. 974 .
6. R. B. Oswald (1973), Dynamic response of aluminium to pulsed energy, deposition in the melt dominated regime, *J. Applied Physics*, **44**(8), p. 3563 .
7. T. Miyazaki and N. Taniguchi (1973), An experimental analysis of penetration mechanism of high power density electron beam into metal, *Sci. Papers Inst. Phys. Chem. Researches*, **1**, p.67 .
8. B. Steverding and F. P. Gibson (1977), Early stages of light interaction with solids at extreme intensities, *J. Physics D: Applied Physics*, **10**(12), p. 1683 .
9. R. M. Schmidt (1978), Temperature dependent spall threshold of four metal alloys, *J. Physics and Chemistry Solids*, **39**(4), p. 375 .
10. R. M. White and R. H. Gobbet (1974), The spalling impulse and velocity of dense, porous and powdered tantalum and lead induced by a pulsed electron beam, *J. Physics D: Applied Physics*, **7**(17), p. 2342.
11. Y. Kondo, M. Hirai, and M. Ueta (1972), *J. Physical Society*, Japan, **33**, p. 151.
12. D. I. Vaisburd, G. I. Gering (1974), Brittle fracture velocity of ionic crystals induced by power pulse irradiation of electron beams, *Fizika Tverdogo Tela (Soviet Physics of Solids (FTT))*, **16**, p. 3178 .
13. I. N. Balichev and D. I. Vaisburd (1975), Two mechanisms of the brittle fracture of ionic crystals by intense electron beams, *Fizika Tverdogo Tela (Soviet Physics of Solids (FTT))*, **17**, p. 1236.
14. D. I. Vaisburd and I. N. Balichev (1975), *Pis'ma v Zhur. Tekhnich. Fiziki (Letters in Journal of Technical Physics (LetZTF))*, **1**, p. 531.
15. G. I. Gering and A. N. Valiaev (1974), *Izvestya Vysch. Uchebn. Zaved. SSSR, Fizika*, (*Notices of Higher Education Colleges, Physics*), **5**, p. 104.
16. I. N. Balichev (1974), *Fracture of Ionic Crystals by Nanosecond Electron Beams of High Intensity*, Doctoral Thesis, Tomsk University Press.
17. I. N. Balichev, D. I. Vaisburd, and G. I. Gering (1976), *Pis'ma a Zhur. Tekhnich.Fiziki (Letters in Journal of Technical Physics (Pis'ma ZTF))*, **2**(7), p. 327 .
18. I. N. Balichev, D. I. Vasburd, and S. B. Maltis (1978), The probability distribution of brittle fracture of ionic crystals irradiated by nanosecond pulses of high-power electron beams, *Doklady Academii Nauk SSSR*

(Soviet Physics -Doklady (DAN)), **241(6)**, p. 1330 .
19. D. I. Vaisburd and I. N. Balichev (1972), *Pis'ma v Zh. Experim. and Teoret. Fiziki (Letters in Journal of Experim. and Teor.Phys.)*, **15**, p. 537.
20. D. I. Vaisburd, I. N. Balichev, and A. N. Valiaev (1976), *Izvestya Acad. Nauk SSSR, Ser. Fizicheskaya (Notices of the USSR Academy of Sciences, Physics)*, **40(11)**, p. 2404 .
21. D.I.Vaisburd, I.N.Balichev, and A.N.Valiaev (1974), *Izvestya Acad. Nauk SSSR, Ser. Fizicheskaya (Notices of the USSR Academy of Sciences, Physics)* **38**, p. 1281 .
22. D. I. Vaisburd, A. N. Valiaev, and G.I.Gering (1975), *Tretye Vsesouznoe Soveshie po Radiats, Fizika i Chimia Ionnykh Cristallov (Physics and Chemistry of Ionic Crystals)*, Riga.
23. A.A.Vorob'ev (1968), *Mechanical and Heat Properties of Alkaline Halogen Crystals*, Vyschaya Shcola Press, Moscow, .
24. D. I. Vaisburd, G. I. Gering, and V. M. Kondrashov (1976), Brittle fracture of glasses affected by pulse electron irradiation of great density beams, *Zh. Tekhnicheskoy Fiziki (J. of Technical Physics)*, **5**, p. 1971.
25. J. Badian and J. Reffi (1968), *Dynamics of Fracture in Glass Resulting from the Effects of Laser Radiation, Deystviye Lazernogo Izlucheniya (Effects of Laser Irradiation)*, Mir Press, Moscow.
26. B. W. Schumacher and R. C. Smith (1974), Rock breakage and rock cutting with focused electron beams, *Advances in Rock Mechanics*, **2**, Part B, Washington .
27. R. T. Avery, and D. Keefe (1975), An electron accelerator for tunneling through hard rock, *IEEE Trans. Nuclear Sci.*, **20**(10), p. 61.
28. R. T. Avery and D. Keefe (1973) , Shattering rock with intense bursts of energy - the electrons, *IEEE Trans.Nuclear Sci.*, **20**(3), p. 1391.
29. A. A. Vorob'ev (1973), *Damage Accumulation, Structure Degradation, and Minerals and Rocks Fracture*, Tomsk University Press .
30. G. A. Mesiatz, A. S. Nasibov, and V. V. Kremnev (1970), *Formation of High-Voltage Nano-Second Pulses*, Energia Press, Moscow .
31. H. Frauenfelder and E. M. Henley (1974), *Subatomic Physics*, Prentice-Hall, New Jersey.
32. C. Kittel (1967), *Introduction to Solid State Physics*, Wiley, NY.
33. W. M. Peffey (1967), A study of stress waves produced by pulsed high-energy radiation, *Applied Physics Letters*, **10**(6), p. 171.
34. G. P. Cherepanov and V. Z. Parton (1972), The mechanics of fracture, Mekhanika v SSSR za 50 let (*Soviet Mechanics for 50 years*), **3**, Nauka Press, Moscow.
35. G. P. Cherepanov (1979), *Mechanics of Brittle Fracture*, McGraw Hill, New York .
36. A. A. Borzykh and G. P. Cherepanov (1980)., Fracture of solids by electron beams, *Problemy Prochnosti (Strength of Materials,(PP))*, **8**, p. 16.
37. A. I. Melker and I. L. Tokmakov (1977), Fracture of solids irradiated by electrons, *Fizika i Chimia Obrabotky Materialov (Physics and Chemistry of Material Treatment)*, **2**, p. 62.
38. A. A.Borzykh and G. P. Cherepanov (1980), On the theory of fracture of solids by high power beams of electrons, *Prikladnaya Matematika i Mekhanica (Applied Mathematics and Mechanics, (PMM))*, **44**, p. 1120.
39. A. A. Borzykh (1979), On the theory of fracture of solids by high power pulsed electron beams, *Mekhanika Tverdogo Tela (Mechanics of Solids (MTT))*, **6**, p. 178 .
40. D. I. Vaisburd and S. I. Kasatkin (1978), On the minimal dimension of a disturbed zone creating a crack in solids, *Izvestya Vysch. Uchebn. Zaved. SSSR, Fizika, (Notices of Higher Education Colleges, Physics)* **7**, p. 142.
41. B. Steverding, C. W. Austin, and A. H. Warkheister (1972), Fracture by superimposing stress waves, *J. Applied Physics*, **43**(7), p. 3217.
42. B. Steverding and S. H. Lehnigk (1971) , Collision of stress pulses with obstacles and dynamics of fracture, *J. Applied Physics*, **42**(8), p. 3231.
43. G. I. Barenblatt and G. P. Cherepanov (1960), Cleavage of brittle bodies, *Prikladnaya Matematica i Mekhanica (Applied Mathematics and Mechanics, (PMM))*, **24**(4).

44. D. R. Curran, D. A. Shockey, and S. Winkler (1970), Crack propagation at supersonic velocities, II, Theoretical model, *Int. J. Fracture Mechanics*, **6**(3), p. 271 .

45. S. R. Winkler, D. A. Shockey, and D. R. Curran (1970), Crack propagation at supersonic velocities, *Int. J. Fracture Mechanics*, **6**(2), p. 151.

46. S. R. Winkler (1973), Supersonic crack propagation and ionic crystals induced by strong laser pulses, in *Dynamic Crack Propagation*, G. Sih (ed.), Noordhoff Publ., Leyden.

47. L. I. Mirkin (1975), *Physical Fundamentals of Material Treatment by Laser Beams*, Moscow State University Press., Moscow.

48. G. P. Cherepanov (1987), *Rock Fracture Mechanics in Drilling*, Nedra Publishers, Moscow.

49. T. Ekobory (1972), *Fracture and Strength Mechanics and Physics of Solids*, Metallurgia Press, Moscow.

50. H. Liebowitz (ed.) (1968 - 1972), *Fracture*, Vol. 1-7 , Academic Press, New York, .

51. G. Sih (ed.) (1973) , *Dynamic Crack Propagation*, Noordhoff Publ., Leyden.

52. A. A. Vedenov and E. P. Velikhov (1962), *Zhurnal Experimentalnoi i Teoreticheskoi Fiziki (Soviet Physics JETP)*, **20**(3), p. 1110.

53. F. C. Perry (1970), Electron beam induced stress waves in solids, *Applied Physics Letters*, **17**(11).

54. F. Platzman and P. Volf (1975), *Volny i Vzaimodeistvia v Plasme Tverdogo Tela (Waves and Interactions in Solid State Plasma)*, Mir Press, Moscow.

55. A. I. Melker, A. I. Michailov, and N. V. Zolotorevsky (1979), Destruction of angarmonic atomic chains, *Fizika Tverdogo Tela (Soviet Physics of Solids (FTT))*, **21**(4), p. 1545.

56. A. I. Melker, S. N. Romanov, and I. L. Tokmakov (1979), ECM modeling of Frenkel defects formed in solids by relativistic electrons, *Fizika i Chimia Obrabotky Materialov (Physics and Chemistry of Material Treatment (PCMT))*, **2**, p. 18.

57. S. I.Molovsky and A. D.Sushkev (1972), *High-Power Electron and Ion Beams*, Energia Press, Leningrad.

58. W. T. Sanders (1972), On the possibility of a supersonic crack in a crystal lattice, *Engineering Fracture Mechanics*, **4**, p. 145 .

59. P. A. Cherenkov (1934), *Doklady Academii Nauk SSSR (Soviet Physics - Doklady (DAN))*, **20**(8), p. 451.

60. P. A. Cherenkov (1938), *Doklady Academii Nauk SSSR (Soviet Physics - Doklady (DAN))*, **22**(9), p. 653.

61. P. A. Cherenkov (1944), Radiation of electrons moving in matter with superluminal velocity, *Trudy Fizich. Inst. Acad.* Nauk SSSR (Proceedings of the Physical Institute of the USSR) **2**(4), p. 3.

62. S. I. Vavilov (1934), On possible causes of blue luminescence of liquids, *Doklady Academii Nauk SSSR (Soviet Physics -Doklady (DAN))*, **2**, p. 457.

63. I. E. Tamm and I. M. Frank (1937) , Coherent radiation of a fast electron in a medium, *Doklady Academii Nauk SSSR (Soviet Physics - Doklady (DAN))*, **14**, p. 107.

64. I. E. Tamm (1939), Radiation emitted by uniformly moving electrons, *J.Physics USSR*, **1**, p. 439.

65. A. A. Borzykh and G. P. Cherepanov (1980), Collective relativistic interactions in electron beams, *Zhurnal Experimentalnoi i Teoreticheskoi Fiziki (Soviet Physics JETP)*, **51**(1), p. 59 .

66. G. P. Cherepanov (1967), On crack propagation in continuous media, *Applied Mathematics and Mechanics (PMM)*, **31**(3), pp. 476-488.

67. G. P. Cherepanov (1977), Invariant Γ-integrals and their applications to mechanics, *Applied Mathematics and Mechanics (PMM)*, **41**(3), pp. 399-412.

68. G. P. Cherepanov (1981), Invariant Γ- Integral, *Engineering Fracture Mechanics*, **14**(1), pp. 39-58.

69. G. P. Cherepanov (1990) , Computation of Invariant Integral at Singular Points, *Computational Methods in the Mechanics of Fracture*, S.Atluri (ed.), Russian Edition, Mir, Moscow , pp. 350-365.

70. L. D. Landau and E. M. Lifshitz (1959), *Electrodynamics of Continuous Media*, Gostekhizdat Press, Moscow .

71. L. D.Landau and E. M.Lifshitz (1975), *Classical Theory of Fields*, Pergamon Press, London.

72. I. I. Goldenblat and S. V. Ul'ianov (1979), *Introduction to the Theory of Relativity*, Nauka Press, Moscow.

73. V. G. Levich (1969), *Course of Theoretical Physics*, **1**, Nauka Press, Moscow.

74. A. A. Sokolov and I. M. Ternov (1974) , *Relativistic Electrons*, Nauka Press, Moscow.

75. B. M. Budak, A. A. Samarsky, and A. N. Tikhonov (1972), *Sbornik Zadach po Matematicheskoi Fizike (Collection of Problems in Mathematical Physics)*, Nauka Press, Moscow.

76. J. Cole and J. Huth (1958), Stresses produced in a half plane by moving loads, *J. Applied Mechanics*, **25**, p. 433.

77. L. M. Milne-Thomson (1960), *Theoretical Hydrodynamics*, St. Martin's Press, New York.

78. G. P. Cherepanov (1993), An introduction to singular integral equations in aerodynamics, Chapter in: S. M. Belotserkovsky and I. K. Lifanov, *Method of Discrete Vortices*, CRC Press, Boca Raton - London, pp. 1-18.

79. J. C. Maxwell (1864), *Treatise on Electricity and Magnetism*, Cambridge University Press, Cambridge.

80. G. B. Dwight (1961), *Table of Integrals and other Formulas*, Collier - MacMillan, New York.

81. N. E. Kochin, I. A. Kibel, and N. V. Roze (1963) , *Theoretical Hydromechanics*, **2**, Nauka Press, Moscow.

82. R. Courant and K. O. Friedrichs (1948), *Supersonic Flow and Shock Waves*, Interscience, New York.

83. G. P. Cherepanov and V. A. Pinsker (1987), On the high-temperature cutting of rocks, *Soviet Mining Science*, **6**.

84. G. P. Cherepanov and V. A. Pinsker (1988), On the theory of cutting materials using a laser beam, *Strength of Materials*, **11**.

85. G. P. Cherepanov and V. A. Pinsker (1989), High-temperature cutting of rocks: 3D Problems, *Soviet Mining Science*, **12**.

86. G. P. Cherepanov (1994), Zero drag of a thin wing moving in an elastic medium at the Rayleigh speed, *J. Applied Mechanics*, **61**(4), pp. 988-991.

CHAPTER 8

FRACTALS IN FRACTURE OF SOLIDS

The revolution in geometry, which has recently led to the notion of the fractional dimension of real world bodies and the creation of fractal geometry, has substantially influenced fracture science. In the present chapter, the methods of the theory of fractals are applied to fracture phenomena. A brief overview of practical fractal geometry and some general work on fractals is given in the Introduction, Section 8.1. In Section 8.2, *Fractal Analysis in Fracture Mechanics*, a description of some experimental measurements of the fractality of fractures is provided for some metals and rocks. In Section 8.3, *Fractal Cracks in Solids*, the methods of fractal geometry are applied to brittle cracks with irregular and complicated structures. In Section 8.4, *Nanofracture*, some basic relations for nanoscale, which can serve as a reference scale in scaling laws, are given. An application of the methods of fractal geometry to fatigue and creep is treated in Section 8.5, *Fatigue and Creep*. Fractal geometry is treated in many books, some of which are indicated in the Introduction. However, for the mastering of the simplest methods, a reading of this chapter suffices; to feel so the reader is encouraged to solve the problems at the end of the chapter.

8.1. Introduction

We start from practical fractal geometry.

Power-law fractals

What are *fractals* and *fractal geometry* from a pragmatic point of view? To explain this elementarily, we consider the simplest mathematical problem of measuring the length between two points in space, and a practical counterpart of this problem, namely, measuring the length of a coastal line between two points.

The length of a mathematical line depends on the path we travel between the two points. The more complicated the path the greater the length of the line. It is easy to invent the paths, the length of which may be however great, even without loops. The construction of these paths includes many turns within a small domain (we can create any number of turns because the thickness of the mathematical line is zero).

The length of a coastal line depends on the scale we use to measure this length. A jogger measures the length by the scale of 1 *m*. A marine microbiologist is interested to know the length on the scale of 10^{-3} *m* for a snail crawling around the coastal line. A *Maxwell's demon* or an infinitesimal man probably using X-ray technique to the effect, would like to measure the same length with an accuracy of 10^{-10} *m*; it is his scale. A cartographer or geographer may be satisfied with measurements on the scale of 10^4 *m* or 10^5 *m*. The general property of these measurements is that length increases when scale decreases. For example, if we suppose that the geodesically shortest distance between two points equals 1000 miles, the measured distance may appear to be many times greater. The smaller the scale, the longer the length, so that the measured jogging distance could be as long as 10,000 miles. It was a striking discovery in the 1960s that, in the range of scales from 1 *m* to 100 *km*, the results of measurements are well approximated by the power function

$$L = C\Delta^{1-d_f} . \tag{8.1}$$

Here L is the length of the coastal line measured in scale Δ, C and d_f are some constants independent of scale and length. If one chooses a certain reference scale Δ_0 , Equation (8.1) can be written as:

$$\frac{L}{L_0} = \left(\frac{\Delta}{\Delta_0}\right)^{1-d_f} . \tag{8.2}$$

Or, after taking the logarithm, we may write

$$log\frac{L}{L_0} = \left(1-d_f\right) log\frac{\Delta}{\Delta_0}. \tag{8.3}$$

Here L_0 is the length in the scale Δ_0.

The dimensionless quantity d_f is referred to as *the metric or fractal dimension* of a coastal line, and the coastal line is called the fractal of dimension d_f . Different coastal lines have different fractal dimensions; in other words, they represent fractals of different dimensions, from $d_f = 1.14$ (Australia) to $d_f = 1.49$ (Norway). The value of $d_f = 1$ is called *the topological dimension;* when $d_f = 1$ the length of the line does not depend on measurement scales and devices (this is the geometry we used before the advent of fractal geometry). All geometrical objects of pre-fractal geometry, namely lines, areas, and volumes, had dimensions 1, 2 and 3 so that only 1D, 2D and 3D objects were recognized. (In the new geometry, these are referred to the topological dimensions).

Many power-law fractals have been discovered not only in geography, but also in mechanics, physics, chemistry, biology and other sciences. Some of them like turbulent fluid flows have been known since the 1930's and even since earlier time (see Section 8.5). However, it was only after Mandelbrot's book[1] that fractality as a general property of the natural world was accepted. Almost all natural objects appeared to be fractals or multifractals. That the only 1D, 2D, 3D and 4D (if time is involved as a specific coordinate) objects were considered in the past, was our simplified geometrical model of the world, and not a law of nature.

A natural object or system can often be described as a power-law fractal with a non-integer dimension using the following equation

$$\frac{P}{P_0} = \left(\frac{\Delta}{\Delta_0}\right)^{d_T - d_f} . \tag{8.4}$$

Here P is a measured property of the object or system in scale Δ, P_0 is the value of the property in a certain reference scale Δ_0, d_T is the topological dimension (equal to 1,2,3 or 4), and d_f is the metric or fractal dimension which usually varies within the following limits:

$$\begin{array}{ll} 0 \le d_f \le 1 & \text{for point-like fractals with } d_T = 0; \\ 1 \le d_f < 1 & \text{for line-like fractals with } d_T = 1; \\ 2 \le d_f < 3 & \text{for area-like fractals with } d_T = 2; \\ 3 \le d_f < 4 & \text{for volume-like fractals with } d_T = 3; \\ 4 \le d_f < 5 & \text{for dynamic 3D objects with } d_T = 4. \end{array} \tag{8.5}$$

The lower bound in Equation (8.5) corresponds to the classical geometry of lines, areas, volumes and dynamic 3D objects.

Experimental studies of the fractal dimensions of numerous natural objects have led to many discoveries. For example, owing to recent astrophysical data, the universe that after Einstein's theory of relativity was seen as a three-dimensional space-time, now appears to have the fractal dimension d_f =2.2; that is, the universe looks like a fractal pancake!

Mathematics has bound together empirical knowledge utilizing strict proofs and notions that were often achieved long before practical fractal geometry appeared. These proofs are treated in the books cited below.

Some remarks

In man-made structures, a variety of nano-, micro-, and macro-defects appear at the production stage that may worsen during the service life of the structure. In geomaterials, joints and faults of various sizes have appeared throughout the geological history of the Earth. Numerous geophysical and fractographic studies[1–6] indicate the fractal characteristics of fractures in solids. Moreover, it has been empirically established[-6] that defects emerging during deformation of initially homogeneous materials gain a hierarchical block structure with specific spatial scales, L_i, obeying the relation, $X = L_{i+1} / L_i = const$, $i = 0, 1,2,...$ The "fractality" manifests itself most strikingly in the following aspects:

(i) the statistical self-similarity of crack configurations, whose metric dimension, d_f does not coincide with the topological dimension, d_T, and usually has the fractional value, so that $d_T < d_H < d_{T+1}$;

(ii) the multifractal character of the spatial distribution of defects, pores, and cracks in solid materials[1–10], and the power law of their size distributions[-4]; and

(iii) the scaling invariance of the pore space volume and crack surface in the broad range of scales.[1–10]

The foundations of fractal solid mechanics have been laid down[1–14] during the past decade. The irreversible deformation and fracture of solids relates to a class of processes in which the macroscopic effects are determined by microscopic-level behaviors[15,16]. For this reason, a reliable prediction of the response of a solid to an external force must be based upon a clear understanding of the mechanics of the processes in nanoscale. It is evident that only the description of these processes within the framework of a single system, taking into account the interrelation of different processes in nano-, micro-, and macroscale, would result in an adequate theory of the deformation and fracture of solids.

From this point of view, nanofracture mechanics[17–22] constructed on a simple fundamental concept in nanoscale, may be extended to micro- and macroscale if one uses the experimental evidence of statistical self-similarity and scaling invariance with respect to cracking in solid materials. The self-similarity of dissipative structures on different scaling levels provides the opportunity for transition from nano- to micro- and macro-levels.

8.2 Fractal Analysis in Fracture Mechanics

The geometrical objects with fractal dimensions called fractals have been known in mathematics for a long time. The Cantor set (Figure 8.1), the Sierpinski triangle (Figure 8.2) and the Koch curve (Figure 8.3) are classic examples of regular, power-law fractals.

Figure 8.1. The Cantor set (fractal dimension, $d_f = ln2/ln3$).

a *b*

Figure 8.2. The Sierpinski triangle as a geometric fractal

k, is proportional, within the framework of self-similarity law, to k^x , where the exponent, x, defines the fractal dimension that relates the mass of the object to its size. In the case of scattering by three-dimensional objects having fractal surfaces, we have $x = 6 - \gamma$, while for porous materials, $x = 7 - \gamma$, where γ is the index specifying the size distribution law of pores:

$$N(r) \sim r^{-\gamma}, \tag{8.6}$$

where $\gamma = d_f - 1$, and r is the pore size.

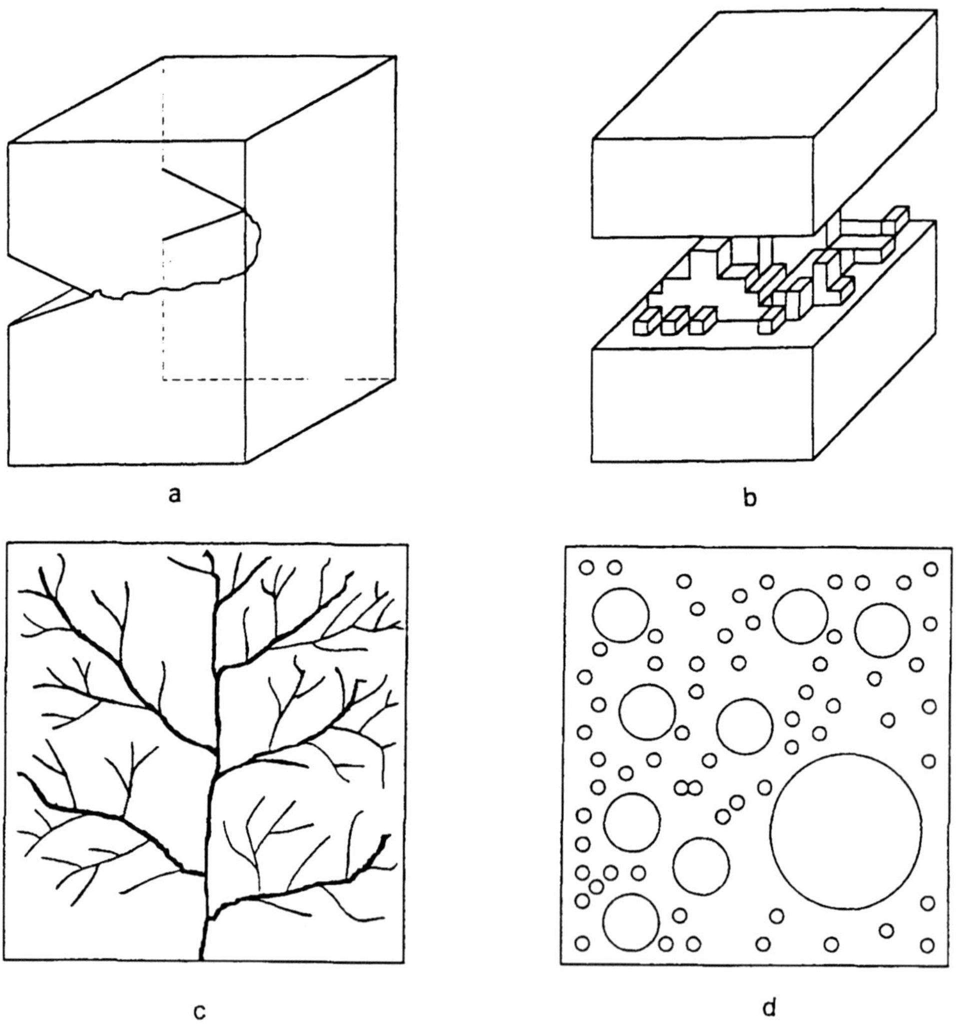

Figure 8.4. Fractal geometry of the fracture of a solid: (a) a crack front showing fractal properties in brittle fracture; (b) a crack surface showing fractal properties in ductile fracture; (c) the crack path showing statistical self-similarity in brittle fracture; (d) the scaling law observed in a crack size distribution by quasibrittle fracture.

Another study method for statistically self-similar objects, the adsorption probe technique, is based on the comparison of adsorption isotherms for different substances in the monolayer region[1,4]. However, the range of molecular sizes of possible adsorbates is quite narrow (from 0.2 to 1 *nm*), whereas the possible use of highly-molecular probes necessitates taking into account the change in the conformation of polymer chains in the adsorption process. The fractal dimension of clusters can be determined from distribution curves. Fractals are typically characterized[13] by the power-law size distribution function *N(r)*; see Equation (8.6). The angle of the slope of straight lines of *ln N* plotted against *ln r* defines the fractal dimension connected with d_f for the crack surface, $d_f = \gamma + 1$. An application of this method is shown in Figures 8.5(a) and 8.5(b) where the size distribution of the number of pores for high-temperature creep in AISI 304 steel is described. Using this method, it has been demonstrated[23] that the fractal dimension of a cluster of pores increases with time.

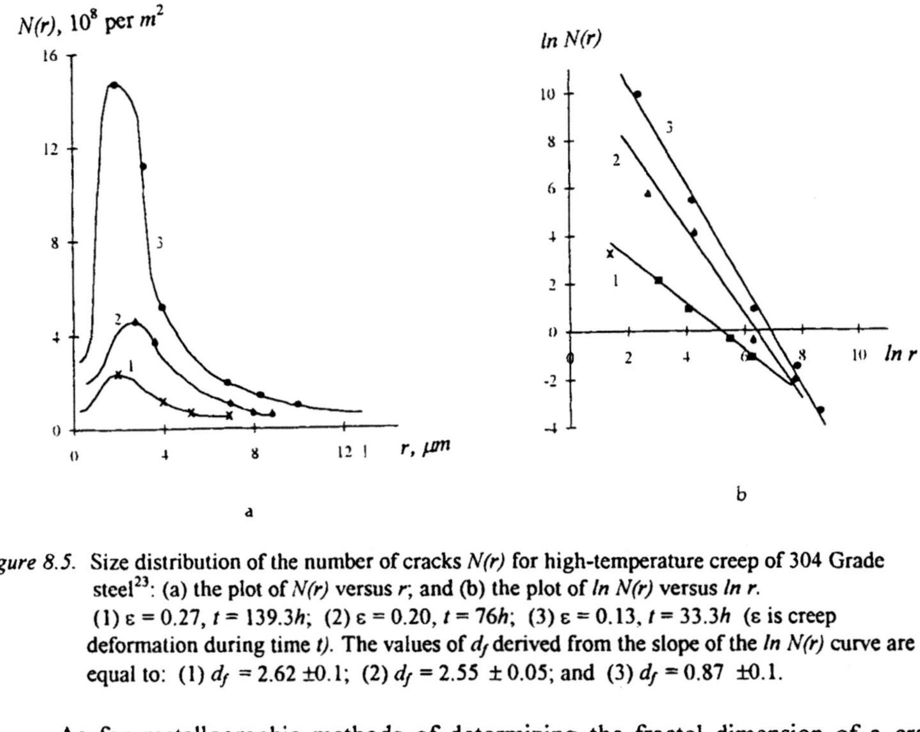

Figure 8.5. Size distribution of the number of cracks *N(r)* for high-temperature creep of 304 Grade steel[23]: (a) the plot of *N(r)* versus *r*; and (b) the plot of *ln N(r)* versus *ln r*. (1) $\varepsilon = 0.27$, $t = 139.3h$; (2) $\varepsilon = 0.20$, $t = 76h$; (3) $\varepsilon = 0.13$, $t = 33.3h$ (ε is creep deformation during time *t*). The values of d_f derived from the slope of the *ln N(r)* curve are equal to: (1) $d_f = 2.62 \pm 0.1$; (2) $d_f = 2.55 \pm 0.05$; and (3) $d_f = 0.87 \pm 0.1$.

As for metallographic methods of determining the fractal dimension of a crack surface, three basic methods compete with one another:

(i) slit island analysis (SIA),
(ii) method of vertical sections (VMS),
(iii) Fourier analysis of the distribution profiles (FAP).

The first method is based on measuring the dependence of the crack surface area, *S*, on the crack perimeter, *P*, when plotted in a logarithmic scale (Figure 8.6) for fractal islands obtained by cutting the crack surface by parallel planes. To do this, a sample with the crack surface is mounted into a conducting bakelite and then ground until many small "islands" become visible on the crack surface. Individual islands are then photographed with high magnification and the photographs are used to measure the area and perimeter of each island (for example, by means of image analysis technique). From the slope of the line (Figure 8.6) drawn through these data points by least square fitting, one can calculate the fractal dimension for each crack surface using the equation, d_f = (slope +1). This method requires rather sophisticated procedures in practical use.

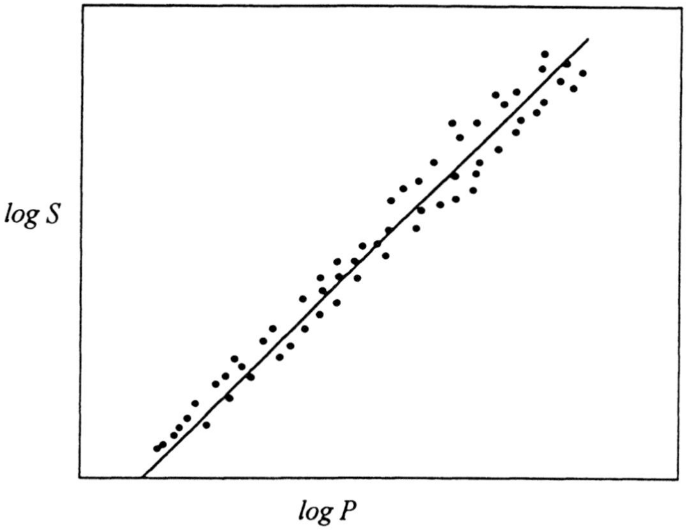

Figure 8.6. A typical *log S* versus *log P* diagram.

The method of vertical sections consists of investigating the relation between the length of the profile of the crack surface and the scale of the measurement. To obtain reliable average figures, the measurements are conducted with respect to several different orientations of the surface profile, obtained by polishing the sections of the crack surface on which a composite or plastic is deposited. It has been established[4] that the dependence of the profile length L on the measurement scale, ε, obeys the asymptotic law of power-law fractals:

$$L(\varepsilon) = C_0 \varepsilon^{-(d_f - 1)}. \tag{8.7}$$

One version of VSM is to measure the ratio, $R_L(\varepsilon)=L(\varepsilon)/L'$, where L' is the length of the profile projection on the plane parallel to the crack surface. This ratio characterizes the surface roughness:

$$R_L(\varepsilon) = C\varepsilon^{-(d_f - 1)}. \tag{8.8}$$

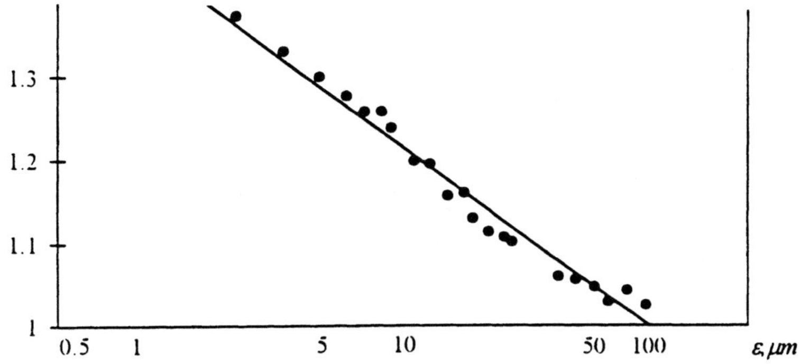

Figure 8.7. The profile roughness parameter R_L versus the measuring unit ε of the crack surface for a specimen of AISI 4340 Grade steel tempered at 700°C, in logarithmic coordinates. (The correlation coefficient is equal to 0.991). The data are taken from reference[23].

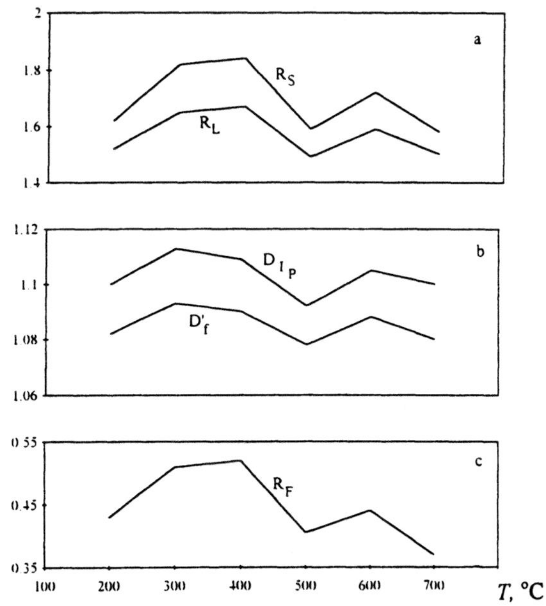

Figure 8.8. Some functions of tempering temperature T for a specimen of AISI 4340 -Grade steel tempered during 1.5 hours, namely: (a) the profile, R_L, and surface, R_S, roughness parameters; (b) the fractal dimension measured by several methods; and (c) the local shape parameter, R_F, of the crack surface. The data are taken from reference[23].

Figure 8.7 shows the dependence of the surface roughness for the specimen made of AISI 4340 steel (tempered at 700°C) on the measurement scale, ε. From the data in Figure 8.8, it can be seen that the fractal dimension of the crack surface, d_f, depends on the tempering temperature of steel, that is, on the initial microstructure.

The value of d_f correlates to the variations of roughness (R_L), shape (R_S), and local configuration (R_F) of the crack surface (Figure 8.8). Here R_S denotes the ratio of the measured area to its projection, and R_F is the ratio of the mean height of the profile peaks to the mean distance between peaks. Evidently, the relation between R_S and R_L can be written in the form,

$$R_S = \frac{4}{\pi}(R_L - 1) + 1. \tag{8.9}$$

Figure 8.9 shows the fractal dimensions of the crack profile and surface in AISI 4340 steel versus the tempering temperature.

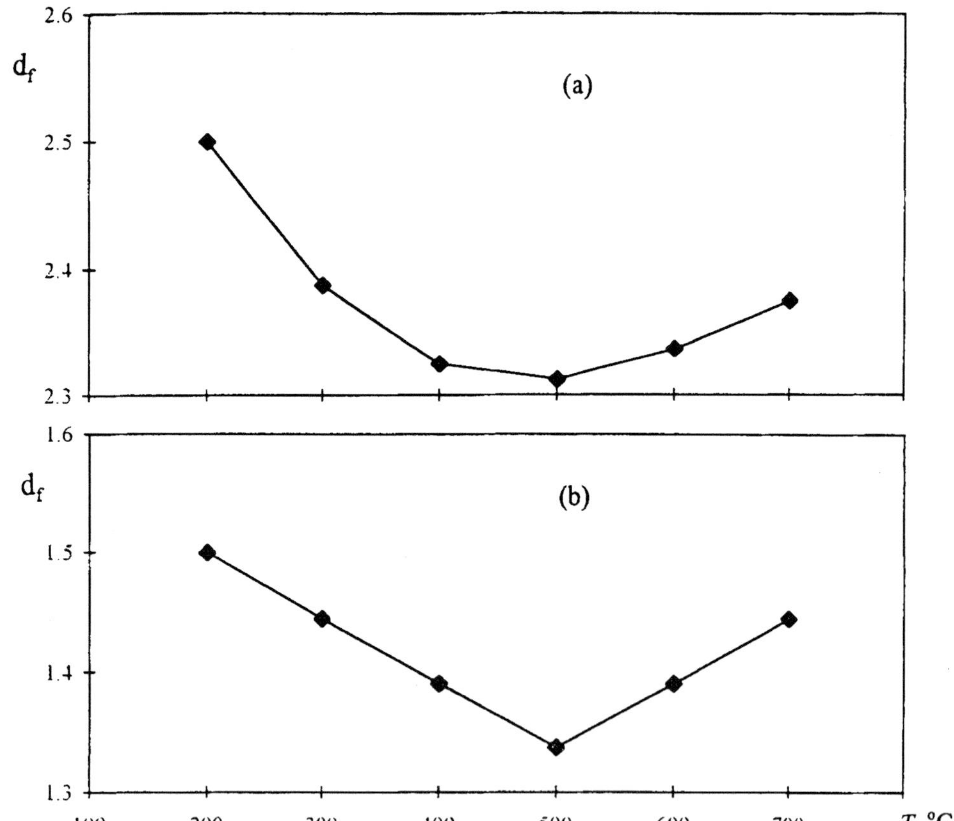

Figure 8.9. The fractal dimension of the surface (a) and profile (b) of cracks in terms of tempering temperature T for specimens of AISI 4340 Grade steel tempered during 1.5 hours. The data are taken from reference[23].

The FAP method for the determination of the fractal dimension of a crack surface is based on an analysis of the Fourier series of the crack surface profile. In this method, the sum of squared amplitudes of the Fourier series, S, is approximated by the power function:

$$S \sim k^{B}. \tag{8.10}$$

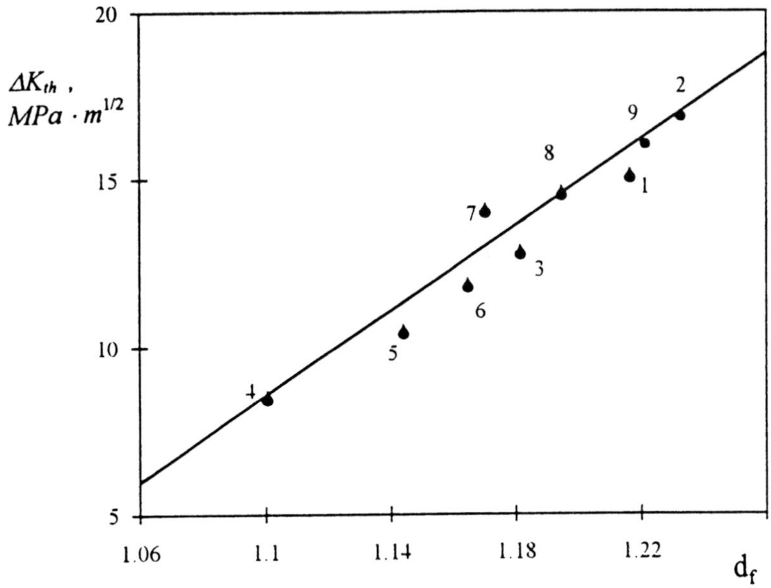

Figure 8.12. Correlation between the fractal dimension, d_f, and the threshold stress intensity factor range amplitude ΔK_{th} for fatigue cracks in a number of steels of different microstructure and heat treatment: (1) needle-shaped martensite; (2) net martensite, (3) granular martensite; (4) rolling steel; (5,6,7,8,9) different types of heat treatment.

Figure 8.12 shows the correlation between the fractal dimension of the fatigue crack profile and the threshold value of the stress intensity factor range, ΔK_{th}.

Rocks

One of the most remarkable exhibitions of fractal fracture mechanics is found in geophysics. Fracturing of geomaterials in the Earth solid shell manifests itself by earthquakes. The earthquake frequency law established by Guttenberg and Richter as far back as 1954 can be expressed in the form[2],

$$lg\, n(M) = a_1 - b_1 M = a - b\, lg\, E. \quad (M \sim lg\, E; \quad M_1 < M < M_2). \tag{8.12}$$

Here $n(M)$ is the number of earthquakes of magnitude M and more, for a long-time interval recorded over a specific region, E is the earthquake energy, and a and b are constants. The constant a determines the average value of seismic activity in a given region, and b is the fractal exponent, which is approximately constant for each region. Equation (8.12) is valid in a wide range of magnitudes, from disastrous earthquakes to the acoustic emission revealed in microcracking[2].

Equation (8.12) is equivalent to the power-law function, $n = CE^{-b}$, which means that an increase in E by the factor λ causes a decrease of n by the factor λ^{-b}:

$$n(\lambda E) = \lambda^{-b} n(E). \tag{8.13}$$

This may be interpreted as a manifestation of the fractal invariance of the earthquake frequency of dimension b in the different scales of energy E. For many seismic regions, the constant b is equal to about 0.9.

Guttenberg and Richter did not explore the spatial limitations of Equation (8.12), assuming the uniform seismicity distribution in each region. However, later studies[2], based on computer analysis of seismic catalogs, have shown the existence of a relation between the number of earthquakes in a certain range of magnitudes and the size of the seismic area for earthquakes with $M > 5$. The modified frequency law has the form[2]:

$$lg\, n(M,L) = a_2 - b\, lg\, E + C\, lg\, L \quad . \tag{8.14}$$

Here a_2 is a constant for a given region ($-3 < a_2 < -5$), b is the fractal dimension in scales of energy E, L is the linear size of the seismoactive region, and C is a spatial fractal dimension ranging, for different regions, from 1 to 2. If earthquake foci are distributed uniformly throughout the seismic area, we have $C = 2$. If earthquake foci are distributed uniformly along a certain line segment, we have $C = 1$. The intermediate value of C characterizes the fractal dimension of earthquake distribution in the area (the spatial fractal). Seismic process is fractal with respect to both energy and area. Moreover, the study of earthquake distribution in time for a fixed region showed that the distribution also obeys a similarity law:

$$lg\, n(\tau) = a_3 + B\;\; lg\, \tau \tag{8.15}$$

where $n(\tau)$ is the number of earthquakes occured for the time span τ, and a_3 and B are constants. Thus, all the major seismicity features obey the similarity (scaling) law. This observation suggests the generalized fractal seismicity law[2]:

$$lg\, n(E,\tau,L) = A - b\; lg\, E + C\, lg\, L + B\, lg\, \tau\,, \tag{8.16}$$

where $n(E, \tau, L)$ is the number of earthquakes of energy E or more that occurred for the period τ in the region of the characteristic linear size L, and A, b, B and C are certain constants. A detailed examination of several seismoactive regions was done by Sadovsky and Pisarenko[2], who estimated these constants in corresponding regions. They divided the maps of these regions into nearly square cells of area ε^2, the minimum ε value being as small as 3 *km*. The frequencies of earthquakes in each cell for a selected period of time were used as sampling estimates of earthquake probability.

The topological features of fields of defects in real solids can be studied using methods of percolation theory[24]. Percolation clusters as typical multifractals were used in simulating some stochastic structures[4,24]. The topological properties of percolation clusters are characterized by a wide spectrum of dimensions linked by the critical indexes that determine the behavior of the system near the percolation threshold[24]. See also references[25-30], which are devoted to the numerical simulation of elastic self-similar structures, that is, elastic fractals.

8.3. Fractal Cracks in Solids

Cracks formed in structural materials and rocks do not resemble the idealized geometric models of cracks with smooth edges which are usually considered in traditional fracture theories. As a rule, the crack front and the crack surface have a quite irregular structure that is characterized by roughness and asperities of different sizes. With the exception of some specific cases, a crack surface and front resemble a chaotic structure. Nevertheless, it has been experimentally established that they possess statistical self-similarity in a wide range of length scales, $\Delta_0 > \Delta > \Delta_m$, where Δ_0 and Δ_m are the greatest and smallest scales which are the upper and lower bounds of the self-similarity[1-4,31-33]. Furthermore, cracking sometimes creates a tree-like fractal structure. Fractality of the crack front and surface, as well as that of tree-like cracking structures are addressed in this section.

Single crack growth

Consider a crack growth with no branching along the front line. Suppose the crack surface created as a result of the growth is a simple fractal obeying the power-law, with a metric dimension $d = d_A$ in the range of length scales, $\Delta_0 > \Delta > \Delta_m$. In the smaller scales $\Delta < \Delta_m$, some plastic deformations and / or more complicated surface structure hold so that, in scales $\Delta > \Delta_m$, the material and crack may be considered to be brittle. Assume that the scale Δ_0 coincides with the scale of the crack length and / or the specimen size (the laboratory scale). Also, we assume that, in the scale Δ_0, the metric dimension d_A is equal to the topological dimension of the crack surface, that is, 2. Hence, due to Equation (8.5), $3 > d_A \geq 2$. In other words, we have

$$A_n = A_0\left(\frac{\Delta_0}{\Delta_n}\right)^{d_A-2} \qquad (m \geq n \geq 0). \tag{8.17}$$

Here, A_n is the area of the crack surface in the scale Δ_n. Suppose the crack surface area has increased by the amount dA_0 in the scale Δ_0 (because of the crack growth). The elastic energy release providing for this area growth is equal to $\Gamma_c\, dA_0$ where Γ_c is the specific energy absorption per unit area measured in the usual laboratory scale, Δ_0. Due to the energy conservation law the same amount of energy must be released if measured in another scale, Δ_n:

$$\Gamma_c\, dA_0 = \Gamma_1\, dA_1 = \ldots = \Gamma_n\, dA_n = \ldots = \gamma\, dA_m. \tag{8.18}$$

Here, Γ_n is the specific energy dissipation per unit area measured in scale Δ_n, $n = 1,2,..,m$. It is reasonable to consider $\Gamma_m = \gamma$ as the efficient surface energy of the material resulting from quasibrittle fracturing. If Δ_m equals the interatomic spacing, the value of γ is the true surface energy. For larger scales, the efficient surface energy can be many times greater because of plastic deformations and / or surface fractality. Using Equations (8.17) and (8.18) we can derive the following scaling laws:

$$\Gamma_c = \gamma\left(\frac{\Delta_0}{\Delta_m}\right)^{d_A-2}, \quad \Gamma_n = \gamma\left(\frac{\Delta_n}{\Delta_m}\right)^{d_A-2} \quad (n = 1,2,\dots,m). \tag{8.19}$$

Equation (8.19) determines the energy dissipation rate at the crack tip by its growth measured in different scales.

Suppose that a material is ideally brittle up to the scale of interatomic spacing a, that is, $\Delta_m = a$, and assume that the power-law with the same metric dimension is valid up to this scale. Probably, these assumptions can approximately describe the properties of ceramics and glass; more tests must be done to find the corresponding metric dimension in these materials. For such materials, we have

$$\Gamma_c = \gamma\left(\frac{\Delta_0}{a}\right)^{d_A-2}, \tag{8.20}$$

where γ is the true surface energy. Equation (8.20) expresses the specific energy dissipation rate on the crack front measured in the laboratory scale Δ_0, in terms of true surface energy and interatomic spacing.

Consider also two simpler cases which are of an academic interest:

(i) A through crack develops in an infinitely thin plate along a fractal curve in a scale Δ_m (similar, for example, to the Koch curve of Figure 8.3).

(ii) The crack front is a fractal line, e.g. a Koch curve, that suffers a translation as a whole in the direction of the crack growth while the crack grows.

We consider these particular cases of a fractal crack surface in more detail:

(i) Suppose that the crack path in a thin sheet is a regular power-law fractal in the range of length scales, $\Delta_0 > \Delta > \Delta_m$ with the metric dimension d_L (where $2 > d_L > 1$ and in this case, $d_T = 1$). In other words, we have

$$L_n = L_0\left(\frac{\Delta_0}{\Delta_n}\right)^{d_L-1} \quad (m \geq n \geq 0). \tag{8.21}$$

Here L_n is the length of the crack path in the scale Δ_n, and L_0 is the crack length in the laboratory scale Δ_0.

Similar to Equations (8.18)-(8.20) we can derive, using same assumptions and notation,

$$\Gamma_c dL_0 = \Gamma_1 dL_1 = \dots = \Gamma_n dL_n = \dots = \gamma\, dL_m, \tag{8.22}$$

$$\Gamma_c = \gamma\left(\frac{\Delta_0}{\Delta_m}\right)^{d_L-1}, \quad \Gamma_n = \gamma\left(\frac{\Delta_n}{\Delta_m}\right)^{d_L-1}, \quad (n = 1,2,\dots,m), \tag{8.23}$$

$$\Gamma_c = \gamma\left(\frac{\Delta_0}{a}\right)^{d_L-1}, \quad (\text{for ideally brittle materials with } \Delta_m = a). \tag{8.24}$$

The equations provide scaling laws for the cascade transfer of the elastic energy flow from the laboratory scale to smaller scales, near a crack tip.

(ii) Suppose that the crack front is a regular power-law fractal, with a metric dimension d_F where $2 > d_F > 1$, in the range of length scales, $\Delta_0 > \Delta > \Delta_m$. The length of such a fractal front is given by

$$W_n = W_0\left(\frac{\Delta_0}{\Delta_n}\right)^{d_F-1}. \tag{8.25}$$

Here, W_n and W_0 are the lengths of the crack front in scales Δ_n and Δ_0 respectively.

If the crack surface is formed by the translation of the fractal front line, the scaling laws corresponding to Equations (8.18)-(8.20) are provided by

$$\Gamma_c dW_0 = \Gamma_1 dW_1 = \ldots = \Gamma_n dW_n = \ldots = \gamma\, dW_m, \tag{8.26}$$

$$\Gamma_c = \gamma\left(\frac{\Delta_0}{\Delta_m}\right)^{d_F-1}, \quad \Gamma_n = \gamma\left(\frac{\Delta_n}{\Delta_m}\right)^{d_F-1}, \quad (n = 1,2,\ldots m), \tag{8.27}$$

$$\Gamma_c = \gamma\left(\frac{\Delta_0}{a}\right)^{d_F-1} \quad (\text{for ideally brittle materials with } \Delta_m = a). \tag{8.28}$$

Equations (8.18)-(8.28) represent the scaling laws of the fracturing of quasibrittle materials cut by a single crack. If we assume that the equation, $dA_n = dL_n\, dW_n$, is statistically valid in the neighborhood of a point on the crack front (where L_n is the length of the crack path in the plane perpendicular to the crack front in scale Δ_0 at the point), then from Equations (8.18) - (8.28) it follows that the metric dimension of the crack surface is equal to the sum of metric dimensions of the crack path and front, that is, $d_A = d_L + d_F$. This important conclusion, however, needs experimental verification.

The stress intensity factor near the fractal crack front depends on the scale of measurement, too. Let us derive the corresponding law of scaling for cracks with fractal surfaces obeying the power law of Equation (8.17). We assume that the crack under study is an opening mode or mode I crack in any scale Δ_n where $n = 0,1,\ldots,m$.

According to the energy conservation law, the release of elastic energy due to the increment of the crack surface area is given by

$$dG = \frac{K_I^2}{\left(1-\nu_0^2\right)E_0} dA_0 = \frac{K_{I1}^2}{\left(1-\nu_1^2\right)E_1} dA_1 = \ldots = \frac{K_{In}^2}{\left(1-\nu_n^2\right)E_n} dA_n . \tag{8.29}$$
$$(n = 1,2,\ldots m)$$

Here, K_I and K_{In} are the stress intensity factors measured in the laboratory scale Δ_0 and scale Δ_n respectively, ν_n and E_n are Poisson's ratio and Young's modulus in Δ_n, and dA_k is the crack surface increment in Δ_k where $k = 0,1,2,\ldots,m$.

Assume that the fractal volume properties of an elastic solid can be neglected, so that for all scales:

$$\nu_0 = \nu_1 = \nu_2 = \ldots \nu_n = \ldots = \nu_m, \quad E_0 = E_1 = E_2 = \ldots E_n = \ldots = E_m . \tag{8.30}$$

Using Equations (8.17), (8.29) and (8.30) we find

$$K_{In} = K_I \left(\frac{\Delta_n}{\Delta_0}\right)^{\frac{d_A - 2}{2}} . \tag{8.31}$$

If the fractality in the elastic solid can be ignored, the asymptotic distribution of stresses, strains and displacements near the front of a single fractal crack in scale Δ_n, is the same as that for the crack in laboratory scale Δ_0, the shape of which coincides with the shape of the fractal crack in scale Δ_n. It is evident because $\Delta_{n+1} \ll \Delta_n$, so that the crack asperities in scale Δ_{n+1} do not influence on the shape of the crack in scale Δ_n. (The conclusion of Cherepanov, et. al.[5] regarding the power-law asymptotic distribution of stresses and strains near the fractal crack front is incorrect). Equation (8.31) makes sense only for statistically mean values because the stress intensity factor K_{In} (in scale Δ_n) depends on the crack surface asperities in all greater scales Δ_k with $k < n$.

Let us compare the smallest elastic scale, Δ_m, with the scale Δ_{m+1} which has an appreciable plastic strain region (which renders K_{m+1} meaningless). Suppose that the ηth part of the elastic energy release rate in scale Δ_m is spent for plastic deformations in scale Δ_{m+1}, and the remaining $(1-\eta)$th part of the elastic energy release rate is spent for the crack growth in scale Δ_{m+1}, i.e.:

$$(1-\eta)\Gamma_m dA_m = \Gamma_{m+1} dA_{m+1} . \tag{8.32}$$

Assume that the crack surface and the energy rate spent for the crack growth obey the power laws in scales Δ_m and Δ_{m+1}, namely

$$\frac{\Gamma_{m+1}}{\Gamma_m} = \left(\frac{\Delta_{m+1}}{\Delta_m}\right)^{d_g} , \quad \frac{A_{m+1}}{A_m} = \left(\frac{\Delta_m}{\Delta_{m+1}}\right)^{d_A^* - 2} \tag{8.33}$$

Here d_E and $d_A^{\bullet}$ are the metric dimensions of energy and area in these scales.

From Equations (8.32) and (8.33) we can derive

$$d_E = d_A^{\bullet} - 2 - \frac{ln(1-\eta)}{ln(\Delta_m / \Delta_{m+1})} \tag{8.34}$$

The energy dimension d_E, as compared to $d_E = d_A - 2$ in scales Δ_0 to Δ_m, increases if $d_A^{\bullet} = d_A$ or the increase of $d_A^{\bullet}$ is less than $\left| ln(1-\eta) / ln(\Delta_m / \Delta_{m+1}) \right|$.

Tree-mode cracking

Let us consider the problem of complicated, tree-shaped cracking of some materials where the ideas of fractal geometry may be very useful (see Figure 8.4c). This mode of fracturing with many crack branches is observed in brittle materials under high dynamic stress, by metal corrosion and by the environmental cracking of some alloys.

We consider the fractal path of the branching crack growth to be analogous to tree growth. However, there are some substantial differences in these two phenomena. The first is that the crack grows along a branching surface while a tree grows along a branching path in space. The other, more important, difference is that a tree, while growing, maximizes the entropy growth rate (because the leaves of the tree try to obtain a maximum amount of solar energy), but the crack minimizes the entropy growth rate (because the crack tries to find the easiest way to get through the material). Therefore, the fractal shape of a tree-like crack is not specific to the fracturing process itself; it is rather the property of a man-made or natural material structure specially organized to maximize the fracturing energy absorption and the entropy rate in order to arrest crack propagation.

So, the fractal branching path of a crack is the intrinsic property of an optimal, well-made material structure which provides the greatest resistance to the crack growth. Ideally-designed structural materials of the future must display only fractal tree-mode cracking paths. A single cracking path characterizes non-optimally designed materials which make up the majority of structural materials today. In dynamic fracturing, fractal tree-mode cracking may be characteristic even of poor materials, because crack speed is limited, but a high dynamic stress activates very small defects, so that this mode cracking can absorb a maximum of the dynamic wave energy. It is reasonable to mention another fractal analogue of tree-mode cracking, namely, the mouth map of the Mississippi and the Volga rivers.

We assume that fractal model of the tree-mode cracking to be characterized by the following properties:

1. The tree-shaped crack grows statistically in one direction, say, along the x axis.
2. At any moment in time the tree-shaped crack possesses the statistical self-similarity described by the power law of a certain fractal dimension in a range of length scales Δ, from Δ_0 to Δ_m ($\Delta_0 > \Delta > \Delta_m$), where Δ_0 is the greatest scale coinciding with the laboratory scale of the whole cracking tree, and Δ_m is the smallest scale.

3. The metric dimension of the fractal and scale Δ_m are material properties.

4. Material deformations are linearly elastic in all scales, from Δ_0 to Δ_m, so that the metric growth of all branches of the cracking tree is brittle in all scales, from Δ_0 to Δ_m.

Due to the latter assumption, plastic deformations are confined within the limits of the scale Δ_{m+1}, and the cracking can continue to the smaller scales with self-similarity of another fractal dimension. In some cases, for example, in failure waves [34] propagating in glass, the cracking occurs up to the scale 0.1 μm. In structurally-ideal brittle materials like metal glass, the cracking in failure waves can occur up to the scale 10 *nm*, which is the mean size of crystalline clusters.

We describe all dissipative mechanisms taking place in the scale Δ_{m+1} and smaller scales using the efficient surface energy, γ, per unit of metric area. The γ is assumed to be a constant of a given material. The elastic energy release rates and energy absorption fluxes in scales Δ, from Δ_0 to Δ_m, as well as the statistically-mean stress intensity factors in these scales are to be found in the framework of the fractal model of the tree-mode cracking. We assume that there are no dead branches on the cracking tree; all branches grow if there is a sufficient supply of elastic energy from an external field of stresses.

We may consider various shapes of the cracking tree, for example, the following ones:

1. All $x = const$ cross-sections of the cracking tree are statistically identical, that is, the *regime of growth* is *steady-state*; and

2. The statistical width of the cracking tree (in the cross-sections $x = const$) is a linear or power-law function of the statistical length of the cracking tree (along the x axis) starting from a certain reference position, for example, an initial single crack trip, or a small defect with size in Δ_{m+1} and smaller scales, or the end of another cracking tree formed earlier (*self-similar linear or power-law regime of growth*).

The regime of growth determined by the material properties and initial conditions influences the development of a cracking tree. The tree-mode cracking is an important mechanism of stress concentration relaxation and stress energy re-distribution. It may serve a way to arrest a single crack or fracturing.

Using the fractal model of the tree-mode cracking we determine the elastic energy release rate in different scales and statistical stress intensity factor, at the growth of the cracking tree. Suppose that there is one crack branch in the scale Δ_0 (single crack), and there are b crack branches in the scale Δ_1. Because of self-similarity, there are b^2 crack branches in the Δ_2 scale, b^3 crack branches in the Δ_3 and b^n crack branches in any scale Δ_n where $n = 1,2,\ldots,m$. From the energy consideration law, it follows that the flux of energy dissipation at any scale is given by

$$\Gamma_c dA_0 = \Gamma_1 b dA_1 = \ldots = \Gamma_n b^n dA_n = \ldots = \Gamma_m b^m dA_m \qquad (8.35)$$

Here Γ_c is the dissipation energy flow rate per unit area in Δ_0, dA_0 is the area increment by crack growth in Δ_0, Γ_n is the dissipation energy flow rate per unit area of a crack branch in Δ_n and dA_n is the corresponding increment of the metric area of the crack branch in Δ_n where $n = 1,2,\ldots, m$. In our fractal model of the tree-mode cracking, $\Gamma_m = \gamma$ is a material constant. Equation (8.35) describes the cascade process of energy distribution on different scales. Suppose also that the crack branch surface is a fractal of metric dimension d_A which is the same for all branches in the given material, namely

$$\frac{A_n}{A_0} = \left(\frac{\Delta_0}{\Delta_n}\right)^{d_A - 2} \qquad (2 < d_A < 3, \quad n = 0,1,2,\dots m) \tag{8.36}$$

Combining Equations (8.35) and (8.36) one can easy derive:

$$\Gamma_n = \gamma\, b^{m-n} \left(\frac{\Delta_n}{\Delta_m}\right)^{d_A - 2} \qquad (n = 1,2,\dots,m;\ k_{Ic} = \sqrt{E\gamma}\,) \tag{8.37}$$

$$\Gamma_c = \gamma\, b^{m} \left(\frac{\Delta_0}{\Delta_m}\right)^{d_A - 2} \qquad K_{Ic} = k_{Ic} b^{\frac{m}{2}} \left(\frac{\Delta_0}{\Delta_m}\right)^{\frac{d_A - 2}{2}} \tag{8.38}$$

The latter equation provides the statistically-mean value of fracture toughness K_{Ic} of a cracking tree in terms of six scaling and physical parameters (k_{Ic}, b, m, Δ_0, Δ_m, and d_A) of the fractal model of the tree-mode cracking. As seen, the efficient fracture toughness of a material can be increased a hundred times by means of a well-organized material structure securing this mechanism of fracturing.

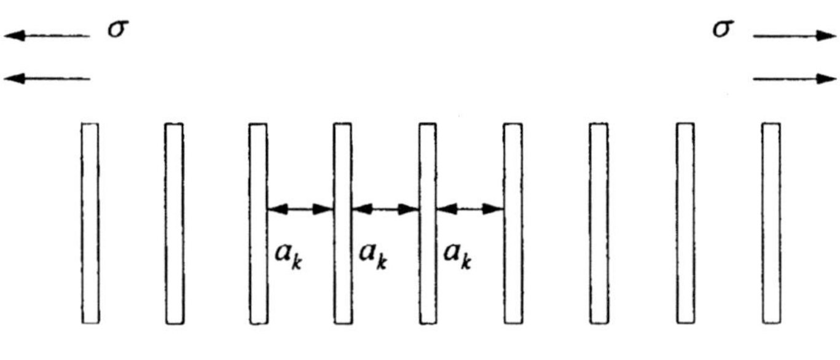

Figure 8.13. A periodic lattice of half-infinite parallel cracks with spacing a_k in the uniform field of tensile stress σ.

Equations (8.35)-(8.38) characterize the frontal part of the cracking tree while it grows. The cracking tree growth occurs because of input of energy of external stress field to the ends of crack branches in the scale Δ_k on the front of the cracking tree. The value of the input is determiend by the structure of frontal crack branches and external stresses.

For a uniform tensile stress field, we estimate the statistically-mean value of stress intensity factors and crack-tip energy flows by using the simple periodic model of half-infinite parallel cracks in the scale Δ_k (Figure 8.13). A simple calculation using the invariant Γ-integral (see Chapter 1) provides the following results[34]:

$$\Gamma_k = \frac{a_k \sigma^2}{2E}(1-\nu^2)(1+2\nu), \qquad k_I = \sigma\sqrt{\left(\nu+\tfrac{1}{2}\right)a_k} \quad . \tag{8.39}$$

Here, σ is the tensile stress, Γ_k and k_I are the energy flow rate and stress intensity factor at the crack-tip respectively, a_k is the lattice spacing (in the scale Δ_k on the front of the cracking tree), and E and ν are Young's modulus and Poisson's ratio.

At the crack growth condition, we have $k_I = k_{Ic}$. From here, we find the statistically-mean value of spacing a_k on the front of the cracking tree in this model:

$$a_k = \frac{4E\gamma}{\sigma^2(1-\nu^2)(1+2\nu)} \quad . \tag{8.40}$$

As seen, the greater the tensile stress, the less spacing between the crack branches in the cracking tree.

The development of the cracking tree along the x axis in terms of loading or time is described by change of the integer index k from 0 to m in the process of the growth. This process can be explained on a simple model of Figure 8.14 depicting a regular mathematical fractal of the cracking tree growth with $b = 2$. The moment of a branching in time coincides with that of the transition from one scale to the next, smaller, one. According to Equation (8.40) and Figure 8.14, the growth of the cracking tree in the initial scale Δ_1 when $k = 1$ proceeds at the stress $\sigma_1 = \eta / \sqrt{a_1}$ where $\eta = k_{Ic}\left(\nu+\tfrac{1}{2}\right)^{-1/2}$ and a_1 is the spacing in the initial scale, Δ_1. In the next scale, Δ_2, when $k = 2$ the cracking tree grows at the greater stress $\sigma_2 = \sigma_1\sqrt{2}$. In the third scale, Δ_3, the cracking tree grows at the stress $\sigma_3 = 2\sigma_1$; and in the fourth scale, Δ_4, at $\sigma_4 = 2\sigma_1\sqrt{2}$. In the kth scale the cracking tree grows at $\sigma_k = 2^{(k-1)/2}\sigma_1$ and so on.

For this fractal, the number of crack branches necessary to absorb the energy of high tensile stress σ_m equals 2^m where:

$$m = 1 + 2\ln\frac{\sigma_m}{\sigma_1} / \ln 2, \qquad a_k = a_1 / 2^{m-1}. \tag{8.41}$$

The index k of the length scale thus plays the role of time in the loading process.

For cracking tree with arbitrary b, the number of crack branches necessary to absorb the energy of high tensile stress σ_k equals b^k where:

$$k = 1 + \frac{2}{\ln b} \ln \frac{\sigma_k}{\sigma_1}, \qquad a_k = a_1 b^{1-k}. \tag{8.42}$$

Remember that $\sigma_1 = \eta / \sqrt{a_1}$ initially.

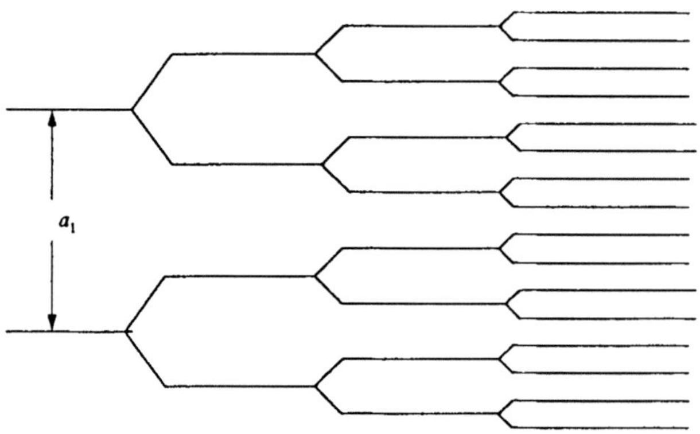

Figure 8.14. A regular mathematical fractal of the cracking tree growth for $N = 2$.

If the external tensile stress σ is a continuous function of time, the stress curve goes through several levels, from $k = 1$ to $k = m$, where: $\sigma_1 < \sigma < \sigma_2$ for $k = 1$, $\sigma_2 < \sigma < \sigma_3$ for $k = 2$, and $\sigma_m < \sigma < \sigma_{m+1}$ for $k = m$, with the corresponding scale fractal of the cracking tree inside each level.

The spatial sizes of the cracking tree in length and width depend on the crack speed on each level. For very high loading rates, the maximum length of the cracking tree can be estimated to be equal to $c_R \Delta t$ where c_R is the limiting speed of the single crack propagation (the Rayleigh speed) and Δt is the time of loading /crack growth.

Tree-mode cracking is an important self-locking mechanism of the crack arrest by the material.

8.4 Nanofracture

Fractal theory can predict fracture and fatigue phenomena in one scale using the corresponding data from another scale. Therefore, it is very important to have reliable data about a fractal in one scale of length, time, energy and so on, which can serve as a reference scale for the approach. The latter is usually the most well-studied laboratory scale of time, length and energy.

Knowledge of reliable data in another scale provides a basis for the estimation of the metric dimension of the power-law fractal. This means that nanofracture mechanics[17-22], which provides a strict approach to an irreversible deformation and fracture in the nanoscale may be used to derive the mechanics of fracture and plasticity in micro- and macro-scales by means of the power-law self-similarity and scaling invariance of fractal cracking. We should not overestimate the possibilities of this approach because of the many dangers of the empirical and statistical nature of the predictions of fractal theory. For example, before an experiment, we do not know whether or not a specific case under study enters into the range between the lower and upper bounds of self-similarity.

However, we cannot neglect this general fractal approach in view of its simplicity. Besides, the information obtained in this way is sometimes unique, especially for a new as yet untested, material. Therefore, we provide here some basic information[17-22] on fracture in nanoscale, that is, roughly in the range from 1 *nm* to 1μm realized inside a grain of a polycrystalline material. For more details concerning the application of nanofracture mechanics to dislocation emission, the reader is referred to Chapter 6.

The brittle versus ductile behavior of a crystalline grain in nanoscale may be specified by a single parameter, called the brittleness number[19]:

$$\eta = \frac{K'_{Ii}}{2\sqrt{G\gamma/(1-\nu)}} = \alpha_1 \frac{\sqrt{\tau_0 a}}{\gamma} \quad . \tag{8.43}$$

Here, K'_{Ii} is the critical value of the stress intensity factor corresponding to the emission of the first pair of dislocations from a mode *I* crack tip (that is, the initiation of plastic deformation), γ is the true surface energy corresponding to the crack growth without any dislocations emitted, G is the shear modulus, ν is Poisson's ratio, a is the interatomic spacing, τ_0 is the Schmidt constant for a given crystal depending on the slip plane, temperature, and purity of the crystal, and α_1 is a number depending on a crystal lattice type (for a cubic lattice $\alpha_1 = 2.236$). The value of K'_{Ii} is equal to:

$$K'_{Ii} = \alpha_2\sqrt{G\tau_0 a/(1-\nu)} \quad . \tag{8.44}$$

Here, α_2 is a number depending on a crystal lattice type (for a cubic lattice, $\alpha_2 = 4.483$).

If $\eta > 1$, a nanocrack tip cannot serve as a source of dislocations, so the fracture will be ideally brittle. If $\eta < 1$, some edge dislocations are being generated from the crack tip under loading, and therefore, a small subcritical growth of the crack occurs. If the η is less than 1 but not very small, the number of emerging dislocations is comparatively small and fracture is quasibrittle.

If

$$\tau_0 L < \alpha_3 a G/(1-\nu), \tag{8.45}$$

the polycrystalline material displays the property of amorfous state, because dislocation emission is impossible in this case. Here, L is the characteristic size of the grain embedded in the material, and α_3 is a number depending on a crystal lattice type (for a cubic lattice, $\alpha_3 = 0.085$). The same Equation (8.45) serves as the condition of superplastic state of a stretched cracked single crystal specimen of linear size L with traction-free sides, because crack growth is impossible in this case.

According to nanofracture mechanics, there is no incubation period for initiating fatigue nanocracks in metal grains[22]. The growth of nanocracks begins under extremely low loads in cyclic loading (K'_{Ii} is the threshold stress intensity factor). The first step on the dl/dN vs. K_I diagram of fatigue nanocrack growth equals one interatomic spacing a per cycle, corresponding to the minimum possible nanocrack growth speed dl/dN (here l is the nanocrack length, and N is the number of cycles). The second step is equal to $2a$ per cycle, the third to $3a$ per cycle, and so on[22]. The minimum fatigue crack growth speeds were observed in thin intercrystalline layers[19].

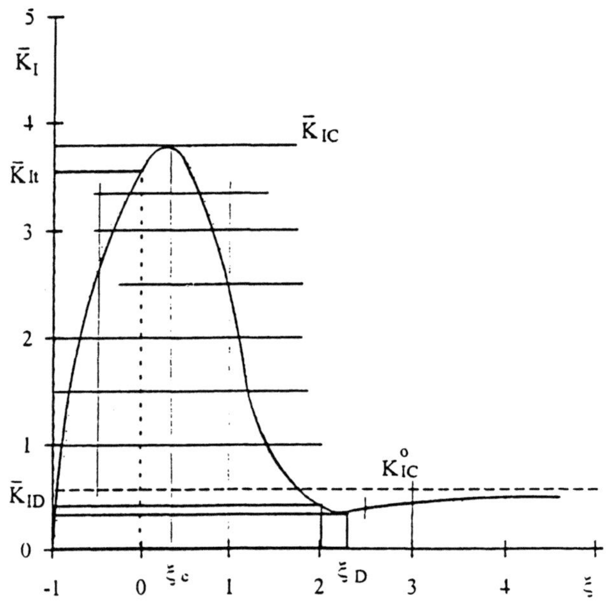

Figure 8.15. Theoretical *R*-curve for a cubic lattice in nanofracture mechanics[22].

The R-curve of a nanocrack specifying nanocrack growth in terms of the stress intensity factor (during loading) is shown in Figure 8.15. We use different scales for $\xi < 0$ and $\xi > 0$, along the horizontal axis so that $\xi = 0$ corresponds to the initiation of a stable brittle nanocrack growth:

$$\xi = -1 + \frac{\Delta l}{a n_i} \quad \text{for } \xi > 0, \tag{8.45a}$$

where Δl is the crack length increment, and

$$\xi = \frac{\sqrt{2}(\Delta l - an_i)}{r_{n_0}}, \quad r_{n_0} = 0.083\frac{aGn_i}{\tau_0(1-\nu)}, \quad n_i = \frac{8.125\gamma}{a\tau_0} \tag{8.46}$$

The characteristic points on the *R*-curve are: the maximum point, $\xi_c = 0.3(\overline{K}_{Ic} = 4.685)$; the minimum point, $\xi_D = \xi_D = 2.3(\overline{K}_{ID} = 0.448)$; the point corresponding to the initiation of a brittle crack, $\xi = 0(\overline{K}_{It} = 4.483)$; and the asymptote $\xi \to \infty(\overline{K}_I = 0.702)$.

At $0 < \xi < \xi_c$ and $\xi > \xi_D$, the nanocrack growth is stable, whereas in the region, $\xi_D > \xi > \xi_c$, the equilibrium is unstable and only dynamical propagation is possible. Thus, we have the following criteria of nanofracture mechanics:

to initiate a brittle nanocrack,

$$K_I > K_{Ii} = 12.78[G\gamma/(1-\nu)]^{1/2} \quad ; \tag{8.47}$$

stable increment of brittle nanocrack length before the onset of instability,

$$\Delta l = \xi_c r_n / \sqrt{2} = 0.146G\gamma/\tau_0^2(1-\nu) \quad ; \tag{8.48}$$

nanofracture toughness,

$$K_{Ic} = 13.35[G\gamma/(1-\nu)]^{1/2} \quad ; \tag{8.49}$$

minimum fracture toughness specifying the arrest condition,

$$K_{ID} = 1.25[G\gamma/(1-\nu)]^{1/2} \quad ; \tag{8.50}$$

nanocrack growth increment before the nanocrack arrest,

$$\Delta l = (\xi_D - \xi_c)\cdot r_{n_0} / \sqrt{2} = G\gamma/\tau_0^2(1-\nu). \tag{8.51}$$

The numerical factors in all these equations correspond to a cubic lattice; they were determined using an approximate approach and will be improved in the future. (See Chapter 6 for some of these improvements). Equations (8.43) to (8.51) may be used as the reference scale data relating to nanoscale.

Nanofracture mechanics[17-22] assumes that the radius of action of atomic cohesion forces is small compared to the interatomic spacing which allows one to neglect the cohesion forces between the opposite sides of the nanocrack almost everywhere, except for a small zone near the crack tip. Probably, this assumption works better for crystals with covalent bonds and metals which have a short-range binding potential, and much worse for crystals with ionic bonds having far-range binding forces. In any case, the direct account of non-linear cohesion forces between the opposite sides of the nanocrack will cause a spontaneous closure of very short nanocracks, so that very short nanocracks locked by the cohesion forces cannot propagate, even for very high stress of the external field[35]. Determination of the critical length of non-propagating nanocracks in terms of different binding potentials is a challenging task to nanofracture mechanics for the future.

8.5 Fatigue and Creep

Time processes make up a large portion of the applications of fractal geometry. In this Section, we consider from this point of view fatigue and creep operation life and fatigue crack growth, which are the most important practical problems of reliability in engineering structures.

General remarks

We consider the time evolution of an arbitrary physical system the state of which is described by N functions (state variables). For an engineering structure made of some materials, the state variables may be external loads or displacements, heat fluxes, irradiation intensity, temperature, concentrations of various chemical components of an environment, and so on. The operation life of the system in time is represented by a path or trajectory in the N-dimensional space of these state variables. For most real systems, the path is a very complicated curve that may be deemed a time fractal obeying the statistical self-similarity or scaling invariance power law in many time scales. This law has the lower and upper bounds of applicability in terms of time scales; we designate them as t_{min} and t_{max} correspondingly.

Time can be measured by the time evolution of the system we study or any other specific systems that exist simultaneously. The operation of some latter systems called *clocks* is periodic in time, to very high preciseness; we can say that they measure the *topological time*, t_T. The clocks include a number of physical systems from atomic clocks, (based, e.g., on Messbauer effect), to mechanical watches, biological clocks and our planetary system. The time-dependent fractal system under consideration has its own time called the *metric or evolutionary time*, t_M ; it is measured using some parameters of the evolution (or marks of aging) of this system. We consider the fractal systems whose metric or evolutionary time is a power law function of topological time,

$$t_M = Ct_T^{-d} \qquad (t_{max} > t_T > t_{min}) \ . \tag{8.52}$$

Here, d is a constant of the fractal system called the *fractal dimension*, and C is a constant. We assume that the metric time t_M coincides with the topological time in either the lower bound scale or upper bound scale (within the limits of reasonable accuracy). From here, Equation (8.52) becomes

$$\frac{t_M}{t_{min}} = \left(\frac{t_{min}}{t_T}\right)^d \quad or \quad \frac{t_M}{t_{max}} = \left(\frac{t_{max}}{t_T}\right)^d \tag{8.53}$$

This form of power law of time self-similarity or scaling invariance is most convenient for practical application. The value of d in (8.52) and (8.53) varies from 0 to 1:

$$0 < d < 1 \tag{8.54}$$

The smaller d is, the closer to a clock the system is.

Any substantial parameter of the system can, in principle, be chosen to be a time meter of metric time of the system under study. Suppose p serves as the metric time of the fractal system. In terms of this parameter, p, the power law, similarly to Equation (8.53), can be written in the form:

$$\frac{p}{p_{min}} = \left(\frac{t_{min}}{t_T}\right)^d \quad \text{or} \quad \frac{p}{p_{max}} = \left(\frac{t_{max}}{t_T}\right)^d \tag{8.55}$$

Here p_{min} and p_{max} are the values of p for $t = t_{min}$ and $t = t_{max}$ correspondingly.

Creep

Suppose a bar is subjected to a stationary tensile load σ. Most solid substances show a measurable creep deformation if the temperature of the bar is greater than one half of the melting temperature. As a result of creep, the bar breaks in a time t_T that essentially depends on σ. The greater the value of σ, the less the value of t_T. Hence, σ can serve as the metric time or time meter of this system corresponding to the topological time of the operation life, t_T, of the bar. Assuming the power law fractality provides:

$$\frac{\sigma}{\sigma_b} = \left(\frac{t_{min}}{t_T}\right)^d \tag{8.56}$$

or

$$t_T = t_{min}\left(\frac{\sigma_b}{\sigma}\right)^{\frac{1}{d}} \tag{8.57}$$

Here σ_b is the ultimate tensile strength in the lower bound time scale, at $t_T = t_{min}$ say 1s or less, and d is the fractal dimension.

Equation (8.57) has become a widely-recognized empirical equation for the operational life of many engineering materials since the last century, when creep of turbine blades turned out to be a serious engineering problem[34,35]. For most metals and alloys, the value of d was measured to vary within the limits of 0.08 to 0.25. The value of t_{max} can reach as high as 10^8 seconds, that is, many years, for some alloys. So, *aposteriori* Equation (8.57) confirms the assumption of the power law fractality of the system under consideration.

Using the first and second invariants of the stress tensor as the controlling parameters allows one to generalize the prediction to arbitrary stress states.

Fatigue

Suppose a bar is subject to a periodic (cyclic) loading that varies from σ_{min} to σ_{max} within every period (cycle) of loading. For an ample class of engineering materials called the elastic-plastic materials, the operational life of the bar depends mainly on $\Delta\sigma = \sigma_{max} - \sigma_{min}$ and may be measured in terms of number of cycles before failure, N. The time variation of load inside a period is less important and may be ignored in many practical applications.

The value of N plays a part of topological time that depends on $\Delta\sigma$, so that the greater $\Delta\sigma$, the less the value of N. Hence, $\Delta\sigma$ can serve as the metric time or time meter of the system. If we assume that $\Delta\sigma$ is a time fractal obeying the power law of self-similarity for various time scales, we derive:

$$\frac{\Delta\sigma}{(\Delta\sigma)_0} = \left(\frac{N_{min}}{N}\right)^d \qquad (N_{min} < N < N_{max}) \tag{8.58}$$

or

$$N = N_{min}\left(\frac{(\Delta\sigma)_0}{\Delta\sigma}\right)^{\frac{1}{d}} \tag{8.59}$$

Here $(\Delta\sigma)_0$ is the value of $\Delta\sigma$ at $N = N_{min}$, that is, at the lower bound of the power law validity, and d is the fractal dimension.

Equation (8.58) has been recognized and popular since Woehler discovered fatigue of metals in the middle of the last century[34,35]. The measurement of d in fatigue tests is a standard requirement for any material operating under variable loadings. The value of d for different metals typically varies from 0.07 to 0.15 and, most frequently, from 0.09 to 0.12. The magnitude of lower and upper bounds for structural alloys can usually be $N_{min}=1\div10^3$ cycles and $N_{max}=10^4\div10^8$ cycles.

For low cycle fatigue, the plastic strain range amplitude, $\Delta\varepsilon_p$, has been shown to be the best controlling parameter of fatigue life (in lieu of $\Delta\sigma$):

$$\Delta\varepsilon_p = \varepsilon_p(\sigma_{max}) - \varepsilon_p(\sigma_{min}) \tag{8.60}$$

Here $\varepsilon_p(\sigma_{max})$ and $\varepsilon_p(\sigma_{min})$ are the plastic strains at $\sigma = \sigma_{max}$ and $\sigma = \sigma_{min}$ determined by the σ -ε diagram of the material in simple loading tests (unloading is usually assumed to be linearly-elastic).

If we assume that $\Delta\varepsilon_p$ is the time meter or metric time of this fractal obeying the power law, we derive:

$$\Delta\varepsilon_p = (\Delta\varepsilon_p)_0 N^{-d} \qquad (1 < N < N_{max}) \tag{8.61}$$

where $(\Delta\varepsilon_p)_0$ is the plastic strain $\Delta\varepsilon_p$ in the lower bound time scale $N = 1$, and d is the fractal dimension.

Equation (8.61) has been popular since the 1950s when Coffin and Manson first suggested it. For most metals and alloys, the d value varies from 0.5 to 0.7.

Fatigue crack growth

For many materials and structures subjected to variable loadings, it has been proven that the measurable fatigue crack growth makes up a significant part of the total fatigue life. Data on fatigue crack growth based on fracture mechanics are much more reliable and statistically much more correlated then those obtained without a consideration of fatigue crack growth.

The fatigue crack growth rate is measured by dl / dN, that is, the statistically mean crack length increment during one cycle of loading. The controlling parameter of the fatigue crack growth has been proven, for many materials and conditions, to be $\Delta K = K_{max} - K_{min}$ where K_{max} and K_{min} are the maximum and minimum values of the stress intensity factor during a loading cycle[34,35].

The value of dl/dN can be considered as a measure or meter of topological length of fatigue crack growth, and ΔK as a measure or meter of metric length of fatigue crack growth. If we assume that the fractal ΔK obeys the power law self-similarity, we derive:

$$\Delta K = (\Delta K)_0 \left(\frac{1}{l_0} \frac{dl}{dN} \right)^d \tag{8.62}$$

or

$$\frac{dl}{dN} = l_0 \left(\frac{\Delta K}{(\Delta K)_0} \right)^{\frac{1}{d}} \tag{8.63}$$

Here, $(\Delta K)_0$ is the value of ΔK in the length scale $dl/dN = l_0$, that is, l_0 is the mean increment of crack length per cycle at $\Delta K = (\Delta K)_0$, and d is the fractal dimension.

Equation (8.63) has been well-known since 1960, when Paris suggested it as a result of the analysis of test data[34,35]. The value of d measured for many engineering materials usually varies from 0.1 to 0.3, and is frequently equal to 0.25, which corresponds to the simplest theory of fatigue crack growth at the constant dissipation energy rate[34,35]. The value of l_0 can vary from 1 *nm* (interatomic distance) to 1 *cm* in metals and alloys. As seen, fatigue and creep of solid matters exhibit the properties of regular power-law fractals in time.

8.6 Conclusion

Reading the chapter and solving the problems leads to several conclusions which follow.

Fractal geometry offers very simple and practical methods for the study of complicated phenomena occurring in many scales of space and time.

The methods allow one to transfer the experimental and theoretical results obtained in a scale of length and time to many other scales. Application of these methods to fatigue and creep allows one to derive the scaling laws earlier obtained empirically.

To use the power function for scaling, some test data are necessary, at least utilizing two different reference scales. As the reference scales in fracture phenomena, the nanoscale and laboratory scale offer advantages.

The complicated shape of the surface and front of a crack, particularly a tree-mode crack, substantially influences the conditions and criteria of the fracturing of solids. These criteria are derived in the Chapter by using the assumption of power law self-similarity, that is, by considering these cracking modes as simple fractals.

8.7 Problems

8.1. Consider a coast line of a country as a fractal. Show how to determine the fractal dimension using the maps of the coast made in two different scales.

8.2. The fractal dimension of the coast line of Australia and Norway are equal to 1.14 and 1.49 respectively. Suppose you plan to jog along the coasts between two towns in Australia and between two towns in Norway. In both cases the distance between the towns equals 200 miles based on map in the global scale (i.e., 10,000 miles). What will your jogging distances be in Australia and Norway?

8.3. Consider a fatigue crack path in a material as a fractal with dimension d_f=1.09. Suppose that the material specimen is subjected to a low cycle load, so that the critical length of the crack in laboratory scale is equal to $l_C = 1cm$ (under the cycle maximum load). Estimate the upper boundary of the life duration of the specimen under the assumption that the crack advances one interatomic distance, $\Delta l = 1nm$, per cycle.

8.4. Derive the equation of the fatigue life of a material in terms of d_f, l_C and Δl under the conditions and assumptions of Problem 8.3.

8.5. Consider an elastic material, say a rock, as a fractal so that

$$\frac{A}{A_0} = \left(\frac{L}{L_0}\right)^{d_f} \qquad (3 > d_f \geq 2) \tag{8.64}$$

Here, A is the area of a material cross-section corresponding to the measurement scale L, d_f is the fractal dimension, and A_0 is the area of a material cross section corresponding to the measurement scale L_0. When d_f = 2, the fractal and metric dimensions coincide. Young's modulus is defined as:

$$E = \frac{F}{A\varepsilon} \tag{8.65}$$

Here, F is the resultant force applied to the material cross section A in the measurement scale L, and ε is the deformation in the same scale.

Show that Young's modulus of the elastic fractal obeys the following scaling law:

$$\frac{E}{E_0} = \left(\frac{L}{L_0}\right)^{2-d_f-d_\varepsilon} \qquad (1 > d_\varepsilon \geq 0) \tag{8.66}$$

Here d_ε is the fractal dimension of strain, and E_0 is Young's modulus of the material in scale L_0. If $d_\varepsilon = 0$, the fractality of strain is negligibly small.

Calculate E in scales $L = 10^5\ m$ and $L = 10^{-8}\ m$, if: $E_0 = 1\ Mbar$ in the laboratory scale $L_0 = 0.1m$. Assume $d_f = 2.15$ and $d_\varepsilon = 0$ (for some rocks).

8.6. Consider creep formation, ε, of a rock under maintained stress σ. The strain rate, $\dot{\varepsilon}$, of creep can be described by the following empirical law,

$$\mu\dot{\varepsilon} = \sigma^m \tag{8.67}$$

where μ is the specific viscosity and m is a material constant ($1 < m < 10$ depending on the material). Show that μ obeys the following scaling law:

$$\frac{\mu}{\mu_0} = \left(\frac{L}{L_0}\right)^{m(2-d_f)} \qquad (d_f \geq 2), \tag{8.68}$$

under the conditions of Problem 8.5, if we ignore the fractality of $\dot{\varepsilon}$.

8.7. As is well known, the Americas are drifting to the west with the speed V=10*cm* per year. Estimate the propelling force, $F = \mu LV$, and the propelling shear stress, $\tau = F/A$, on the bottom of the Americas' lithosphere, where A is the area of Americas, L is the north-south distance ($L = 2\cdot10^7\ m$, $A = 10^{13}\ m^2$), and μ is the linear viscosity corresponding to $m = 1$ in Problem 8.6. For μ_0 in laboratory scale $L_0 = 1\ cm$, use the Nabarro-Herring equation:

$$\mu_0 = \frac{2\Omega D}{kT\Delta^2}, \tag{8.69}$$

Here k is Boltzmann's constant, Δ is the grain size, Ω is the atomic volume of the geomaterial on the bottom of the lithosphere, and D is equal to:

$$D = D_0\, exp\left(-\frac{\Delta H}{kT}\right) \tag{8.70}$$

where ΔH is the enthalpy part of the activation energy, and D_0 is the temperature - independent diffusivity of the geomaterial on the bottom of the lithosphere. Put $\Omega = a^3$, $a = 0.5\ nm$, $D_0 = 1.5\cdot10^3\ cm^2sec^{-1}$, $\Delta H = 4\ eV$ (for Si), $T = 500\ K$, and $\Delta = 1\mu m$ for the estimate.

8.8. In the calculations of interatomic potential of a solid lattice, Young's modulus is used based on measurements of a single crystal deformation. Show that the grain size of a polycrystalline material, Δ, provides the natural lower bound of the scale range, in which power law fractality holds. Derive the equation for the Young's modulus of a polycrystalline fractal for larger scales, L,

$$E = E_0(L/\Delta)^{2-d_f-d_e}, \tag{8.71}$$

where E_0 is Young's modulus for a single crystal.

8.9. Consider an ideally brittle material with a crack; the area of the crack surface is a fractal obeying the self-similarity power law with fractal dimension $d_f \geq 2$. Show that the efficient surface energy of the fractal material, γ, in scale L is equal to

$$\frac{\gamma}{\gamma_0} = \left(\frac{L}{L_0}\right)^{2-d_f} \tag{8.72}$$

where γ_0 relates to reference scale L_0.

8.10. The strength, σ_B, of a specimen of a brittle material with a crack of characteristic length, l, is equal to

$$\sigma_B = \eta\sqrt{\frac{E\gamma}{(1-\nu^2)l}} \tag{8.73}$$

where η is a coefficient depending on the shape of the specimen and crack, and the crack position. Based on Problems 8.5 and 8.8, find the scaling law of σ_B for an elastic fractal solid with a fractal crack surface. Assume that l is a power law fractal with fractal dimension d_l.

8.11. The strength, σ_F, of a cracked specimen of an elastic-perfectly plastic material can be expressed by the following equation:

$$\sigma_F = \eta \sqrt{\frac{E\gamma_{\bullet}}{(1-\nu^2)(l+\delta)}}, \tag{8.74}$$

where

$$\delta = \frac{E\gamma_{\bullet}\eta^2}{(1-\nu^2)\sigma_Y^2} \tag{8.75}$$

Here, σ_Y is the yield stress, and $\gamma_{\bullet}$ is an efficient surface energy defined by fracture toughness, K_{Ic},

$$\gamma_{\bullet} = \frac{(1-\nu^2)K_{Ic}^2}{E} \tag{8.76}$$

Other designations coincide with those of Problem 8.9. Derive the scaling law of σ_F by assuming that E, $\gamma_{\bullet}$ and l are power law fractals.

8.12 Show that the efficient solution to Problems 8.1-8.10 may be achieved only under the following assumptions:

(i) the power law of self-similarity is valid for scales under consideration,

(ii) the fractality is small, so that the metric (fractal) dimension is slightly different from the corresponding topological dimensions (the latter is always a whole number).

8.13. Isotropic turbulence of water starts from the Kolmogorov scale of viscous-inertial transition,

$$L_k = (\nu^3/\varepsilon)^{1/4}, \tag{8.77}$$

which is the lower bound of the power law turbulence governed by the Kolmogorov - Oboukhov law of scaling,

$$U = (\varepsilon L)^{1/3}, \tag{8.78}$$

where

$$\varepsilon = 2\nu e_{ij} e_{ij}, \quad e_{ij} = \frac{1}{2}(u_{i,j} + u_{j,i}) \tag{8.79}$$

Here ν is the kinematic viscosity, L is the characteristic linear scale of the flow under consideration, U is the characteristic turbulent velocity scale, ε is the characteristic scale of turbulence energy dissipation (e_{ij} is the strain rate tensor).

The upper bound of the power law turbulence by Kolmogorov - Oboukhov is caused by natural geometrical limits or other physical phenomena. Find the upper bound, L_C, induced by local fractures and bubble formation in water (the cavitation-inertial transition). The critical negative pressure of rupture of a liquid is defined by linear scale (bubble size) and γ / k_v, where γ is the surface tension and k_v is the volume compressibility of the liquid. At the cavitation-inertial transition, the inertial force of turbulent motion, $\rho U^2 L^2$ is equal to the critical cavitation force of negative pressure, $(\gamma k_v^{-1} L^{-1})^{1/2} L^2$, where ρ is density.

Answer:

$$L_C = \left(\frac{\gamma}{\rho k_v}\right)^{3/7} \varepsilon^{-4/7}$$

References

1. B. B. Mandelbrot (1984), *The Fractal Geometry of Nature*, Freeman, New York.

2. M. A. Sadovsky and V. F Pisarenko (1991), *Seismic Process in Block Medium* [in Russian], Nauka, Moscow.

3. B. M. Smirnov (1991), *Physics of Fractal Clusters*, [in Russian], Nauka, Moscow.

4. J. Feder (1991), *Fractals*, Plenum Press, New York .

5. G. P. Cherepanov, A.S. Balankin and V. S. Ivanova (1995), Fractal fracture mechanics, *Engineerin Fracture Mechanics*, **51**(6), pp. 997-1035.

6. E. Hornbogen (1989), Fractals in microstructure of metals, *Int. Met. Reviews*, **34**(5), pp.277-296.

7. A. S. Balankin (1992), Fractal mechanics of deformed medium and topology of solid fracture, *Doklady Ross.Akad. Nauk*, **322** (5), pp. 869-874.

8. Z. Q. Mu and C. W. Lung (1988), Studies on the fractal dimension and fracture toughness of steel, *J.Physics* D, **21**(5), pp. 848-854.

9. J. J. Mecholsky, D.E. Passoja, and K. S. Feinberg-Rigel (1989), Quantitative analysis of brittle fracture surfaces using fractal geometry, *J. Amer. Ceramic Soc.*, **72**(1), pp. 60-65.

10. R. W. Cahn (1989), Fractal dimension and fracture, *Nature*, **388**, pp. 201-202.

11. A. R. Rosenfield (1987), Fractal mechanics, Ser. Met., 21(11), pp. 1359-1361.

12. L. E. Richards and B. D. Dempsey (1988), Fractal characterization of fractured surface in Ti-4.5 Al-5.0 Mo-1.5 Cr alloy (Coronas), *Scripta Metallurgica.*, **22**(5), pp. 687-689.

13. A. B. Mosolov (1991), Cracks with fractal surface, *Doklady Akad. Nauk SSSR*, **319**(4), pp. 840-844.

14. M. Matsushita (1985), Fractal viewpoint of fracture and accretion, *J. Phys. Soc.* Japan, **54** (3), pp. 857- 860.

15. H. Haken (1988), *Information and Self-Organization: A Macroscopic Approach to Complex Systems*, Springer, Berlin.

16. G. Nicolis and I. Prigogine (1989), *Exploring Complexity*, Freeman, New York.

17. G. P. Cherepanov (1994), Nanofracture mechanics approach to disocation generation and fracturing, *Applied Mechanics Review*, **47**, part 2, pp. S326-330 [Invited paper at the 12th U.S. National Congress of Applied Mechanics].

18. G. P. Cherepanov (1993), Some novel approaches in the mechanics of composites, *Composite Materials and Structures*, C. W. Bert, V. Birman, and D. Hui (eds.), ASME Press, pp. 1-12 [Invited paper at the 1993 Winter ASME Meeting].

19. G. P. Cherepanov (1988), Initiation of microcracks and dislocations, *Soviet Applied Mechanics [American Edition]*, **23**(12), pp. 1165-1176.

20. G. P. Cherepanov (1988), Microcrack growth under monotonic loading, *Soviet Applied Mechanics [American Edition]*, **24**(4), pp. 396-409.

21. G. P. Cherepanov (1989), Closure of microcracks during unloading and nucleation of reverse dislocations, *Soviet Applied Mechanics [American Edition]*, **24**(7), pp. 635-648.
22. G. P. Cherepanov and A. G. Cherepanov (1987), Nanocrack nucleation by condensation of pores, *Soviet Materials Science [American Edition]*, Plenum Press, (6), pp. 17-21.
23. E. E. Underwood and K. Banerji (1986), Fractals in fractography, *Materials Science and Engineering*, **80**, pp. 1-14.
24. I. M. Sokolov (1986), Dimensions and other geometric critical exponents in percolation theory, *Uspekhi Fiz. Nauk [Advances inPhysical Sciences]*, **15**(2), pp. 221-248.
25. D. Bergman and E. Duering (1986), Universal Poisson's ratios in two-dimensional random network of rigid and nonrigid bonds, *Physics Reviev* B, **34**(11), pp. 8199-8201.
26. D. Bergman (1986), Elastic moduli near percolation in a two-dimensional random network of rigid and nonrigid bonds, *Physics Reviev* B, **33**(3), pp. 2013-2016.
27. S. Feng and P. N. Sen (1989), *Physics Reviev Letters*, **52**, pp. 216-218.
28. P. G. De Gennes (1979), *Scaling Concepts in Polymer Physics*, Cornell University Press, London.
29. I. Webman (1985), The elasticity of fractal structures, *Proc. 6th Trieste Int. Symp. on Fractals in Physics, ICTP*, Trieste, Italy, July 9-12, 1985; L. Pietronero and E. Tosatti (eds.), Elsevier, pp. 488-497.
30. M. Muthukamar (1984), *Physics Reviev Letters*, **52**(3), pp. 212-214.
31. E. Louis, F. Guinea, and F. Flores (1985), Fractal nature of cracks, *Proc. 6th Trieste Int. Symp. on Fractals in Physics, ICTP*, Trieste, Italy, July 9-12; L. Pietronero and E. Tosatti (eds.), Elsevier, pp. 487-497.
32. P. Meakin, G. Li, L. M. Sander, E. Louis, and F. Guinea (1986), A simple two-dimensional model for crack propagation, *J. Physics* A, **22**(8.9), pp. 1393-1403.
33. P. H. Dauskardt, F. Houbensak, and R. D. Ritchie (1990), On interpretation of the fractal character of fracture surfaces, *Acta Metallurgica Materialia*, **38**(2), pp. 143-159.
34. G. P. Cherepanov (1974), *Mekhanika Khrupkogo Razrushenia*, Moscow, Nauka; English edition (1979), *Mechanics of Brittle Fracture*, R. de Wit and W. C. Cooley (eds.), McGrow Hill, New York.
35. *Fracture: A Topical Encyclopedia of Current Knowledge* (1966), G. P. Cherepanov (ed.). Krieger Publ., Melbourne, USA.
36. *Scale-Size Effects in the Failure Mechanisms of Materials and Structures* (1996), A. Carpinteri (ed.), E&FN Spon, London.

SUBJECT INDEX

Mechanics

SOLID MECHANICS AND ITS APPLICATIONS

Series Editor: G.M.L. Gladwell

Aims and Scope of the Series

The fundamental questions arising in mechanics are: *Why?, How?,* and *How much?* The aim of this series is to provide lucid accounts written by authoritative researchers giving vision and insight in answering these questions on the subject of mechanics as it relates to solids. The scope of the series covers the entire spectrum of solid mechanics. Thus it includes the foundation of mechanics; variational formulations; computational mechanics; statics, kinematics and dynamics of rigid and elastic bodies; vibrations of solids and structures; dynamical systems and chaos; the theories of elasticity, plasticity and viscoelasticity; composite materials; rods, beams, shells and membranes; structural control and stability; soils, rocks and geomechanics; fracture; tribology; experimental mechanics; biomechanics and machine design.

1. R.T. Haftka, Z. Gürdal and M.P. Kamat: *Elements of Structural Optimization.* 2nd rev.ed., 1990 ISBN 0-7923-0608-2
2. J.J. Kalker: *Three-Dimensional Elastic Bodies in Rolling Contact.* 1990 ISBN 0-7923-0712-7
3. P. Karasudhi: *Foundations of Solid Mechanics.* 1991 ISBN 0-7923-0772-0
4. *Not published*
5. *Not published.*
6. J.F. Doyle: *Static and Dynamic Analysis of Structures.* With an Emphasis on Mechanics and Computer Matrix Methods. 1991 ISBN 0-7923-1124-8; Pb 0-7923-1208-2
7. O.O. Ochoa and J.N. Reddy: *Finite Element Analysis of Composite Laminates.* ISBN 0-7923-1125-6
8. M.H. Aliabadi and D.P. Rooke: *Numerical Fracture Mechanics.* ISBN 0-7923-1175-2
9. J. Angeles and C.S. López-Cajún: *Optimization of Cam Mechanisms.* 1991 ISBN 0-7923-1355-0
10. D.E. Grierson, A. Franchi and P. Riva (eds.): *Progress in Structural Engineering.* 1991 ISBN 0-7923-1396-8
11. R.T. Haftka and Z. Gürdal: *Elements of Structural Optimization.* 3rd rev. and exp. ed. 1992 ISBN 0-7923-1504-9; Pb 0-7923-1505-7
12. J.R. Barber: *Elasticity.* 1992 ISBN 0-7923-1609-6; Pb 0-7923-1610-X
13. H.S. Tzou and G.L. Anderson (eds.): *Intelligent Structural Systems.* 1992 ISBN 0-7923-1920-6
14. E.E. Gdoutos: *Fracture Mechanics.* An Introduction. 1993 ISBN 0-7923-1932-X
15. J.P. Ward: *Solid Mechanics.* An Introduction. 1992 ISBN 0-7923-1949-4
16. M. Farshad: *Design and Analysis of Shell Structures.* 1992 ISBN 0-7923-1950-8
17. H.S. Tzou and T. Fukuda (eds.): *Precision Sensors, Actuators and Systems.* 1992 ISBN 0-7923-2015-8
18. J.R. Vinson: *The Behavior of Shells Composed of Isotropic and Composite Materials.* 1993 ISBN 0-7923-2113-8
19. H.S. Tzou: *Piezoelectric Shells.* Distributed Sensing and Control of Continua. 1993 ISBN 0-7923-2186-3

Kluwer Academic Publishers – Dordrecht / Boston / London

Mechanics

SOLID MECHANICS AND ITS APPLICATIONS

Series Editor: G.M.L. Gladwell

20. W. Schiehlen (ed.): *Advanced Multibody System Dynamics*. Simulation and Software Tools. 1993 ISBN 0-7923-2192-8
21. C.-W. Lee: *Vibration Analysis of Rotors*. 1993 ISBN 0-7923-2300-9
22. D.R. Smith: *An Introduction to Continuum Mechanics*. 1993 ISBN 0-7923-2454-4
23. G.M.L. Gladwell: *Inverse Problems in Scattering*. An Introduction. 1993 ISBN 0-7923-2478-1
24. G. Prathap: *The Finite Element Method in Structural Mechanics*. 1993 ISBN 0-7923-2492-7
25. J. Herskovits (ed.): *Advances in Structural Optimization*. 1995 ISBN 0-7923-2510-9
26. M.A. González-Palacios and J. Angeles: *Cam Synthesis*. 1993 ISBN 0-7923-2536-2
27. W.S. Hall: *The Boundary Element Method*. 1993 ISBN 0-7923-2580-X
28. J. Angeles, G. Hommel and P. Kovács (eds.): *Computational Kinematics*. 1993 ISBN 0-7923-2585-0
29. A. Curnier: *Computational Methods in Solid Mechanics*. 1994 ISBN 0-7923-2761-6
30. D.A. Hills and D. Nowell: *Mechanics of Fretting Fatigue*. 1994 ISBN 0-7923-2866-3
31. B. Tabarrok and F.P.J. Rimrott: *Variational Methods and Complementary Formulations in Dynamics*. 1994 ISBN 0-7923-2923-6
32. E.H. Dowell (ed.), E.F. Crawley, H.C. Curtiss Jr., D.A. Peters, R. H. Scanlan and F. Sisto: *A Modern Course in Aeroelasticity*. Third Revised and Enlarged Edition. 1995 ISBN 0-7923-2788-8; Pb: 0-7923-2789-6
33. A. Preumont: *Random Vibration and Spectral Analysis*. 1994 ISBN 0-7923-3036-6
34. J.N. Reddy (ed.): *Mechanics of Composite Materials*. Selected works of Nicholas J. Pagano. 1994 ISBN 0-7923-3041-2
35. A.P.S. Selvadurai (ed.): *Mechanics of Poroelastic Media*. 1996 ISBN 0-7923-3329-2
36. Z. Mróz, D. Weichert, S. Dorosz (eds.): *Inelastic Behaviour of Structures under Variable Loads*. 1995 ISBN 0-7923-3397-7
37. R. Pyrz (ed.): *IUTAM Symposium on Microstructure-Property Interactions in Composite Materials*. Proceedings of the IUTAM Symposium held in Aalborg, Denmark. 1995 ISBN 0-7923-3427-2
38. M.I. Friswell and J.E. Mottershead: *Finite Element Model Updating in Structural Dynamics*. 1995 ISBN 0-7923-3431-0
39. D.F. Parker and A.H. England (eds.): *IUTAM Symposium on Anisotropy, Inhomogeneity and Nonlinearity in Solid Mechanics*. Proceedings of the IUTAM Symposium held in Nottingham, U.K. 1995 ISBN 0-7923-3594-5
40. J.-P. Merlet and B. Ravani (eds.): *Computational Kinematics '95*. 1995 ISBN 0-7923-3673-9
41. L.P. Lebedev, I.I. Vorovich and G.M.L. Gladwell: *Functional Analysis*. Applications in Mechanics and Inverse Problems. 1996 ISBN 0-7923-3849-9
42. J. Menčik: *Mechanics of Components with Treated or Coated Surfaces*. 1996 ISBN 0-7923-3700-X
43. D. Bestle and W. Schiehlen (eds.): *IUTAM Symposium on Optimization of Mechanical Systems*. Proceedings of the IUTAM Symposium held in Stuttgart, Germany. 1996 ISBN 0-7923-3830-8

Kluwer Academic Publishers – Dordrecht / Boston / London

Mechanics

SOLID MECHANICS AND ITS APPLICATIONS

Series Editor: G.M.L. Gladwell

44. D.A. Hills, P.A. Kelly, D.N. Dai and A.M. Korsunsky: *Solution of Crack Problems.* The Distributed Dislocation Technique. 1996 ISBN 0-7923-3848-0
45. V.A. Squire, R.J. Hosking, A.D. Kerr and P.J. Langhorne: *Moving Loads on Ice Plates.* 1996 ISBN 0-7923-3953-3
46. A. Pineau and A. Zaoui (eds.): *IUTAM Symposium on Micromechanics of Plasticity and Damage of Multiphase Materials.* Proceedings of the IUTAM Symposium held in Sèvres, Paris, France. 1996 ISBN 0-7923-4188-0
47. A. Naess and S. Krenk (eds.): *IUTAM Symposium on Advances in Nonlinear Stochastic Mechanics.* Proceedings of the IUTAM Symposium held in Trondheim, Norway. 1996 ISBN 0-7923-4193-7
48. D. Ieşan and A. Scalia: *Thermoelastic Deformations.* 1996 ISBN 0-7923-4230-5
49. J. R. Willis (ed.): *IUTAM Symposium on Nonlinear Analysis of Fracture.* Proceedings of the IUTAM Symposium held in Cambridge, U.K. 1997 ISBN 0-7923-4378-6
50. A. Preumont: *Vibration Control of Active Structures.* An Introduction. 1997 ISBN 0-7923-4392-1
51. G.P. Cherepanov: *Methods of Fracture Mechanics: Solid Matter Physics.* 1997 ISBN 0-7923-4408-1

Kluwer Academic Publishers – Dordrecht / Boston / London

Mechanics

FLUID MECHANICS AND ITS APPLICATIONS

Series Editor: R. Moreau

Aims and Scope of the Series

The purpose of this series is to focus on subjects in which fluid mechanics plays a fundamental role. As well as the more traditional applications of aeronautics, hydraulics, heat and mass transfer etc., books will be published dealing with topics which are currently in a state of rapid development, such as turbulence, suspensions and multiphase fluids, super and hypersonic flows and numerical modelling techniques. It is a widely held view that it is the interdisciplinary subjects that will receive intense scientific attention, bringing them to the forefront of technological advancement. Fluids have the ability to transport matter and its properties as well as transmit force, therefore fluid mechanics is a subject that is particularly open to cross fertilisation with other sciences and disciplines of engineering. The subject of fluid mechanics will be highly relevant in domains such as chemical, metallurgical, biological and ecological engineering. This series is particularly open to such new multidisciplinary domains.

1. M. Lesieur: *Turbulence in Fluids*. 2nd rev. ed., 1990 ISBN 0-7923-0645-7
2. O. Métais and M. Lesieur (eds.): *Turbulence and Coherent Structures*. 1991 ISBN 0-7923-0646-5
3. R. Moreau: *Magnetohydrodynamics*. 1990 ISBN 0-7923-0937-5
4. E. Coustols (ed.): *Turbulence Control by Passive Means*. 1990 ISBN 0-7923-1020-9
5. A.A. Borissov (ed.): *Dynamic Structure of Detonation in Gaseous and Dispersed Media*. 1991 ISBN 0-7923-1340-2
6. K.-S. Choi (ed.): *Recent Developments in Turbulence Management*. 1991 ISBN 0-7923-1477-8
7. E.P. Evans and B. Coulbeck (eds.): *Pipeline Systems*. 1992 ISBN 0-7923-1668-1
8. B. Nau (ed.): *Fluid Sealing*. 1992 ISBN 0-7923-1669-X
9. T.K.S. Murthy (ed.): *Computational Methods in Hypersonic Aerodynamics*. 1992 ISBN 0-7923-1673-8
10. R. King (ed.): *Fluid Mechanics of Mixing*. Modelling, Operations and Experimental Techniques. 1992 ISBN 0-7923-1720-3
11. Z. Han and X. Yin: *Shock Dynamics*. 1993 ISBN 0-7923-1746-7
12. L. Svarovsky and M.T. Thew (eds.): *Hydroclones*. Analysis and Applications. 1992 ISBN 0-7923-1876-5
13. A. Lichtarowicz (ed.): *Jet Cutting Technology*. 1992 ISBN 0-7923-1979-6
14. F.T.M. Nieuwstadt (ed.): *Flow Visualization and Image Analysis*. 1993 ISBN 0-7923-1994-X
15. A.J. Saul (ed.): *Floods and Flood Management*. 1992 ISBN 0-7923-2078-6
16. D.E. Ashpis, T.B. Gatski and R. Hirsh (eds.): *Instabilities and Turbulence in Engineering Flows*. 1993 ISBN 0-7923-2161-8
17. R.S. Azad: *The Atmospheric Boundary Layer for Engineers*. 1993 ISBN 0-7923-2187-1
18. F.T.M. Nieuwstadt (ed.): *Advances in Turbulence IV*. 1993 ISBN 0-7923-2282-7
19. K.K. Prasad (ed.): *Further Developments in Turbulence Management*. 1993 ISBN 0-7923-2291-6
20. Y.A. Tatarchenko: *Shaped Crystal Growth*. 1993 ISBN 0-7923-2419-6

Kluwer Academic Publishers – Dordrecht / Boston / London

Mechanics

FLUID MECHANICS AND ITS APPLICATIONS

Series Editor: R. Moreau

21. J.P. Bonnet and M.N. Glauser (eds.): *Eddy Structure Identification in Free Turbulent Shear Flows.* 1993 ISBN 0-7923-2449-8
22. R.S. Srivastava: *Interaction of Shock Waves.* 1994 ISBN 0-7923-2920-1
23. J.R. Blake, J.M. Boulton-Stone and N.H. Thomas (eds.): *Bubble Dynamics and Interface Phenomena.* 1994 ISBN 0-7923-3008-0
24. R. Benzi (ed.): *Advances in Turbulence V.* 1995 ISBN 0-7923-3032-3
25. B.I. Rabinovich, V.G. Lebedev and A.I. Mytarev: *Vortex Processes and Solid Body Dynamics.* The Dynamic Problems of Spacecrafts and Magnetic Levitation Systems. 1994 ISBN 0-7923-3092-7
26. P.R. Voke, L. Kleiser and J.-P. Chollet (eds.): *Direct and Large-Eddy Simulation I.* Selected papers from the First ERCOFTAC Workshop on Direct and Large-Eddy Simulation. 1994 ISBN 0-7923-3106-0
27. J.A. Sparenberg: *Hydrodynamic Propulsion and its Optimization.* Analytic Theory. 1995 ISBN 0-7923-3201-6
28. J.F. Dijksman and G.D.C. Kuiken (eds.): *IUTAM Symposium on Numerical Simulation of Non-Isothermal Flow of Viscoelastic Liquids.* Proceedings of an IUTAM Symposium held in Kerkrade, The Netherlands. 1995 ISBN 0-7923-3262-8
29. B.M. Boubnov and G.S. Golitsyn: *Convection in Rotating Fluids.* 1995 ISBN 0-7923-3371-3
30. S.I. Green (ed.): *Fluid Vortices.* 1995 ISBN 0-7923-3376-4
31. S. Morioka and L. van Wijngaarden (eds.): *IUTAM Symposium on Waves in Liquid/Gas and Liquid/Vapour Two-Phase Systems.* 1995 ISBN 0-7923-3424-8
32. A. Gyr and H.-W. Bewersdorff: *Drag Reduction of Turbulent Flows by Additives.* 1995 ISBN 0-7923-3485-X
33. Y.P. Golovachov: *Numerical Simulation of Viscous Shock Layer Flows.* 1995 ISBN 0-7923-3626-7
34. J. Grue, B. Gjevik and J.E. Weber (eds.): *Waves and Nonlinear Processes in Hydrodynamics.* 1996 ISBN 0-7923-4031-0
35. P.W. Duck and P. Hall (eds.): *IUTAM Symposium on Nonlinear Instability and Transition in Three-Dimensional Boundary Layers.* 1996 ISBN 0-7923-4079-5
36. S. Gavrilakis, L. Machiels and P.A. Monkewitz (eds.): *Advances in Turbulence VI.* Proceedings of the 6th European Turbulence Conference. 1996 ISBN 0-7923-4132-5
37. K. Gersten (ed.): *IUTAM Symposium on Asymptotic Methods for Turbulent Shear Flows at High Reynolds Numbers.* Proceedings of the IUTAM Symposium held in Bochum, Germany. 1996 ISBN 0-7923-4138-4
38. J. Verhás: *Thermodynamics and Rheology.* 1997 ISBN 0-7923-4251-8
39. M. Champion and B. Deshaies (eds.): *IUTAM Symposium on Combustion in Supersonic Flows.* Proceedings of the IUTAM Symposium held in Poitiers, France. 1997 ISBN 0-7923-4313-1

Kluwer Academic Publishers – Dordrecht / Boston / London

Mechanics

From 1990, books on the subject of *mechanics* will be published under two series:
***FLUID* MECHANICS AND ITS APPLICATIONS**
Series Editor: R.J. Moreau
***SOLID* MECHANICS AND ITS APPLICATIONS**
Series Editor: G.M.L. Gladwell

Prior to 1990, the books listed below were published in the respective series indicated below.

MECHANICS: DYNAMICAL SYSTEMS
Editors: L. Meirovitch and G.Æ. Oravas

1. E.H. Dowell: *Aeroelasticity of Plates and Shells.* 1975 ISBN 90-286-0404-9
2. D.G.B. Edelen: *Lagrangian Mechanics of Nonconservative Nonholonomic Systems.* 1977 ISBN 90-286-0077-9
3. J.L. Junkins: *An Introduction to Optimal Estimation of Dynamical Systems.* 1978 ISBN 90-286-0067-1
4. E.H. Dowell (ed.), H.C. Curtiss Jr., R.H. Scanlan and F. Sisto: *A Modern Course in Aeroelasticity.* *Revised and enlarged edition see under Volume 11*
5. L. Meirovitch: *Computational Methods in Structural Dynamics.* 1980 ISBN 90-286-0580-0
6. B. Skalmierski and A. Tylikowski: *Stochastic Processes in Dynamics.* Revised and enlarged translation. 1982 ISBN 90-247-2686-7
7. P.C. Müller and W.O. Schiehlen: *Linear Vibrations.* A Theoretical Treatment of Multi-degree-of-freedom Vibrating Systems. 1985 ISBN 90-247-2983-1
8. Gh. Buzdugan, E. Mihăilescu and M. Radeş: *Vibration Measurement.* 1986 ISBN 90-247-3111-9
9. G.M.L. Gladwell: *Inverse Problems in Vibration.* 1987 ISBN 90-247-3408-8
10. G.I. Schuëller and M. Shinozuka: *Stochastic Methods in Structural Dynamics.* 1987 ISBN 90-247-3611-0
11. E.H. Dowell (ed.), H.C. Curtiss Jr., R.H. Scanlan and F. Sisto: *A Modern Course in Aeroelasticity.* Second revised and enlarged edition (of Volume 4). 1989 ISBN Hb 0-7923-0062-9; Pb 0-7923-0185-4
12. W. Szemplińska-Stupnicka: *The Behavior of Nonlinear Vibrating Systems.* Volume I: Fundamental Concepts and Methods: Applications to Single-Degree-of-Freedom Systems. 1990 ISBN 0-7923-0368-7
13. W. Szemplińska-Stupnicka: *The Behavior of Nonlinear Vibrating Systems.* Volume II: Advanced Concepts and Applications to Multi-Degree-of-Freedom Systems. 1990 ISBN 0-7923-0369-5
Set ISBN (Vols. 12–13) 0-7923-0370-9

MECHANICS OF STRUCTURAL SYSTEMS
Editors: J.S. Przemieniecki and G.Æ. Oravas

1. L. Frýba: *Vibration of Solids and Structures under Moving Loads.* 1970 ISBN 90-01-32420-2
2. K. Marguerre and K. Wölfel: *Mechanics of Vibration.* 1979 ISBN 90-286-0086-8

Mechanics

3. E.B. Magrab: *Vibrations of Elastic Structural Members.* 1979 ISBN 90-286-0207-0
4. R.T. Haftka and M.P. Kamat: *Elements of Structural Optimization.* 1985
   Revised and enlarged edition see under Solid Mechanics and Its Applications, Volume 1
5. J.R. Vinson and R.L. Sierakowski: *The Behavior of Structures Composed of Composite Materials.* 1986 ISBN Hb 90-247-3125-9; Pb 90-247-3578-5
6. B.E. Gatewood: *Virtual Principles in Aircraft Structures.* Volume 1: Analysis. 1989
   ISBN 90-247-3754-0
7. B.E. Gatewood: *Virtual Principles in Aircraft Structures.* Volume 2: Design, Plates, Finite Elements. 1989 ISBN 90-247-3755-9
   Set (Gatewood 1 + 2) ISBN 90-247-3753-2

MECHANICS OF ELASTIC AND INELASTIC SOLIDS

Editors: S. Nemat-Nasser and G.Æ. Oravas

---

1. G.M.L. Gladwell: *Contact Problems in the Classical Theory of Elasticity.* 1980
   ISBN Hb 90-286-0440-5; Pb 90-286-0760-9
2. G. Wempner: *Mechanics of Solids with Applications to Thin Bodies.* 1981
   ISBN 90-286-0880-X
3. T. Mura: *Micromechanics of Defects in Solids.* 2nd revised edition, 1987
   ISBN 90-247-3343-X
4. R.G. Payton: *Elastic Wave Propagation in Transversely Isotropic Media.* 1983
   ISBN 90-247-2843-6
5. S. Nemat-Nasser, H. Abé and S. Hirakawa (eds.): *Hydraulic Fracturing and Geothermal Energy.* 1983 ISBN 90-247-2855-X
6. S. Nemat-Nasser, R.J. Asaro and G.A. Hegemier (eds.): *Theoretical Foundation for Large-scale Computations of Nonlinear Material Behavior.* 1984 ISBN 90-247-3092-9
7. N. Cristescu: *Rock Rheology.* 1988 ISBN 90-247-3660-9
8. G.I.N. Rozvany: *Structural Design via Optimality Criteria.* The Prager Approach to Structural Optimization. 1989 ISBN 90-247-3613-7

MECHANICS OF SURFACE STRUCTURES

Editors: W.A. Nash and G.Æ. Oravas

---

1. P. Seide: *Small Elastic Deformations of Thin Shells.* 1975 ISBN 90-286-0064-7
2. V. Panc: *Theories of Elastic Plates.* 1975 ISBN 90-286-0104-X
3. J.L. Nowinski: *Theory of Thermoelasticity with Applications.* 1978
   ISBN 90-286-0457-X
4. S. Łukasiewicz: *Local Loads in Plates and Shells.* 1979 ISBN 90-286-0047-7
5. C. Fiřt: *Statics, Formfinding and Dynamics of Air-supported Membrane Structures.* 1983 ISBN 90-247-2672-7
6. Y. Kai-yuan (ed.): *Progress in Applied Mechanics.* The Chien Wei-zang Anniversary Volume. 1987 ISBN 90-247-3249-2
7. R. Negruţiu: *Elastic Analysis of Slab Structures.* 1987 ISBN 90-247-3367-7
8. J.R. Vinson: *The Behavior of Thin Walled Structures.* Beams, Plates, and Shells. 1988
   ISBN Hb 90-247-3663-3; Pb 90-247-3664-1

Mechanics

MECHANICS OF FLUIDS AND TRANSPORT PROCESSES
Editors: R.J. Moreau and G.Æ. Oravas

1. J. Happel and H. Brenner: *Low Reynolds Number Hydrodynamics.* With Special Applications to Particular Media. 1983 ISBN Hb 90-01-37115-9; Pb 90-247-2877-0
2. S. Zahorski: *Mechanics of Viscoelastic Fluids.* 1982 ISBN 90-247-2687-5
3. J.A. Sparenberg: *Elements of Hydrodynamics Propulsion.* 1984 ISBN 90-247-2871-1
4. B.K. Shivamoggi: *Theoretical Fluid Dynamics.* 1984 ISBN 90-247-2999-8
5. R. Timman, A.J. Hermans and G.C. Hsiao: *Water Waves and Ship Hydrodynamics.* An Introduction. 1985 ISBN 90-247-3218-2
6. M. Lesieur: *Turbulence in Fluids.* Stochastic and Numerical Modelling. 1987 ISBN 90-247-3470-3
7. L.A. Lliboutry: *Very Slow Flows of Solids.* Basics of Modeling in Geodynamics and Glaciology. 1987 ISBN 90-247-3482-7
8. B.K. Shivamoggi: *Introduction to Nonlinear Fluid-Plasma Waves.* 1988 ISBN 90-247-3662-5
9. V. Bojarevičs, Ya. Freibergs, E.I. Shilova and E.V. Shcherbinin: *Electrically Induced Vortical Flows.* 1989 ISBN 90-247-3712-5
10. J. Lielpeteris and R. Moreau (eds.): *Liquid Metal Magnetohydrodynamics.* 1989 ISBN 0-7923-0344-X

MECHANICS OF ELASTIC STABILITY
Editors: H. Leipholz and G.Æ. Oravas

1. H. Leipholz: *Theory of Elasticity.* 1974 ISBN 90-286-0193-7
2. L. Librescu: *Elastostatics and Kinetics of Aniosotropic and Heterogeneous Shell-type Structures.* 1975 ISBN 90-286-0035-3
3. C.L. Dym: *Stability Theory and Its Applications to Structural Mechanics.* 1974 ISBN 90-286-0094-9
4. K. Huseyin: *Nonlinear Theory of Elastic Stability.* 1975 ISBN 90-286-0344-1
5. H. Leipholz: *Direct Variational Methods and Eigenvalue Problems in Engineering.* 1977 ISBN 90-286-0106-6
6. K. Huseyin: *Vibrations and Stability of Multiple Parameter Systems.* 1978 ISBN 90-286-0136-8
7. H. Leipholz: *Stability of Elastic Systems.* 1980 ISBN 90-286-0050-7
8. V.V. Bolotin: *Random Vibrations of Elastic Systems.* 1984 ISBN 90-247-2981-5
9. D. Bushnell: *Computerized Buckling Analysis of Shells.* 1985 ISBN 90-247-3099-6
10. L.M. Kachanov: *Introduction to Continuum Damage Mechanics.* 1986 ISBN 90-247-3319-7
11. H.H.E. Leipholz and M. Abdel-Rohman: *Control of Structures.* 1986 ISBN 90-247-3321-9
12. H.E. Lindberg and A.L. Florence: *Dynamic Pulse Buckling.* Theory and Experiment. 1987 ISBN 90-247-3566-1
13. A. Gajewski and M. Zyczkowski: *Optimal Structural Design under Stability Constraints.* 1988 ISBN 90-247-3612-9

Printed in the United States
86706LV00001B/68/A

9 780792 344087